The Economics of Creativity

Understanding the economic implication of creative individuals and firms is at the heart of the new economy and of related fields such as the economics of knowledge, the economics of science and innovation management. This book brings together a panel of theoretical and empirical contributions which address the generation of creative ideas and their transformation into products and services by firms or universities, as well as the interplay of those organizations in networks and markets.

The word 'creativity' has been used a great deal recently in relation to efforts to recover from the global financial crisis and re-launch economic activity. Little has been added to explain how and why an economic approach of creativity is useful and necessary. It is useful to understand how the most creative people work and think, and how to foster their creative productivity. It is useful to understand how organizations integrate and exploit creative ideas. It is useful to understand how market mechanisms can handle creativity and how policies must be adapted. It is necessary in the light of the recent economic crises that made innovation, invention and creativity the basis of a new industrialization and fuel for a new economic development.

This new book assesses the economic impact of creativity, defining the term and then going on to explore theoretically and practically the economic consequences of creativity through a range of themes including: creativity and evolutionary theories of technological change; creativity and organizational learning; creativity and technological policy; and creativity and economics of networks. This volume offers a rich source of inspiration and ideas for the pursuit of research which merges economic tradition and management perspectives.

Thierry Burger-Helmchen is a researcher at BETA Strasbourg and Dean of the Department of Economics and Management at the University of Strasbourg, France.

Routledge Studies in Global Competition
Edited by John Cantwell, Rutgers, the State University of New Jersey, USA and David Mowery, University of California, Berkeley, USA

1. Japanese Firms in Europe
Edited by Frédérique Sachwald

2. Technological Innovation, Multinational Corporations and New International Competitiveness
The case of intermediate countries
Edited by José Molero

3. Global Competition and the Labour Market
Nigel Driffield

4. The Source of Capital Goods Innovation
The role of user firms in Japan and Korea
Kong-Rae Lee

5. Climates of Global Competition
Maria Bengtsson

6. Multinational Enterprises and Technological Spillovers
Tommaso Perez

7. Governance of International Strategic Alliances Technology and transaction costs
Joanne E. Oxley

8. Strategy in Emerging Markets
Telecommunications establishments in Europe
Anders Pehrsson

9. Going Multinational
The Korean experience of direct investment
Edited by Frédérique Sachwald

10. Multinational Firms and Impacts on Employment, Trade and Technology
New perspectives for a new century
Edited by Robert E. Lipsey and Jean-Louis Mucchielli

11. Multinational Firms
The global–local dilemma
Edited by John H. Dunning and Jean-Louis Mucchielli

12. MIT and the rise of entrepreneurial science
Henry Etzkowitz

13. Technological Resources and the Logic of Corporate Diversification
Brian Silverman

14. The Economics of Innovation, New Technologies and Structural Change
Cristiano Antonelli

15. European Union Direct Investment in China
Characteristics, challenges and perspectives
Daniel Van Den Bulcke, Haiyan Zhang and Maria do Céu Esteves

16. Biotechnology in Comparative Perspective
Edited by Gerhard Fuchs

17. Technological Change and Economic Performance
Albert L. Link and Donald S. Siegel

18. Multinational Corporations and European Regional Systems of Innovation
John Cantwell and Simona Iammarino

19. Knowledge and Innovation in Regional Industry
An entrepreneurial coalition
Roel Rutten

20. Local Industrial Clusters
Existence, emergence and evolution
Thomas Brenner

21. The Emerging Industrial Structure of the Wider Europe
Edited by Francis McGowen, Slavo Radosevic and Nick Von Tunzelmann

22. Entrepreneurship
A new perspective
Thomas Grebel

23. Evaluating Public Research Institutions
The U.S. advanced technology program's intramural research initiative
Albert N. Link and John T. Scott

24. Location and Competition
Edited by Steven Brakman and Harry Garretsen

25. Entrepreneurship and Dynamics in the Knowledge Economy
Edited by Charlie Karlsson, Börje Johansson and Roger R. Stough

26. Evolution and Design of Institutions
Edited by Christian Schubert and Georg von Wangenheim

27. The Changing Economic Geography of Globalization
Reinventing space
Edited by Giovanna Vertova

28. Economics of the Firm
Analysis, evolution and history
Edited by Michael Dietrich

29. Innovation, Technology and Hypercompetition
Hans Gottinger

30. Mergers and Acquisitions in Asia
A global perspective
Roger Y. W. Tang and Ali M. Metwalli

31. Competitiveness of New Industries
Institutional framework and learning in information technology in Japan, the U.S and Germany
Edited Cornelia Storz and Andreas Moerke

32. Entry and Post-Entry Performance of Newborn Firms
Marco Vivarelli

33. Changes in Regional Firm Founding Activities
A theoretical explanation and empirical evidence
Dirk Fornahl

34. Risk Appraisal and Venture Capital in High Technology New Ventures
Gavin C. Reid and Julia A. Smith

35. Competing for Knowledge
Creating, connecting and growing
Robert Huggins and Hiro Izushi

36. Corporate Governance, Finance and the Technological Advantage of Nations
Andrew Tylecote and Francesca Visintin

37. Dynamic Capabilities Between Firm Organisation and Local Systems of Production
Edited by Riccardo Leoncini and Sandro Montresor

38. Localised Technological Change
Towards the economics of complexity
Cristiano Antonelli

39. Knowledge Economies
Innovation, organization and location
Wilfred Dolfsma

40. Governance and Innovation
Maria Brouwer

41. Public Policy for Regional Development
Edited by Jorge Martinez-Vazquez and François Vaillancourt

42. Evolutionary Economic Geography
Location of production and the European Union
Miroslav Jovanovic

43. Broadband Economics Lessons from Japan
Takanori Ida

44. Targeting Regional Economic Development
Edited by Stephan J. Goetz, Steven C. Deller and Thomas R. Harris

45. Innovation, Knowledge and Power
Theodora Asimakou

46. Creativity, Innovation and the Cultural Economy
Edited by Andy C. Pratt and Paul Jeffcutt

47. Co-opetition Strategy
Giovanni Battista Dagnino and Elena Rocco

48. Knowledge Intensive Entrepreneurship and Innovation Systems
Evidence from Europe
Edited by Franco Malerba

49. Innovation in Complex Social Systems
Edited by Petra Ahrweiler

50. Internationalization, Technological Change and the Theory of the Firm
Edited by Nicola De Liso and Riccardo Leoncini

51. Territory, Specialization and Globalization in European Manufacturing
Helena Marques and Francisco Puig

52. Institutional Diversity and Innovation
Continuing and emerging patterns in Japan and China
Cornelia Storz and Sebastian Schäfer

53. Innovation and Economic Crisis
Daniele Archibugi and Andrea Filippetti

54. The Communications Industries in the Era of Convergence
Catherine Mulligan

55. Innovation, Technology and Knowledge
Charlie Karlsson, Borje Johansson and Roger R. Stough

56. Evolution of Competition Laws and their Enforcement
Pradeep S. Mehta

57. The Economics of Structural Change in Knowledge
Francesco Quatraro

58. Economic Geography and the Unequal Development of Regions
Jean-Claude Prager and Jacques-François Thisse

59. Social Networks, Innovation and the Knowledge Economy
Edited by Isabel Salavisa and Margarida Fontes

60. The Economics of Creativity
Ideas, firms and markets
Edited by Thierry Burger-Helmchen

Bureau d'Economie Théorique et Appliquée (BETA)

Bureau d'Economie Théorique et Appliquée (BETA) is a research laboratory of the University of Strasbourg, University of Lorraine and the French National Centre for Scientific Research CNRS. It was created in 1972 and it became an 'associate' member of CNRS in 1985. BETA's activities cover a wide range of topics dealing with basic as well as applied scientific research in the fields of economics and management science. Historically, the laboratory developed its research programme along several research directions based on micro- and macro-economic theories, and is heir to a local tradition of history of economic thought. It has also developed a number of focus areas, often resulting from the fruitful interaction of 'theoretical' approaches and 'applied' approaches – as expressed in its name – such as the economics of innovation, management of technology and organizations, environmental economics, socio-economic approaches of education, training and employment and, more recently, historical economics and cliometrics. The result has been the creation of collective competences from years of practice with different methodologies: modelling techniques, econometrics, methods of technology evaluation and foresight, experimental economics, to mention only a few. In July 2012 BETA numbered some 120 staff members with permanent positions (mainly professors and assistant professors of the university, researchers, administrative and technical personnel) and over 80 non-permanent staff (associate members, PhD students and post–docs).

Since 2010 several developments have taken place in the domain of economics and management of creativity. In particular, BETA is now successfully supported by three local companies: SALM, Socomec and Cabinet Voirin, through a Chair in Management of Creativity. It has allowed BETA and the Faculty of Economics and Management to organize a Fall School in Management of Creativity and structure an Academy of Management of Creativity and Innovation.

Bureau d'économie théorique et appliquée (BETA)
UMR 7522

The Economics of Creativity

Ideas, firms and markets

Edited by
Thierry Burger-Helmchen

Routledge
Taylor & Francis Group
LONDON AND NEW YORK

First published 2013
by Routledge
2 Park Square, Milton Park, Abingdon, Oxon OX14 4RN

Simultaneously published in the USA and Canada
by Routledge
711 Third Avenue, New York, NY 10017

Routledge is an imprint of the Taylor & Francis Group, an informa business

British Library Cataloguing in Publication Data

A catalogue record for this book is available from the British Library

Library of Congress Cataloging in Publication Data
The economics of creativity : ideas, firms and markets / edited by Thierry Burger-Helmchen.
p. cm.
Includes bibliographical references and index.
1. Creative ability in business. 2. Intellectual capital. 3. New products. 4. Diffusion of innovations–Economic aspects. 5. Technological innovations–Economic aspects.
I. Burger-Helmchen, Thierry.
HD53.E26 2013
338'.064–dc23
2012029764

ISBN: 978-0-415-52529-9 (hbk)
ISBN: 978-0-203-07341-4 (ebk)

Typeset in Times New Roman by
Newgen Imaging Systems

Contents

Bureau d'Economie Théorique et Appliquée (BETA) vi
List of figures xii
List of tables and boxes xiv
Notes on contributors xvi
Preface xxiv
Acknowledgements xxvi

PART I
Economics of creativity, ideas and policies 1

1 Exploring the convergence of the rationales underlying the policies for science, technology and innovation and for creative industries 3
LAURENT BACH

2 A strategic management of forgetting 20
GILLES LAMBERT

3 Product lines innovation: on the use of technical laws of evolution 41
CHRISTOPHE LERCH, ERIC SCHENK AND DENIS CAVALLUCCI

4 Tools and implications of managing knowledge and creativity 58
CLAUDE GUITTARD, CLAUDE PARAPONARIS, JEAN-LOUIS ERMINE AND PASCAL LIÈVRE

5 Innovation, territories and creativity: some reflections about usual and less usual instruments for innovation-driven regional policies 78
EMMANUEL MULLER, JEAN-ALAIN HÉRAUD AND ANDREA ZENKER

6 Governing communities of experts 89
JEAN-PHILIPPE BOOTZ, OLIVIER DUPOUËT, FRANCIS KERN AND ERIC SCHENK

7 Science based entrepreneurship through the lens of sectoral and national innovation systems: French academic start-ups developing hydrogen and fuel cell technologies 110
ARMAN AVADIKYAN

PART II
Economics of creativity, firms and organizations 129

8 Managing creativity in the firm: the fuzzy front end of innovation and dynamic capabilities 131
PATRICK COHENDET, JEAN-FRANÇOIS HARVEY AND LAURENT SIMON

9 Appropriation strategies and endogenous technological regime: towards a dynamic theory of the role of patents 151
JÉRÔME FREYERMUTH, PATRICK COHENDET AND JULIEN PÉNIN

10 Estimating the impact of a potential process innovation and the optimal strategy of licensing 167
WEIFANG LOU

11 Strategic alliances and knowledge flows between firms 192
MÜGE ÖZMAN AND PATRICK LLERENA

12 Deal values for acquisitions by larger incumbents in high-technology industries: the influence of non-linearities and moderation effects 208
MARCUS WAGNER

13 Marshalling resources and dynamic capabilities to overcome innovative barriers and foster creativity in the service sector 226
CAROLINE HUSSLER, LAURENCE FRANK, MATTHIEU BUNEL AND PATRICK EPARVIER

14 Research, teaching and commercial activities: the behaviour of laboratories in an entrepreneurial university 252
MIREILLE MATT AND VÉRONIQUE SCHAEFFER

PART III
Economics of creativity, networks and markets 269

15 The public economics of creativity: economies of scope in technological infrastructure 271
MOSHE JUSTMAN

16 Fits and misfits: technological matching and R&D networks 281
ROBIN COWAN, NICOLAS JONARD AND BULAT SANDITOV

17 Market occupation in a circular market: product innovation and/or patenting 309
MURAT YILDIZOĞLU AND EHUD ZUSCOVITCH

18 Creativity and uncertainty in the act of work: the contribution of the viability theory 327
FRANCIS MUNIER

19 What makes movies tick: ivory tower insights, studio views and research directions 339
S. ABRAHAM RAVID

Index 351

Figures

3.1	Overview of the laws of evolution	51
3.2	Laws of evolution for computer keyboards	55
4.1	The TEEXMA tool for technical knowledge management	67
4.2	The three approaches: knowledge/competencies/processes	69
4.3	Identifying the critical knowledge in organic farming	70
4.4	Knowledge macroscope in MASK I	71
4.5	Indirect and direct knowledge transfer, according to Nonaka	74
8.1	The three steps of the FFE phase	136
9.1	The management of patents: a wide spectrum of uses	152
9.2	Combining Callon and Zuscovitch analysis to understand the dynamics of the technological regime	161
9.3	The evolution of technological regime	163
9.4	Patenting strategies in a dynamic setting	164
10.1	Evolution of camera characteristics (January 2003–May 2005)	175
11.1	Our hypothesis	198
11.2	Sectoral distribution of alliances in the study	199
11.3	Country distribution of the dyads covered in the study	200
11.4	Duration and learning	200
13.1	Total number of perceived barriers by sector	234
13.2	Barriers to innovation: whole sample	236
13.3	Barriers to innovation: innovative firms only	237
13.4	Barriers to innovation: firms with abandoned innovation project only	238
13.5	Constraints faced and innovation introduced in the service sector	239
13.6	Impact of public support on the service sector	240
14.1	Eigenvalues and percentage of explained variability	260
14.2	Projection of the variables on the plan 1×2	260
14.3	Projection of labs in the first factorial plan	261
16.1	Density of the equilibrium network	291
16.2	Expected transitivity and rescaled expected transitivity	293

16.3 Frequency distribution of rescaled distance 296
16.4 Ratio of rescaled transitivity to rescaled average path length as a function of network density 297
16.5 Degree distributions for different parameter values 301
16.6 Comparing competition with no competition: ratios of path length and transitivity over the parameter space 303
16.7 Comparing competition to no competition case: ratio of the small world ratio over the parameter space 304
17.1 Optimal number of varieties in monopoly 316
17.2 Circular city and entry 316
17.3 Duopoly profits and variety 320
17.4 Variety proliferation and entry blockading 321
17.5 Entry blockading: incumbent's profits 321
17.6 Gross profits: patent or not? Two cases, one result... 322
17.7 Circular city and entry 324
18.1 Viable evolution of creative acts 333
18.2 Creative evolution and crisis of viability 334
18.3 Heavy solution and non-creativity 335

Tables and boxes

Tables

1.1 Basic principles of standard approaches in STI policy and corresponding examples in CCI policy 10
2.1 Forgetting in some companies 37
3.1 Families and descent lines according to Deforge 45
3.2 Paths of evolution 47
4.1 Characteristics of the three types of knowledge 64
4.2 Three types of approach for cognitive resources mapping 68
4.3 The different models in MASK I 71
4.4 The two types of community 73
4.5 Entrepreneur and manager roles 73
5.1 The proposed framework 81
5.2 A synoptic comparison of the three types of instruments 83
5.3 Elements of a counter-manifesto for rethinking local innovation-supporting policy 86
6.1 The three faces of the expert 92
8.1 The fuzzy front end of innovation and dynamic capabilities 145
9.1 Technological regime and firms' optimal patenting strategies 155
10.1 Model-level total sales statistics 176
10.2 Demand estimates 177
10.3 Cost estimates 178
10.4 Change in producer surplus with one adopter 179
10.5 Size of digital cameras in cubic inch by brand 180
10.6 Change in consumer surplus with one adopter 181
10.7 Welfare change measure with multiple adopters 182
10.8 Average price changes by adopter(s) 183
10.A1 Change in producer surplus with one adopter of 'metal body' cameras 187
10.A2 Change in consumer surplus with one adopter of 'metal body' cameras 188

10.A3 Welfare change due to 'metal body' cameras with multiple adopters 189
11.1 Descriptive statistics 202
11.2 Some results of the OLS regressions 204
12.1 Base model for all sectors jointly, dependent variable: average annual value of acquisitions 214
12.2 Extended model for all sectors jointly, dependent variable: average annual value of acquisitions 215
12.3 Base model for the EDA industry, dependent variable: average annual value of acquisitions 216
12.4 Extended model for the EDA industry, dependent variable: average annual value of acquisitions 217
12.5 Base model for the biotechnology industry, dependent variable: average annual value of acquisitions 218
12.6 Extended model for the biotechnology industry, dependent variable: average annual value of acquisitions 219
12.7 Base model for the semiconductor industry, dependent variable: average annual value of acquisitions 220
12.8 Extended model for the semiconductor industry, dependent variable: average annual value of acquisitions 221
12.9 Summary of results for average annual value 223
12.A1 Summary statistics and correlations 223
13.1 Innovative projects by sector: success and failure 242
13.2 Innovation impediments as perceived by service sector firms depending on manpower qualification 243
13.3 Innovation impediments as perceived by firms depending on the scope of their market 245
14.1 Definition of the variables 258
14.2 Descriptive statistics of the variables 259
14.3 Correlation coefficients between our variables 259
14.4 Variables defining the three components 261
14.5 Means of the variables for each class 262
14.6 Characteristics of each class 262
16.1 Constructing a longest shortest path 295
19.1 Percentage of nominations and awards in different categories for the Oscar winning film over 40 years (1971–2010) and Razzies from their inception, 1981–2010 343

Boxes

13.1 Innovation styles in service industries 227
13.2 Using CIS data to study innovation in the service sector 233

Contributors

Arman Avadikyan has a PhD in Economics. He is a research engineer at BETA (Bureau d'Economie Théorique et Appliquée, UMR 7522 – CNRS), a research unit of the University of Strasbourg, University of Lorraine and the CNRS (France). His current research work focuses on innovation management, sustainable technologies and S&T policy.

Laurent Bach, PhD, is Associate Professor in Economics at the University of Strasbourg. He specializes in the economics of innovation, in particular in the analysis of science, technology and innovation policies, in the evaluation of their economic impact and in science–industry relations. He is regularly involved in field/contract based research for various international and national entities (EC, ESA, OECD, French Ministry of Research, Canadian Space Agency, Japanese MITI, etc.).

Jean-Philippe Bootz is Associate Professor at the University of Franche-Comté and researcher at the EM Strasbourg Business School (Humanis). His research focuses on strategic foresight, knowledge management and human resources management.

Matthieu Bunel, PhD, is Associate Professor at the University of Caen – Centre d'études de l'emploi.

Thierry Burger-Helmchen (PhD in Management Science, EM Strasbourg, University of Strasbourg), is a researcher at BETA (Bureau d'Economie Théorique et Appliquée, UMR 7522 – CNRS) and Dean of the Department of Economics and Management at the University of Strasbourg. His research topics are entrepreneurship and innovation management. He analyses the economic impact of creativity notably in the video game industry. He teaches Strategy of SMEs and Managerial Economics at the University of Strasbourg.

Denis Cavallucci is Associate Professor at INSA Strasbourg, and a researcher at LGECO.

Patrick Cohendet is Professor of Economics at HEC Montréal and was previously a professor at the University of Strasbourg. His research interests include economics of innovation, technology management, knowledge management, theory of the firm and economics of creativity. He is the author or co-author of 15 books and over 60 articles in refereed journals. He has conducted a series of economic studies on innovation and economics of knowledge (measurement of spin-offs, evaluation of the economic benefits of R&D projects, evaluation of technology transfer, etc.). These studies were carried out by his research laboratory BETA (Bureau d'Economie Théorique et Appliquée, UMR 7522 – CNRS) and Mosaic at HEC Montréal for different European and North American organizations such as the European Commission, the EU, OECD, Council of Europe and the Canadian Space Agency.

Robin Cowan is Professor of Management at BETA (Bureau d'Economie Théorique et Appliquée, UMR 7522 – CNRS) and Professor of Economics at Maastricht University and UNU-MERIT. He received his PhD from Stanford University. His research focuses largely on the economics of networks, the psychology of organizations and the economics and dynamics of knowledge.

Olivier Dupouët is Professor at BEM Management School. He teaches Management Information Systems. He worked for three years as a consultant in the area of information systems for the banking industry. His missions included the conception of software for operational risk management at Ixis CIB and the development of software for risk computation in compliance with the Basel II framework. He holds a PhD in Economics from the University of Strasbourg. His research interests lie in organizational learning, knowledge management and organizational design. He has published articles in journals such as *Journal of Economic Behavior and Organization*, *Research Policy* and *Journal of Management and Governance*.

Patrick Eparvier, PhD, is Senior Consultant – Technopolis Group France.

Jean-Louis Ermine has a PhD in Fundamental Mathematics (Paris University Denis Diderot) and a National Research Director Title in Computer Science (Bordeaux University). After 15 years at the university, he worked in the French Atomic Energy Commission (CEA) as a knowledge management (KM) expert for over 10 years. Since 2003, he has been Professor at the national public organisation Institut Mines-Télécom, where he is now Director for Innovation at Telecom Business School. He has written 5 books and more than 100 articles in peer-reviewed journals and conferences. He has been President of the French Knowledge Management Club since 1999, an association rallying many French speaking companies, and of the French speaking academic association for KM since 2012. He has been project leader or advisor in numerous research or industrial KM projects in public or private companies in France and abroad.

Laurence Frank is Professor at ISC School of Management in Paris and a researcher at BETA (Bureau d'Economie Théorique et Appliquée, UMR 7522 – CNRS). Her research encompasses competitive strategies in the airport industry, public reforms and business models founded on public-private partnerships.

Jérôme Freyermuth is a PhD candidate at BETA (Bureau d'Economie Théorique et Appliquée, UMR 7522 – CNRS). He is also a junior consultant at VOIRIN Consultants. His research interests are the economics of innovation, in particular the analysis of appropriation strategies and the use of information technologies in organizations.

Claude Guittard has a PhD in Business Studies from the University of Strasbourg. He works as a researcher at BETA (Bureau d'Economie Théorique et Appliquée, UMR 7522 – CNRS) and is Associate Professor at the Department of Economics and Management at the University of Strasbourg. He is in charge of the Chair IT – Globalization from EM Strasbourg. His work tries to answer the general question 'how to use IT to manage knowledge'. From this perspective he has written several articles about knowledge creation and transfer on the Internet and the role of community of practice in IT entrepreneurship or, more recently, about crowdsourcing, or the definition of IT experts.

Jean-François Harvey is a PhD candidate at HEC Montréal and SKEMA Business School (joint supervision). He holds a Joseph-Armand Bombardier CGS Doctoral Scholarship from the Social Sciences and Humanities Research Council (SSHRC) and his research deals with organizational learning, with a particular emphasis on the design and development of networks and communities of practice to foster knowledge creation. His research has appeared in *Journal of Knowledge Management* and *Gestion*.

Jean-Alain Héraud is Professor of Economics at the University of Strasbourg. He was head of the Department of Economics and Management from 2008 to 2012. A member of BETA (Bureau d'Economie Théorique et Appliquée, UMR 7522 – CNRS), he specializes in economics of innovation and science and technology policy and has worked also in specific fields like regional development, training and employment, energy systems and technology foresight. In recent years he has organized collective research on multi-level and multi-actor governance of science and technology, in particular within European research programmes. In France, he has been involved in a national experience on regional systems of innovation, leading to the design of regional innovation policies. He is presently active in the cross-border Upper-Rhine region for studies and expertise in the field of creative and knowledge based development of territories.

Caroline Hussler is currently Associate Professor at UTBM (University of Technology of Belfort-Montbéliard, France) and a research fellow at BETA (Bureau d'Economie Théorique et Appliquée, UMR 7522 – CNRS). Her main research interests lie in the determinants of territorial innovation and the stakes of social network building. Her works have been published in major international journals such as *Research Policy*, *International Journal of Technology Management*, *Technological Forecasting and Social Change*, *Regional Studies* and *Economics of Innovation and New Technology*.

Nicolas Jonard is Associate Professor of Economics and Management at the University of Luxembourg. He received his PhD from the University of Strasbourg. His present research focuses on the architecture and performance of innovation networks, the psychology of organizations and the behaviour of interacting agents systems.

Moshe Justman is based at the Department of Economics, Ben-Gurion University of the Negev, Beersheba, Israel, and Melbourne Institute for Applied Economic and Social Research, University of Melbourne.

Francis Kern is Professor in Economics at the Faculty of Economics and Management, University of Strasbourg. His research topics are located in the economy and management of innovation and knowledge. He specialises in the analysis and management of knowledge communities in developing countries, notably in projects involving ecotourism, agroecology and pharmacopoeia.

Gilles Lambert is Professor at the École de Management Strasbourg of the University of Strasbourg where he teaches Corporate Finance, Entrepreneurship and Strategy. He is a researcher at BETA (Bureau d'Economie Théorique et Appliquée, UMR 7522 – CNRS). He holds a degree in Management Sciences from University Louis Pasteur, and is working on real options, corporate governance, innovation management and organizational change.

Christophe Lerch is Associate Professor in Management at BETA (Bureau d'Economie Théorique et Appliquée, UMR 7522 – CNRS). His general research focus is on the monitoring of business processes. He devotes particular attention to semi and unstructured forms of problem solving at work in design processes.

Pascal Lièvre holds a PhD in Production Economics from the University of Lyon-II. He is Professor (HDR, University of Aix-Marseille-II) in Management Science at Clermont University and ESC Clermont Graduate School of Management. He is Director of Research at the Centre de Recherche Clermontois de Gestion et de Management (CRCGM, EA 3849). Since 2000, he has been in charge of a research programme on the management of extreme situations dealing mainly with polar expeditions.

He works on expansion of knowledge (knowing, knowledge, scientific knowledge) in projects.

Patrick Llerena is Professor of Economics at the University of Strasbourg and a researcher at BETA (Bureau d'Economie Théorique et Appliquée, UMR 7522 – CNRS). He is the author and co-author of many articles on the economics of innovation and the economics of science.

Weifang Lou, PhD Economics, is a lecturer in Economics and Finance at La Trobe University. Her research lies in the area of empirical IO in investigating the economic behaviour of consumers, firms and industries. Weifang Lou's current work focuses on modelling demand and supply dynamics in the new durables product markets, and the strategic choice decisions of firms in the banking industry and other financial institutions.

Mireille Matt is Directeur de Recherche (Research Professor) at INRA (National Institute of Agricultural Research). She is conducting her research at Grenoble Applied Economics Lab, a joint research unit between INRA and University Pierre Mendès France (Grenoble). She has a PhD in Economics from the University of Strasbourg where she started her career as Associate Professor. Her research interests are in economics of innovation, economics of knowledge, in technology policy and in evaluation of large R&D programmes. Her work is published in *Research Policy*, *Journal of Technology Transfer*, *International Journal of Entrepreneurship and Innovation Management*, *Information Economics and Policy*, *Technovation*, *International Journal of Technology Management* and *Scientometrics*.

Emmanuel Muller is an economist specializing in innovation and knowledge economics and has been active for 20 years in these fields in France and Germany. Since 2009, he has coordinated the evoREG initiative devoted to innovation policy and research in the Upper Rhine and associating BETA (Bureau d'Economie Théorique et Appliquée, UMR 7522 – CNRS) and the Fraunhofer Institute for Systems and Innovation Research ISI (Karlsruhe, Germany). Apart from teaching and publishing in the fields of innovation economics, management and policy he performs contract research mainly on the issues of regional innovation systems, knowledge-intensive business services (KIBS) and creative economics.

Francis Munier is Associate Professor at the Faculty of Economics and Management of the University of Strasbourg and a member of BETA (Bureau d'Economie Théorique et Appliquée, UMR 7522 – CNRS). His research focuses on the economics of innovation and knowledge, with an original mathematical application which allows some modelling in a radical and non-deterministic environment.

Müge Özman is Associate Professor at the Télécom École de Management, Institut-Mines Télécom, and a researcher at BETA (Bureau d'Economie Théorique et Appliquée, UMR 7522 – CNRS). She has published many articles on the economics of networks and economics of innovation.

Claude Paraponaris holds a PhD in Industrial Economics from the University of Mediterranée (France) and a national Research Director Title in Management Science, and he is Professor at the University of Paris Est. He worked for the National Centre for Scientific Research (CNRS) for 8 years in Aix-en-Provence (France). He was an expert for the EU, focusing on science–industry relations. He has written 2 books and more than 50 articles in peer-reviewed journals and conferences. He is the vice-president of the French Speaking Congress in Knowledge Management (GeCSO). His research interests include innovation and knowledge management, multinational information systems, shared creativity, epistemic communities and network effects.

Julien Pénin is a researcher at BETA (Bureau d'Economie Théorique et Appliquée, UMR 7522 – CNRS). He has authored and co-authored many publications related to the economics and management of patents, open innovation and open source. Julien Pénin teaches Economics and Management of Innovation at the Department of Economics and Management at the University of Strasbourg.

S. Abraham (Avri) Ravid is the Sy Syms Professor of Finance at Yeshiva University, New York. Previous appointments and visits include Rutgers University, Chicago Booth, Wharton, Cornell, Yale, NYU, Columbia University, UCLA and Haifa University, Israel. He has more than 40 refereed publications, in major journals in finance, economics, management and marketing, as well as book chapters, monographs and other publications. He has won several research awards and his research has been covered by press around the world. His current research interests include corporate finance (mergers, bankruptcy and corporate debt), contracting theory and venture capital as well as the entertainment industry.

Bulat Sanditov is an Associate Professor at Télécom École de Management, Institut Mines-Télécom. He received his PhD in Economics from Maastricht University. He was a postdoctoral fellow at Bocconi University, Maastricht University and a visiting researcher at BETA (Bureau d'Economie Théorique et Appliquée, UMR 7522 – CNRS). His research interests include university–industry linkages, structure and dynamics of knowledge networks, path-dependency in collective learning and the effects of social structure on consumers' behaviour.

Véronique Schaeffer is Associate Professor at the Department of Economics and Management at the University of Strasbourg and a researcher at

BETA (Bureau d'Economie Théorique et Appliquée, UMR 7522 – CNRS). Her research interests are entrepreneurship, economics of science and evaluation.

Eric Schenk is Associate Professor in Management at INSA Graduate School of Science and Technology of Strasbourg and an associate researcher at BETA (Bureau d'Economie Théorique et Appliquée, UMR 7522 – CNRS). His research focuses on innovation management and the questions related to competences. His recent work deals with the management of experts and of creativity.

Laurent Simon, PhD, is Associate Professor in Management at HEC Montréal. His research activities aim at developing a better understanding of the design and management of collaborative, creative and innovative processes in organizations, with a specific focus on techno-creative industries (video games and media), artistic organizations, cultural industries and professional services. He co-chairs Mosaic, a research group dedicated to the study of the management of creation in an innovation society, a partnership with business organizations (Bell, Ubisoft Montreal, National Bank, Hydro-Québec, Cirque du Soleil and more).

Marcus Wagner is Professor at the Julius-Maximilians University of Würzburg where he holds the Chair in Entrepreneurship and Management. He is an associate member of BETA (Bureau d'Economie Théorique et Appliquée, UMR 7522 – CNRS), where he was Marie Curie Fellow from 2006 to 2008. Prior to this, he was Associate Professor at the Chair in Technology and Innovation Management, Munich University of Technology and a senior manager in the semiconductor industry. His research interests are entrepreneurship, innovation management and strategic management.

Murat Yıldızoğlu is Professor of Economics at Bordeaux University. He is a member and former director of the joint research unit GREThA (UMR CNRS 5113). His past positions were at the universities of Strasbourg and Aix-Marseille. He was a member of BETA (Bureau d'Economie Théorique et Appliquée, UMR 7522 – CNRS) from 1989 to 2000 and he continues to be an associated member. His recent research is in evolutionary computational economics and focuses on agents' learning and economic dynamics.

Andrea Zenker is a geographer, specializing in innovation economics and regional innovation. She works as a researcher and project manager at the Fraunhofer Institute for Systems and Innovation Research ISI in Karlsruhe (Germany), Competence Center 'Policy and Regions'. She received her PhD in 2007 from the University of Strasbourg, Faculty for Geography and Development. Andrea Zenker is engaged in contract research in innovation economics, regional innovation systems and strategies. Her recent research and publication activities focus on creativity, knowledge-intensive business services, German-French research cooperation and innovation systems.

Preface

Economics of creativity

One of many debates that can be read about when studying creativity is the definition of what creativity really is. This debate is misplaced for at least two reasons.

First there are a handful of well-known and accepted definitions of what creativity is in art, sociology, economic and managerial literature. Claiming that the field is blurred and not interesting because of that is being blind to what is going on in many prolific and important domains in economics and management (Vesa, 2012).For example, creativity is defined by the 2008 *Creative Economy Report* (UNCTAD, 2008) and several other public and internal agencies. The report states:

> 'Creativity' … refers to the formulation of new ideas and to the application of these ideas to produce original works of art and cultural products, functional creations, scientific inventions and technological innovations. There is thus an economic aspect to creativity, observable in the way it contributes to entrepreneurship, fosters innovation, enhances productivity and promotes economic growth. 'Creativity' is associated with originality, imagination, inspiration, ingenuity and inventiveness. It is an inner characteristic of individuals to be imaginative and to express ideas; associated with knowledge, these ideas are the essence of intellectual capital. The twenty-first century has seen a growing understanding of the interface between creativity, culture and economics, the rationale behind the emerging concept of the 'creative economy'.
>
> (UNCTAD, 2008: 3)

Many important concepts used in strategy, economics and management have much weaker definitions and are not easy to measure. For example, dynamic capabilities in strategy and transaction costs in economics can be defined in many ways and are difficult to measure.

Second, claiming that the measure of 'creativity' is difficult is not the core problem: the real point of interest is to measure the economic consequence of creativity. For this it is necessary to disentangle creativity from other concepts, in order to know precisely what is studied. When you have to introduce 'economics of creativity' the difficulty comes from the term 'economics' not 'creativity'. Galenson (2010: 351) notes:

> it is important to recognize that economists' failure to study individuals has prevented them from understanding the sources of the contributions of the most productive people in our society. Breaking this disciplinary taboo may now allow us not only to understand, but perhaps also to increase, the creativity of these remarkable individuals, and to help others to follow them.

Galenson is probably on the right track but misses a part of the picture. Towse (2010: 464) takes into account that economists cannot hope to fully explain creativity but they can create and test theories that provide considerable empirical evidence.

It might seem that the economic influence of creativity depends on creative people, but this is not the whole picture. Firms, organizations, networks and markets play a major, yet understudied, role in how creative ideas, products and services emerge and are built, distributed and measured. Therefore this book is divided into sections reflecting these important elements of the economics of creativity.

Thierry Burger-Helmchen,
Professor of Management Science,
BETA,
University of Strasbourg
July 2012

References

Galenson, D.W. (2010). 'Understanding Creativity', *Journal of Applied Economics*, **13**(2): 351–62.

Towse, R. (2010). 'Creativity, Copyright and the Creative Industries Paradigm', *Kykos*, **63**(3): 461–78.

UNCTAD (2008). *Creative Economy Report*, Geneva, UNCTAD.

Vesa, P. (2012). 'Entrepreneurial Creativity as Discovery and Exploitation of Business Opportunities', *Entrepreneurship – Creativity and Innovative Business Models*, Thierry Burger-Helmchen (ed.), InTech, Available from: www.intechopen.com/books/entrepreneurship-creativity-and-innovative-business-models/entrepreneurial-creativity-as-discovery-and-exploitation-of-business-opportunities (accessed July 2012).

Acknowledgements

I would like to thank the 38 authors of the 19 articles in this book. They answered with enthusiasm the call I sent out some time ago: the first reason for this initiative was to pay tribute to our deceased colleague Ehud Zuscovitch who had been an inspiration for many of us across generations and whose ideas are still being taken up and debated. Coming from 8 countries and 22 different institutions the contributors collaborated and evaluated each other's works with great professionalism. Hosting a workshop in Strasbourg on 19 and 20 April 2012 was also a very pleasant moment and insightful for all participants whose works were improved during these two days of discussion.

We also acknowledge the support of Ms Anne Leitzgen (SALM), M. Ivan Steyert (Socomec) and M. Frédéric Creplet (Cabinet Voirin) to our activities on creativity.

My thanks to Natalie Tomlinson, Penny Harper and Emily Kindleysides at Routledge for supporting the production of this book.

Last, but not least, my most sincere thanks go to Monique Flasaquier who did an incredible job (as usual) in helping me polish and finalize this book.

Part I

Economics of creativity, ideas and policies

1 Exploring the convergence of the rationales underlying the policies for science, technology and innovation and for creative industries

Laurent Bach

Ehud Zuscovitch was probably as good as an economist as he would have been as a rock 'n' roll performer.

Anonymous observer, late twentieth century

Three characteristics of Ehud's professional activities were particularly striking for me. One was his ability to provide 'elliptic' and simple metaphors to illustrate his arguments; the second was his so often used metaphor (again) of research activities as 'chewing a bone like a dog'; the third was his writing ('German philosopher') style, extremely dense and articulated (one has seldom seen such a density of concepts per paragraph). Sorry then if, because of a poor attempt to mimic Ehud in those respects, the present article uses obscure and seemingly Planet Mars-originating examples, if some analyses are seemingly those of a very clumsy young dog, and if the text is often far more indigestible than any of Ehud's favourite Lebanon cooking specialities.

Introduction

In the recent period, there have been apparent similarities in the arguments put forward by the advocates of the so-called creative sectors and by the scholars and specialists of economics of innovation. System approaches, learning cities and clusters, innovative milieus, creation and diffusion processes, etc. seem to constitute a new commonly shared vocabulary and conceptual arsenal. This chapter proposes a policy-oriented examination of the possible convergence between the economics of Science, Technology and Innovation (STI) and of Cultural and Creative Industries (CCI).

STI policies have been profoundly renewed in the last decades, adding (or partially substituting) heterodox approaches to the old time 'market failures' lines of argument. In the case of CCI, endless debates have been centred on the definition of creative industries, and this has also blurred the frontiers and the scope of policies towards supporting them. But a growing use is made of arguments more oriented towards the importance of such sectors in generating innovation and growth than towards the classical public good and heritage arguments. Thus, is there a real shift in rationales for CCI policies, and is this shift similar to the one observed in the case of STI policies? This chapter then explores the rationales (and the representations that are lying behind those rationales) on which policy for creative industries is or may be built, using a comparison with STI policy which serves us as a reference point.

In the first part, we indicate briefly what we mean by STI policy and by CCI policy, based in particular on the possibility of defining the type of goods involved. In the second part, we present a simple grid of analysis of policy objectives that seems common to STI and CCI policies. These objectives include those of a particular economic nature, the rationales of which will be the subject of the following parts. We will quickly review the case of STI policies. Then we will focus on CCI policy analysis. In particular, we will propose a slight change in the perspective of analysis that could set the ground for devising a more coherent and integrated background for the design of STI and CCI policies. We conclude by identifying some obstacles to the exploration of this research avenue.

The scope of STI and CCI policies: a simple definition

As regards STI policies, we simply refer to 'what government [or other public authorities] have done and could do to promote the production, diffusion and use of scientific and technical knowledge in order to realize national objectives [or objectives defined at other levels of public authorities]' (from Lundvall and Borras, 2005, p. 599, to which we add a generalization to other policy levels). The focus could be put on the production of scientific knowledge, on the advancement and commercialization of sectoral technological knowledge and on the overall innovative performance of the economy.

We simply propose at first instance to label CCI policy what governments or other public authorities do to promote the production, diffusion and consumption of cultural and creative goods and services in order to meet the national objectives or objectives defined at other levels of public authorities.

This raises immediately the issue of the nature of cultural and creative goods, to which the long debate about the scope of arts, cultural industries and creative industries is of course related (see for instance Galloway and Dunlop, 2007; KEA, 2006; O'Connor, 2007; or UNCTAD, 2008, and the corresponding main references dating back to Adorno and Horkheimer's seminal 1940s contribution). We suggest referring here to the following list of criteria put forward by Galloway and Dunlop (2007) to identify cultural (including arts) and creative goods, that provides a quite general interpretative framework:

1) The creativity required to produce them.
2) The possibility of relating IPR (intellectual property rights) to them.
3) The fact that they convey symbolic messages and cultural content (aesthetic, spiritual, social, symbolic, historical, authenticity considerations,

contribution of the activities to community understanding of cultural identity …), which confer on them some cultural value (and even an essential role in communication freedom and democracy support) and thus make them more than simply utilitarian, insofar as they serve in addition some larger communicative purposes.
4) Their functional or user value.
5) The characteristics of their production process.

Using these criteria and briefly summarizing, one can identify three main different types of classification and delineation that have been applied to goods and/or sectors.[1]

One consists in mainly taking into account the first two criteria, and then ranging in the category of 'creative industries' all activities that produce the corresponding goods, without further clear categorization. This is more or less the approach of the Australian and UK DCMS (Department for Culture, Media and Sport) tradition, which has strongly been put in question for having left aside what supposedly makes cultural goods different from other goods.

A second one consists in taking into account the first four criteria, but using some of these to introduce a finer categorization. In this perspective,[2] criteria three and four are used to operate a sort of hierarchy between sectors, sometimes following a so-called 'concentric circle approach' in the vein of initial works by D. Throsby (KEA, 2006; Throsby, 2010; Work Foundation, 2007): the ones dealing with the most 'cultural intensive content' goods are labelled 'cultural industries', and form a sub-set of a larger sector called 'creative industries' dealing with the whole range of 'cultural content intensive goods'. In this approach, it is sometimes mentioned (in Throsby's argument at least), that the cultural content supposedly varies in inverse proportion to the user/functional content, which would preserve the inherent specificities of 'pure arts'. A sub-classification can then be proposed within the cultural industries as well as within the rest of the creative industries.

According to the third approach, the third criterion would be refined and used in a more abrupt way. Roughly speaking, high and serious arts (non-industrially produced and distributed), with the highest cultural content, would be excluded from creative and even cultural industries (or left at the periphery), while all other products respecting the other characteristics would be included, because they are symbolic texts (i.e. objects, artefacts, events, e.g. TV programmes, films, recordings, books, etc. that are meaningful) conveyed by various media, industrially produced, disseminated and consumed. Then, the fifth criterion is also used in combination.

We neglect the third classification, and consider the first two ones without so far differentiating between them, hence the term of creative and cultural industries.

The general objectives of STI and CCI policies

Far be it for us the attempt to be comprehensive in this matter, but it seems that a first parallel can be drawn between the objectives targeted by these two types of policies. Thus, Pavitt and Walker (1976) summarize the objectives of STI policy in four main points:

- to ensure the independence, security and prestige of a nation;
- to promote growth and economic development in particular through improved competitiveness;
- to promote social development in the broad sense (including dimensions such as education, health, employment, quality of life, environment …);
- to favour progress of scientific and technical knowledge as such, and their diffusion through time and space.

It is tempting to find the correspondence between this typology and those proposed for example by Mulcahy (2006) and Pratt (2005) according to which the objectives of CCI policy are organized along the following lines (slightly reorganized for sake of clarity of presentation but without changing the meaning):

- culture as glorification – 'ideological-political: nationalist' for Pratt;
- cultural utilitarianism, economic impact of culture, attractiveness (creative cities) – 'economic' for Pratt;
- democratization of culture (aesthetic enlightenment, enhanced dignity, educational development, general citizenry) and cultural democracy (equal opportunity for citizens to be culturally active) – 'ideological-political: humanist, aesthetic' for Pratt;
- culture per se.

The authors therefore adopt a broad view, and integrate a particular economic role of CCI policies. We find some of these overall objectives, although ordered differently, in the presentation that Throsby (2010) offers of 'arts and culture' objectives of cultural policy:

- defining cultural identity;
- producing and consuming creative arts, by favouring excellence and quality, by promoting innovation and by securing access;
- celebrating diversity;
- ensuring continuity, through preservation of heritage and education.

Besides, Throsby precisely considers economic performance and development as an 'economic' objective of CCI policy, as well as the 'classic' objectives of economic policy (efficiency, equity, growth, full employment, price stability, external balance).[3]

There would therefore be economic aims to STI policies and CCI policies. These are the rationales associated with these objectives we will now discuss.

The renewal of economic rationales for STI policy

The evolution of rationales put forward by economists for justifying STI policy has been extensively analysed (Bach and Matt, 2005; Lundvall and Borras, 2005; Metcalfe and Georghiou, 1998; Steinmueller, 2010). At the risk of over-simplification, it can be sketched using a distinction between the standard or neo-classic approach and the 'evolutionary structuralist' approach (to use the expression of Lundvall and Borras). Let us very briefly summarize the main points.

The standard approach to STI policy rationales

The standard analysis of STI policy rationales is based on the central argument of the supposed existence of market failures, related to (i) the uncertainty inherent to R&D and innovation, (ii) the existence of indivisibilities in the production of knowledge and (iii) the properties of public good (non-rivalry and non-exclusiveness) of scientific and technological knowledge, resulting in particular phenomena of externalities (knowledge externalities, but also strategic externalities and network externalities). Against the objectives listed above, the main articulation is as follows. Private agents are not encouraged enough to undertake R&D and innovation likely to generate the benefits outlined above, and then governments are called upon to intervene. Especially, it is not just because a particular good raises a public good problem that its production and distribution are supported, but because this particular good is useful in one way or another to society.

Basically, State intervention aims at reducing uncertainty, bearing part of or all the risks and/or costs, or providing compensation to private investment in these activities. The basic forms of intervention (from which multiple concrete policy instruments can be derived) then consist in increasing the information available to actors, in playing the role of (*ex ante* funding, *ex post* reward) the supply and / or the demand, in facilitating cooperation (between users and producers, between producers) and in establishing property rights on the outputs.

In the STI policy area, it is common to emphasize the close proximity between this so-called standard approach and two pillars underpinning the policy rationales: knowledge assimilated to a public good (or even as an informational good) and a linear model of innovation.

A great consistency unifies these elements of analysis. Due to specialization and differentiation of roles, in classical representations, upstream of the model essentially resides a world governed by its own laws, in which the actor is essentially individual: the researcher, working in the world of open science, and gifted with a creative genius. Specialization and differentiation also correspond to the intensity of the problems of market failure and therefore to the intensity and shape of state interventions. Simply put, this intensity would be higher upstream than downstream. Correspondingly, the frontiers between science policy, technology policy and innovation policy (including diffusion) are then quite straightforward in these approaches, as they are articulated along the sequences of the model.

Lastly, for some major investments in STI such as large research infrastructures, the argument of the economic impact through the multiplier effect of spending is also sometimes presented (though rather advanced as an argument of 'second tier' coming after an argument about the nature of to-be-supported STI activities).

The evolutionist structuralist approach to STI policy rationales

The rationales for STI policies have been largely completed and renewed by the contributions of the economics of innovation and the different schools of thought that have irrigated this branch of economics during the last three or four decades.[4] As a noticeable consequence, the frontiers between science, technology and innovation policies tend to be blurred. Three major influences can be distinguished, which share many common roots but are each focused on one or more specific dimensions.

The evolutionary approach has led to the identification of selection failures, advising the policy maker to avoid (or depart from) situations of lock-in by maintaining sufficient diversity, to ensure the relevance of selection methods or to foster paradigm transition. Systems and networks approaches, enabling the identification of 'system failures', laid the foundation of (or allowed the re-thinking of) a range of actions to foster proximity (especially locally), to enrich, complete and connect networks of all kinds, and to promote consistency and coordination of objectives and actions within the networks and over time. Approaches based on the economics of knowledge, focusing on the processes of creation, codification, capitalization, sharing of and access to different forms of knowledge are only beginning to irrigate the field(s) of policy design, for example through the roles of knowledge communities and of the collective nature of cognitive processes.

About economic rationales for policy for cultural and creative industries

It is clear that the paradigms on which policies are based are not a priori structured in the same manner as the STI policy, and especially are not organized exactly like an opposition (or separation of fields of relevance) between a standard vision and an alternative vision. Still, let us try to find parallels and to follow a logical analysis in order to identify a few common points.

A standard approach to CCI policy rationales?

First, there are of course the arguments of existence of market failures in the CCI field. Ridge *et al.* (2007) explicitly detailed e.g. externalities, lack of information, quality of choice, coordination failure, public goods, transaction costs. Artistic, cultural goods or services would have characteristics of public goods (or sometimes of 'merit goods' following Musgrave typology), which are also reinforced by the wider distribution of digital technologies in the process (and media) of creation and dissemination of CCI goods. The low cost of reproduction of certain goods (again reinforced by digitization) further accentuates the lack of incentive for these goods. There is a risk of sub-allocation of resources to these activities and therefore sub-optimal production, as compared to the supposed 'needs' of the society, including the long-term usefulness for future generations (heritage) who are also integrated into the non-economic arguments.

Other aspects of market failure are also often put forward, even if their relevance greatly varies from one CCI activity to the other. The production would be carried out in an environment largely uncertain: in terms of outcome and process of production and as regards demand response in the case of production of a new good. The demand characteristics are also at play here ('nobody knows' principle, no saturation and rational addiction, infinite variety, A list/B list and long-tail phenomenon, network externalities, fashion, mimetism, often collective consumption, etc., see for instance Caves, 2002). Production would also sometimes exhibit indivisibilities (producing a blockbuster film, producing an opera …).

A particular aspect, the famous 'cost disease' introduced by Baumol and Bowen (1966) has long been a pillar of the rationale for state intervention, although related to a quite limited set of cultural goods, i.e. live performance. According to this argument, live performance cannot achieve efficiency gain in time or labour or any resources, and because of labour cost increase, this live performance becomes more and more inefficient from an economic standpoint, thus requires subsidies (needs to be subsidized).

In all these cases, the relationship between the CCI and the rest of the economy is the one sketched by Potts and Cunningham (2008) in their 'welfare' model, in which the economy drives creative industries through transfers of resources. In other terms, they consume more resources than they produce because of 'cost disease', or they do not produce because of other aspects. Producing such goods is welfare enhancing, and may even have a broader cultural/heritage impact, but it raises market failures. Thus welfare subsidies as well as the whole range of neo-classical analysis-based policy tools are required.

This state intervention is ultimately supposed to be largely made on the same principles, although the design of CCI policies only partly pushes the logic of intervention in the same directions as did the design of STI policies (Table 1.1).

Table 1.1 Basic principles of standard approaches in STI policy and corresponding examples in CCI policy (in italics are possible interpretations of existing or past actions)

STI policy	*CCI policy*
Increasing information	Information on cultural supply, education of demand, definition of professional standards governing the arts market
Supply side: producing Supply side: funding	Public theatre company or TV channels, *creators, musicians or authors of the King or any local authority (Molière, Lully and Bach)?* 'Residences' or scholarships, repayable advances (audiovisual work, movies)
Demand side: ordering Demand side: funding	Public museums purchases Tax deductions on the artistic patronage, discount rate access to exhibitions
Fostering cooperation	*'Open Days' inviting the public to visit artists' workshops as cooperation producer-user? Creation of live performances combining musicians, writers and visual artists as cooperation between 'competitors'?*
Establishing IPR	Copyrights, moral rights, fair use, 'private copying', neighbouring rights, resale rights, collective management of rights

Let us turn now to the two 'pillars' associated with the standard approach. First, the public good nature of CCI goods can be argued from a slightly different perspective, and all the so-called cultural goods do not exhibit the same public good properties, if they ever do:

- Some are partly rival: e.g. live performance or a great meal designed and prepared by a three-star cook.
- Technology is largely influencing the properties of the cultural goods: digitization is the topical case for a broad range of cultural goods.
- Some goods present different levels of rivalry and exclusivism while they are based on the same original piece of work: music again or theatre are perfect examples (live performance, recorded live performance, studio recording, music score or theatre play …).
- In addition, different sources of market failures apply differently from one type of CCI good to the other, and even from one distribution stage to the other of a given CCI good. To put it in an even more complex way, some CCI goods (video games, movies, some live performances …) are made from the combination of other basic CCI goods (the latter being inputs) which do not exhibit the same public good properties.

Second, it can be argued that, to some extent, the role of the linear model could find its equivalent in the value chain frequently used to analyse the functioning of the CCI. For example, Throsby (2010) very simply breaks it down into Production–Distribution–Consumption, and indicates clearly that it can serve as a guide in identifying points of state intervention and evaluation of their impact. Pratt (2007) offers a somewhat more detailed analysis of the 'upstream' (Creation–Making–Dissemination–Exhibition–Reception–Consumption). From our perspective, we would be inclined to break the sequence into Creation, Publication and Distribution, Dissemination and, finally, Consumption; these steps would then play the role of invention (or the production of scientific knowledge and / or technology), innovation (that is to say, the first provision to 'clients' or 'users'), distribution and consumption.

As in the STI world, there is often the individual artist as the central figure upstream of the model: he evolves in the arts world often compared with an 'underground', and he is also of course gifted with a creative genius. And as in STI, these individuals can be grouped by 'schools', but it is individuals who acquire recognition and posterity.

In addition to the market failures issue, a second line of argument inspired by a standard economic analysis is to justify the CCI policy based on the multiplier effect through spending mechanisms that some activities may have on other activities. The organization of a festival, the presence of concert halls or theatre, museum or architectural elements, the activity of an important core business of a particular industry (film, edition …) generate cascading purchases, thus with an essentially local economic impact. Hence there is also an argument for government support if the economic impacts are considered as significant. As in the case of STI, there is nothing in this matter that reflects a truly distinctive nature of CCI (with the possible exception of the structure of expenditure that may be specific). As underlined by Potts and Cunningham (2008) when proposing to refer to their 'competition' model for such an approach, the creative industries are just another industry, requiring standard industry policy, but not necessarily something specific. However, the high level of this effect (related to issues of volume or faster development) could argue in favour of a preferential involvement in CCI.

A renewal of economic rationales of CCI policies

Economic analysis of CCI has already borrowed heavily from the evolutionary school(s) to explain the dynamics of emergence and evolution of artistic movements or specific industrial configurations. The music industry offers significant examples of studies relying on such approaches to investigate the evolution of genres, business models or the entire industry (Geels, 2007; Lampell *et al.*, 2008; Tschmuck, 2006). But even if the policies have to address the issue of paradigm shifts (digitization is a particularly relevant case), they do not deal specifically with the tools and concepts of evolutionary analysis. Similarly, the maintenance of diversity is of course at the heart of a number of policies, but it is the cultural diversity in itself (past, present and future) that is valued from the perspective of heritage and multicultural respect, not in order to maintain the adaptability and then to avoid situations of lock-ins that would be negative. In other words, any lock-in (understood as a cultural form) appears as negative per se. And a priori 'mutations' are encouraged in the same perspective.

As in the case of STI, one of the main changes in the arguments used to justify the CCI policies has been the incorporation of concepts from the systemic and network approaches, in particular applied to a local level. They are used in the case of support and development of creative cities and other cultural clusters: a significant presence of activities and creative people supports the exchanges and the dynamics of a given geographical area, and is conducive to economic development that strengthens and regenerates by pulling other activities and creative people. This avenue has been widely explored at least since the work of Landry (2000) and Florida (2002).

The inclusion of these approaches in the rationales of CCI policy was actually previously achieved as part of what was analysed by many as a crucial policy shift from traditional cultural policies to creative industries operated in the UK in the mid-1990s, represented especially by the work of the DCMS (Hesmondhalgh, 2007; O'Connor, 2007; Pratt, 2007). According to these authors, there was a change from sectoral boundaries (or rather an extension of the traditional cultural sector, as we have seen above) together with a modification of the rationales. Political discourse tends to ignore the traditional approach partly (funding for the creation and implementation of conservation and cultural heritage, supposedly combined with an elitist vision and an overburdened tax-payer), to take over the arguments very close to those put forward in defence of state intervention in innovation. It is therefore based on the one hand on the existence of spillover effects associated with the faster growth rate of CCI relative to other industries (differential observable when the perimeter of selected sectors is itself modified in a way that one may qualify as ad hoc and opportunistic), and on the other hand on these systemic effects at the local level. Furthermore, the role of actors and factors believed to be determinants in the context of innovation policies (SMEs, entrepreneurship, etc.) is largely taken over by the new policy for CCI. On this ground the use of public funding is justified, as well as the highlighting of the role of IPR to protect all stakeholders involved in the processes.

This role of creative industries seems to us perfectly in line with the 'growth' model proposed by Potts and Cunningham (2008), in which the creative industries drive the economy through higher rates of growth (as coal and steel did in the late 1900s, for instance), and thanks to their role in adopting and absorbing new ideas and technologies.

It therefore appears that CCI is increasingly treated as industries subject to STI policy, and that CCI policy would therefore be increasingly confused with STI policies. This position is also reinforced by an assumption that the fast-growing CCI would also be inherently innovative industries, but whose activities would be under-estimated or ignored in Frascati/Oslo Manuals-based surveys and studies, like those service activities to which they would then be considered very close (Cunningham, 2008; Potts, 2009).

Towards a reappraisal of the convergence of rationales?

It seems to us that achieving such a shift, which can be attractive in terms of political marketing, leads to a weakening of the coherence of rationales.

First, if one wishes to keep and strengthen the use of IPR for the sake of protecting economic income (and not so much protecting moral rights), it means that one assumes that one is dealing with a matter of public good. Why then deliberately weaken the weight of the other tools dealing with the same matter?

A second and related point is, what is the limit of the scope of CCI if one no longer takes into consideration something that is symbolic or cultural content? And then on what basis should the 'heritage' aspect be kept in the fields of CCI policy? This issue indicates, as Menger (2010, p. 286, my translation) suggests, a growing dichotomy between two types of cultural policy, one type dedicated to the heritage, the other type dedicated to arts and practices and to creative industry 'now placed under the influence of technological innovations and their power of creative destruction'.

Third, the mechanisms by which creative goods become tradable goods, and especially the possible specificity of creative industries in this respect does not seem analysed in a fully articulated way with the changing CCI paradigm. However, some work on the anatomy of the Creative Cities trace routes, showing how, in localized areas, these phenomena involve an ecology of individuals, firms and communities interacting at different levels, and coordinated by different mechanisms (Cohendet *et al.*, 2010). These studies then touch upon what is related to our last point: the confrontation with the STI policy rationales incline us to think that in the field of CCI policy we lack something like the equivalent of the consideration of knowledge and knowledge processes in the analytic construction in the field of STI policy. Indeed, it is this dimension that somehow ensured consistency of the standard approach building as well as of the evolutionary structuralist building.

One possible way might be to reinstate one of the ideas underlying the typology of CCI explicitly proposed by authors such as Throsby: they consider that the arts are at the heart of the creative industries and gradually as one moves away from the heart, the part value of cultural goods (which mirror the creativity in content and symbolic message content) is reduced in favour of strictly economic value.

The content in symbolic and cultural value of the good would be more or less pronounced in a multitude of goods, and even of an increasing share of goods if we follow Potts (2009) when he emphasizes the classic reference made at least since the 1990s to a 'culturalization' and 'aesthetization' of economic life. Then, some goods (and activities that develop them) would have a more or less important artistic/cultural content, just as some goods (and sectors that have developed them) would have a more or less intense scientific content.

Following this argument would therefore lead to stopping trying to isolate the creative industries without actually proposing an indisputable classification, and to opposing other sectors that are more or less based on STI activities. It comes to fully take into account the possible continuum of configurations characterized by the content of scientific and technical knowledge and by the cultural and symbolic content of the good produced, and the possible proximity of some of the processes that give these goods such content. Exploring this direction further could lead to a homogeneous view between STI and CCI, in which:

- core creative arts (basic research) are at the heart of the creative (standard) industries: they produce 'creative bricks' (fundamental scientific knowledge);
- elements of 'cultural/artistic content', i.e. works, forms, styles, attitudes, ways of understanding the world or events (elements of 'scientific/technical content', i.e. knowledge) have a value on their own …
- … but are also combined, embodied in various proportions in goods, services, devices and artefacts, together with other types of knowledge and managerial/ organisational skills;
- the generation and processing of new creative bricks (new knowledge) are to be analysed using an interactive model exhibiting the same properties as the 'interactive model of innovation', including:
    - possibility of generation of some forms of cultural/artistic bricks (knowledge) at any location in the model;
    - feed-back and overlapping phases;
    - possibility of creating new goods without requiring new artistic creation (new fundamental research) by tapping into an idiosyncratic cultural (knowledge) base and recombining elements;
    - collective (more often than purely 'genius individual'-based) processes of creation and processing, with a key role played by various creative (knowledge) communities.

From a policy perspective, this made it possible for instance to reinterpret, (such) as in the STI field, the role of IPR in some creative industries (video games and music which combine in varying degrees many 'creative elements') not only as a balance between ownership and distribution, but also as a guarantor of the possibility of creating in the future and as a tool of coordination between actors (Bach *et al.*, 2010).

This representation seems partly in line with the proposals of Potts, on products ('creative industries, like science and technology, also produce outputs that are inputs in the innovation process' (Potts, 2009, p. 139)), and those of Cunningham (2008) rather on processes (whereby creative industries 'is a coordinating mechanism, or – in John Hartley's terms – an enabling social technology for distributed innovation' (Cunningham, 2008, p. 3)), or more generally with the 'innovation' model of these two authors (Potts and Cunningham, 2008) in which the creative industries may evolve and operate in the economy through their role as an element in the innovation system of the whole economy. As stated by the authors (p. 10): 'Specifically, this is the same model as proposed for the effect of science, education and technology in the national system of innovation approach.'

However, it may be possible to push the argument one step further, and consider the possibility of abolishing the distinction between the two interactive models (the one of the creative industries and the one of the standard industries), by considering the increasing share of sectors for which the two interactive models are largely, not only connected, but may also be more and more merging; digitization and the Internet being crucial factors here. Those sectors are using, combining, re-organizing both knowledge and creative bricks and including them in various 'proportions' into marketable products and services.

This would provide a homogeneous ground on which STI and CCI policies design could gain in coherence and deploy, possibly in an integrated way through generic or meta-policies, a range of tools that would be based on largely similar concepts and approaches. It may also open the way for an extensive view on creativity, playing somewhere between cultural/artistic bricks (knowledge) and innovation, and thus helping re-examine the possibility of creativity-focused policies.

Conclusion

This chapter is an attempt to compare the STI policy rationales and CCI policy rationales; we have tried to reassemble the puzzle of what might constitute a common platform, on which, using similar concepts or at least some neighbouring ones, it would be possible to identify numbers of failures and propose some remedies.

However, reasoning by analogy has reached its limits at least for now because of four important differences, which determine the field and/or the direction of actions in the policy framework.

The first relates to the cumulative aspect, and the notion of progress that can be attached to it. Concerning the STI world, even if one can call into question the ethical dimension of scientific progress, there is progress in the sense that on the one hand new knowledge challenges, makes obsolete or completes 'old' knowledge, while opening channels to guide the production of future knowledge, and on the other hand this accumulation of knowledge improves our understanding of the world and its workings. In the CCI field, the second of these two aspects relates to diversification rather than improvement. If we are to understand, appropriate, develop the art of today, the history of arts provides us with some keys, maybe not essential but illuminating certain aspects. If we are to understand, appropriate, make changes in the state of knowledge today, it is not necessarily required to know the theories advanced by the great Bolos of Mendes in the first or second century BC.

The second difference refers to the direction of scientific and technical progress. In science and technology, beyond a preoccupation with short-term response to market needs, the need is increasingly emphasized to address global society (environmental issues, ageing, etc.). As noted above, maintaining the diversity of potential scientific and technical answers (or instead focusing on a particular route) is from this perspective a policy strategy, but not necessarily a policy objective in itself. In the field of CCI, the answer is less clear as regards creation: Should the State focus on a particular route (i.e. a specific form of art, expression, etc.) or should it support diversity for the sake of diversity? A common response may be to find out whether we reason in a completely uncertain future in the sense of Knight. So, the maintenance of diversity is in itself a goal and a strategy for STI as for CCI.

The third difference concerns a practice sometimes called amateur practice, and this is not a trivial and marginal issue. If education and dissemination contribute to the consumption as well as to the preservation and the development of potential future creation, they also allow such practices. Can we equate this with another type of consumption only? Or to a non-market re-creation? It also helps in the transmission itself and more generally in the functioning of communities which are probably the reservoir and the fabric of the creators of tomorrow. Furthermore the transition between amateur performance and economic value is also sometimes part of the valorization process (the arts markets in the central squares of our cities, for example). Is there any equivalent in the sphere of STI? Sunday DIY and Lépine contest? The pleasure of playing with the 'Mr Chemistry set' of one's young nephew? Competition between engineering schools for the most energy efficient car? Or participation in crowd-sourcing calls launched by big companies on social networks?

Finally, a fundamental issue in STI was whether or not to codify knowledge and make it explicit and independent of the context that generated it and of the context that uses it. Moreover the issue of the incentive of agents to process knowledge to this end was a central theme. In a way, it is almost the entire edifice of the standard approach on the one hand and of the heterodox approach on the other which are based on different conceptions on the subject, as these conceptions determine the heterogeneity of agents over time (path-dependence) and space, the conditions for sharing and dissemination of knowledge, etc. As regards CCI, we mentioned a few arguments in the last part of the chapter. But it is probably on this issue, and on the corresponding one of knowledge processes, that our limited culture in the analysis of creative processes so far hinders the reflection proposed in this paper.

Notes

1 Among the numerous other typologies, one could also mention the WIPO approach focusing on the IPR aspect only, or the original UNESCO one that focused mainly on the second and third aspects.
2 UNCTAD (2008) also adopts the first four criteria, but proposes another way of categorization between heritage, arts, media and functional creation that we will not detail here.
3 Other less general classifications could fit in the one presented in the text; e.g. Braun and Lavanga (2007) distinguish between cultural objectives (diversity, quality and distribution) and economic objectives (innovation, entrepreneurship, export, investment, clustering and economic growth).
4 Obviously, the opposition here is in part artificially summarized, without mentioning the issue of 'policy mix', at the level of the rationales as well as at the one of the tools actually put in place.

References

Bach, L. and Matt, M. (2005) 'From economic foundations to S&T policy tools: a comparative analysis of the dominant paradigms', in M. Matt and P. Llerena (eds) *Innovation Policy in a Knowledge Based Economy: Theories and Practises*, Berlin: Springer Verlag, pp. 17–45.

Bach, L., Cohendet, P., Penin, J. and Simon, L. (2010) 'Creative industries and the IPR dilemma between appropriation and creation: some insights from the videogame and music industries', *Management International*, 14(3): 59–72.

Baumol, J. B. and Bowen, W. G. (1966) *Performing Arts: The Economic Dilemma*, New York: The Twentieth Century Fund.

Braun, E. and Lavanga, M. (2007) *An International Comparative Quick Scan of National Policies for Creative Industries*, Report of the EURICUR-Erasmus University of Rotterdam for the Ministry of Education, Culture and Science of the Netherlands.

Caves, R. E. (2002) *Creative Industries: Contracts Between Arts and Commerce*, Cambridge, MA: Harvard University Press.

Cohendet, P., Grandadam, D. and Simon, L. (2010) 'The anatomy of the creative city', *Industry & Innovation*, 17(1): 91–111.

Cunningham, S. D. (2008) 'The parallel universes of innovation policy and creative industries policy, and the growing convergence between them, or "so you want us to stop trying to find the cure for cancer and start funding reality TV?" ', *Cultural Science*, 1(1), available at http://eprints.qut.edu.au/18123/1/c18123.pdf (last accessed 13 September 2012).

Florida, R. (2002) *The Rise of the Creative Class*, New York: Basic Books.

Galloway, S. and Dunlop, S. (2007) 'A critique of definitions of the cultural and creative industries in public policy', *International Journal of Cultural Policy*, 13(1): 17–31.

Geels, F. W. (2007) 'Analysing the breakthrough of rock'n'roll (1930–1970) – multi-regime interaction and reconfiguration in the multi-level perspective', *Technological Forecasting & Social Change*, 74: 1411–31.

Hesmondhalgh, D. (2007) *The Cultural Industries*, second edition, London: Sage.

Kea (2006) 'The economy of culture in Europe', Study prepared for the European Commission (DG for Education and Culture), available at www.keanet.eu (last accessed 13 September 2012).

Lampel, J., Bhalla, A. and Pushkar J. (2008) 'Intellectual property rights and industry evolution: the case of the recorded music industry', Working Paper DIME, No. 51, April 2008, available at www.dime-eu.org/working-papers/wp14?page=4 (last accessed 13 September 2012).

Landry, C. (2000) *The Creative City: A Toolkit for Urban Innovators*, Earthscan: London.

Lundvall, B.-Å. and Borras, S. (2005) 'Science, technology, and innovation policy', in J. Fagerberg, D. C. Mowery and R. R. Nelson (eds) *The Oxford Handbook of Innovation*, Oxford: Oxford University Press, pp. 599–631.

Menger, P.-M. (2010) 'Les politiques culturelles en Europe: modèles et évolutions', in P. Poirrier (ed.) *Politiques et pratiques de la culture, La Documentation Française*, Paris: Coll. Les Notices, pp. 277–86.

Metcalfe, J. and Georghiou, L. (1998) 'Equilibrium and evolutionary foundations of technology policy', *OECD STI Review* 22: 75–100.

Mulcahy, K. V. (2006) 'Cultural policy: definitions and theoretical approaches', *The Journal of Arts Management, Law, and Society*, 35(4): 319–30.

Muller, E., Baier, E., Blanrue, F. *et al.* (2009) *Regards croisés sur la culture: l'innovation et la créativité en Alsace*, Rapport pour la Région Alsace, Juin.

O'Connor, J. (2007) *The Cultural and Creative Industries: A Review of the Literature*, Creative Partnership Series. London: Art Council England.

Pavitt, K. and Walker, W. (1976) 'Government policies towards industrial innovation: a review', *Research Policy*, 5(1): 11–97.

Potts, J. (2009) 'Creative industry and innovation policy: introduction', *Innovation Management, Policy and Practice*, 11(2): 138–47.

Potts, J. and Cunningham, S. (2008) 'Four models of the creative industries', *International Journal of Cultural Policy*, 14(3): 233–48.

Pratt, A. C. (2005) 'Cultural industries and public policy', *International Journal of Cultural Policy*, 11(1): 31–44.

Pratt, A. C. (2007) 'The state of the cultural economy: the rise of the cultural economy and the challenges to cultural policy making', LSE Research Online, available at http://eprints.lse.ac.uk/20707/ (last accessed 13 September 2012).

Pratt, A. C. (2008) 'Creative cities: the cultural industries and the creative class', *Geografiska Annaler: Series B, Human Geography*, 90(2): 107–17.

Ridge, M., O' Flaherty, D., Caldwell-Nichols, A., Bradley, R. and Howell, C. (2007) *A Framework for Evaluating Cultural Policy Investment*, Report prepared by Frontier Economics Ltd for DCMS (Department for Culture, Media and Sport, UK).

Steinmueller, W. E. (2010) 'Economics of technology policy', in B. H. Hall and N. Rosenberg (eds) *Handbook of the Economics of Innovation*, Vol. II, Amsterdam: North-Holland, pp. 1181–1218.

Tschmuck, P. (2006) *Creativity and Innovation in the Music Industry*, Dordrecht: Springer.

Throsby, D. (2010) *The Economics of Cultural Policy*, New York: Cambridge University Press.

UNCTAD (2008) 'Creative economy report 2008 – The challenge of assessing the creative economy: towards informed policy-making', UNCTAD/DITC/2008/2. ISBN 978-0-9816619-0-2, Paris.

Work Foundation (2007) *Staying Ahead: The Economic Performance of the UK's Creative Industries*. London: The Work Foundation.

2 A strategic management of forgetting[1]

Gilles Lambert

> We can treat memory as an enemy. The rules of consistency and rationality require a technology of memory. For most purposes, good memories make good choices. But the ability to forget, or overlook, is also useful. If I do not know what I did yesterday or what other people in the organization are doing today, I can act within the system of reason and still do things that are foolish.
>
> March (1988: 263)

Even if it is impossible to forget Ehud Zuscovitch for those who had near relations with him, I am sure that he agreed with the above mentioned idea which was discussed with him now two decades ago: the necessity to forget. Evidently, we do not mean forgetting as an abyssal slip of memory of organizations but as an assumed effort of *destruction-creation*, a beloved concept to the indefatigable and obsessional researcher that Ehud was. Considering forgetting as a strategic fad of management is equivalent to accepting a position of researcher made of doubt and humility more than monolithic certainty considering that the vocation of Science is the search for the truth. This was the strength of Ehud, as it was also true for the researcher he came across in Strasbourg in 1992 during his long quest, J.G. March. Interviewed on the reasons for choosing such a 'harebrained' image for the Garbage Can model of decision, March justified this choice in the following terms:

> I think one should be very serious about his work but not serious about himself and so serious about your ideas that you spend the rest of your life defending them. And one of the ways of protecting yourself against that is to give silly labels to the ideas so that you don't glorify them beyond what they are … This was a way of giving it a label that made it susceptible to abandonment.

However, this model of rationality has proved robust since, after more than 20 years, it continues to inspire empirical and theoretical developments in management science. These were some learning premises that I learnt during my association with both of them. According to me, they contain the main rudiments for the functioning of organizations: the reason for a sustained memory and the unreasonableness of forgetting. In his strenuous work on routines on the one hand, and, on the other hand, in his constant call to disobedience, to subversion vis-à-vis established rules, Ehud Zuscovitch was touching, maybe unconsciously, two central dimensions in our organizations. Indeed, two constructs seem to be central in the work of this researcher: the phenomenon of routine creation and the meso-economic analysis of innovation providing a wide influence of social aspects in the process of innovation such as the rule of actors and management. In that sense, Ehud Zuscovitch was a non-conventional economist including early strong dimensions of management in the theoretical developments.

Born into an Ashkenazi family, Ehud Zuscovitch had been forced to forget the mark of the destructive foolishness of Nazism associated with an inhuman administration of life. In such situations, it became essential to forget, to escape, to be able to build again. For Einstein all theories are necessarily wrong, but their importance is to be useful at some moments. We have to challenge a theory if a better one appears. This non-dogmatic flexibility fits perfectly with the attitude of Ehud Zuscovitch (1984) as a researcher. Around this parable of forgetfulness, we propose to develop a prolongation of this academic posture into the field of change in organizations.

The literature on the phenomena of forgetting reveals two important phases. The concept first appears as belonging to insufficient memory that falters and loses information. Feigenbaum and Simon (1961) provide another mechanism of forgetting from a physical phenomenon of learning that takes place without loss of information. Forgetting, concomitant with learning, for these authors comes from the loss of access to causal links (associations) in a context of growth in the number of associations to remember. For these scholars, learning and forgetting are related; the second concept is finally consubstantial with the first. Our problem is to introduce an extra dimension to this phenomenon: the will of individuals. It is expressed in the organizational phenomenon through managerial strategies seeking to introduce change without disrupting the bureaucratic balance obtained. Based on business cases such questions raise in learning and forgetting, we propose to explore this new dimension. The managerial logic that guides organizations requires that we look at forgetting on the collective level. We will see how the ability to change, a factor performance of companies, depends strongly on this managerial logic enforced by March and that implies preparing strategically to forget concepts, procedures, routines in case more powerful elements of knowledge occur.

Status of forgetting in the managerial literature

In his summary of the phenomena of learning and unlearning in organizations, Hedberg (1981) advanced two forms of unlearning. The first corresponds to an abrupt phenomenon which is simultaneous unlearning relearning. According to Postman and Underwood (1973), this example shows that this mechanism is at work when a causal outcome of the observation (B always follows A) is called into question after a new contradictory observation (C follows always A). The first causality is suddenly forgotten in a process of immediate replacement. In this perspective, unlearning is at the same time the action of unlearning. Forgetting and learning are the manifestations of the same phenomenon. But one can also consider, as Simon did (1971), that learning is a gradual phenomenon. Unlearning is separated in its conduct of relearning. No longer are these two sides of the same coin, but two separate phenomena that take place gradually by dint of contradictory empirical observations. This design is compatible with many theories linking memory with a dynamic allocation process based on the observation frequency (Feigenbaum, 1970).

This phenomenon of forgetting, or rather the so-called principles and rules of operation in organizations, seem to be located at the heart of the problem of learning in business structures. Indeed, the analogy with natural selection models of evolutionary inspiration, especially with developments in the context of population ecology proposed by Hannan and Freeman (1977), emphasizes the idea that organizational inertia of discontinuous processes of adaptation explains the life and death of social constructions. However, understanding these organizational delays requires explaining the internal mechanisms of adaptation to the model that population ecology lacks, and the phenomena of inertia can then be analysed as a time for the completion of effects of learning and experience specific to the organization (Burgelman, 1991).

Individuals need memory to make rational decisions (Simon, 1997). Similarity, familiarization and meaningfulness impact strongly the phenomenon of learning (Feigenbaum and Simon, 1964). Memory is the repository of routines and thus of premises which influence decisions and decision-making, making them consistent with purpose and environment. A lack of memory leads to inconsistency, disorder and irrational behaviour (Simon, 1997). What remains in a memory is critical for an organization's behaviour. For example, that a memory element is active or dormant can make a big difference. When an element is dormant, we can talk of forgetting. Research on these issues suggests that forgetting is never final (De Holan and Phillips, 2004). Individuals forget only temporarily, subject to circumstances and to contingencies to which they, or the organization, are subjected.

Forgetting routines has even more significant implications for an organization's behaviour. First, routines are closely related to learning (Ancori, 1992; Cyert and March, 1992). In changing routines, organizations build decision premises (Simon, 1997), and in so doing try to introduce adjustments and modifications suggested by experience. This means that an individual makes decisions in response to both new and old impulses. Forgetting affects both. According to theory, unlearning routines takes time and resources sometimes in greater amounts than learning (Hedberg, 1981). Impulses coming from old routines tend to dominate because recalling them consumes less energy. In fact, it is hard to forget old routines. In contrast, emerging routine impulses are imperfectly understood and, to become operational, require a greater level of attention and energy. It is easy to forget emerging routines.

Learning is also experimenting. Routine impulses are objects or rules proposed to actors' judgement and strategy. To use Feldman and Pentland's vernacular (2003), they are 'ostensible', a little as structures would be. What actors do with ostensible routines belongs to the creative part of behaviour, and takes into account both the manifest content of routines, and the meaning that they acquire in action. As a result, they are translated into what Feldman and Pentland call 'performative routines'. Forgetting is considerably affected by the energy required to recall and by the information, knowledge and interests of the actors involved (Simon, 1997).

When top managers undertake a change, they believe or hope that they have control over cause and effect relationships. They cannot stand the idea that forgetting may introduce uncertainties in the decision-making process and thus in the achievement of change. They are tempted to do whatever it takes to keep it under control (De Holan and Phillips, 2004). In this article, we intend to explore the management of forgetting. To avoid falling prey to the complexities of physical and psychological mechanisms of forgetting, we adopt a deliberately 'Simonian' decision-making approach. We intend first to describe the process by which decisions are influenced, referring back to March (1991), Simon (1997) and their co-researchers at Carnegie-Mellon. Then we discuss forgetting to clarify the concept and appreciate its effect on decision-making. This will lead us to a process theory of forgetting, which will be illustrated and made more precise using three detailed empirical cases. The first case describes a balanced management of forgetting to allow decision-making coherent both in space and in time. The second case reports on a situation where the management of forgetting leads to periodic cleansing of the memory, obliging the organization to keep reinventing its strategic routines. The third case emphasizes the rule of routines as mechanisms of forgetting but also creation of new learning. The last case focuses on the renewal phenomenon of knowledge at stake inside the routine itself. These cases illustrate the theory proposed and clarify it further by discussing how forgetting can be managed. In conclusion, implications and recommendations for research and practice of strategic management are proposed.

Another dynamic of forgetting is in action in organizations based on the bureaucratic models and compared with the logic of change. Instead of mothballing some organizational efforts through the routinization of procedures learned at length, this dynamic is to loosen some more or less enduring coupling mechanisms inherent in the bureaucratic structure in order to respond to a situation not routinely. However, we believe that the demonstration of such a phenomenon may provide an alternative to shedding light on the change process derived from the model of Nelson and Winter, built mainly in the technical sphere, insofar as it is only at the social structure of the company (principle of authority, resource allocation, human resources management) that adaptation will be conducted. Moreover, the use of organizational flexibilities obtained through further decentralization of the principles of authority completes the previous model which limits the dynamics of organizations to the development of routines, and, at the same time, withdraws all non-routine processes from the scope of the analysis.

The examples we propose to study here are the many attempts in the industrial field at correcting the Taylorist model. Despite a backdrop of growing uncertainty and increasing technological complexity, this model is still based on modes of organization in force on different continents. If it has not really been replaced, it is however subject to constant rearrangements, the common point of which focuses around the principles of bureaucratic hierarchy. The study of industrial organization provides us with many examples of such releases, reporting hierarchical relationships by establishing coordination mechanisms with respect to the horizontal organization. These are lateral liaison mechanisms updated by Galbraith (1973) in aeronautics and in the form of 'liaison' between independent departments, the 'intervention group' or 'Project Team'. Their goal temporarily is to meet non-recurrent problems facing the organization, or more sustainable forms such as the matrix structure. Such patches are at work in manufacturing industries across the relatively widespread practice now of 'quality circles', 'value analysis' or 'design review', whose focal point seems to be the need for a moment to get out of paths of routine activity to develop solutions from informal structures as the backbone of the formal business can generate. In this case, forgetting requires the unlearning of certain rules of bureaucratic functioning, unable to meet the new situation by giving some players in the organization the opportunity to develop solutions on intuitive grounds or within tacit knowledge.

The organizational inertia as a manifestation of the phenomenon of formal routine

Routine in theories of organization

Generally models for dynamic adaptation of the company derived from evolutionary theories outline paths where the ideal adjustment to changes in the environment is almost continuously in accordance with a biological organism. From the perspective of dynamic flexibility, for example, Cohendet and Llerena (1992) describe a model of responsiveness of firms reflecting 'an ability to respond continuously, over time, to changes in the environment'.

The same ideal of continuous adaptation is present in the representations of organizations that Quinn (1978) uses in his incrementalist approach based on the experience of managers, or DiMaggio and Powell (1983) in the neo-institutionalist stream. A against the current, and somewhat provocative in its infancy, research conducted as part of population studies supports a different organizational logic of evolution (Hannan and Freeman, 1977 and 1984). Organizational inertia is present as the main source of business survival.

This position is prima facie a theoretical alternative to the evolutionary theory. The arguments of this analysis are based in part on the existence of external pressures from the environment. Based on statistical studies on various US economic sectors such as catering, the semiconductor industry, the press or telecommunications, mortality rates were observed over long periods (Carroll, 1984). Rather than a form of continuous adaptation, these studies reveal organizational entities that disappear or survive in response to significant changes in the environment. The structural inertia of organizations that survive is analysed according to the authors as a defence system for completing the high degree of reliability of their performances through collective action (e.g. acquisition and maintenance of a certain level of quality products). For Hannan and Freeman, such objectives can be achieved in relatively flat organizational structures. The improvement of the outputs of a company depends directly on its ability to maintain stable and unchanged structures. The work of Freeman (1990) on the semiconductor industry shows that the mortality rate has increased in this sector as a result of a change strategy.

Looking closely at the model of population ecology is not totally to disagree with the evolutionary paradigm. Indeed, these studies do not exclude the change of internal business structures, but they focus on observations of macroeconomic accounting reflecting more discontinuous forms of adaptation, directly from the inertial force at stake in the organizations. According to the authors, the changes appear as uncommon, or extremely rare, in some organizations. The main factors leading to organizational inertia are internal and come from the irreversibility of the choice of equipment, the specific skills required by human technology, the limited information available concerning the activity of the firm, reducing the diagnostic capabilities of the members of the executive or internal policies of resource allocation between different departments.

Another phenomenon reinforces these arguments. This is the notion of imprint of organizations that addresses over time the strong imprint of the circumstances obtained at the time of their creation. This hypothesis, advanced in the work of Stinchcombe (1965) and later repeated by Aldrich (1979), justifies another source of inertia to change. Throughout its life, the company attempts to preserve unchanged the resources acquired, primarily because they have proved themselves in terms of organizational arrangement, providing a competitive advantage over other forms of arrangement, and because the organization is itself a place of emergence of interest groups, rooted in tradition or ideology. All these elements are all anchors that tend to isolate the business pressures of the environment (Tucker *et al.*, 1990). Empirical studies conducted by Hannan and Freeman also show that organizational inertia increases with the size and age of the company, reinforcing on the one hand the hypothesis of structural stability that large companies are better able to achieve given the amount of resources available internally, and second, the role played by the creation date of the organization reflecting the depth of the imprint founder.

Finally, this view is consistent with the first current social Darwinists who saw progress in the social field as the result of the survival of organizations best suited to the environment at any given time, not in terms of capacity control by the human factor (Perrow, 1986). Indeed, if one follows the logic of ecological models, social change proceeds first by a change in behaviour (intentional or not according to Perrow), followed by a phase of natural selection in which the most efficient organizations survive. In this dynamic, the last stage is characterized by retention mechanisms where organizational inertia allows the system to maintain itself and reproduce its business efficiently. However, as noted by Dunne *et al.* (1988) and Winter (1990), the analogy with the biological model to explain the progress in the social field gives an exaggerated picture of the movements of variation and selection that drive organizations. To the extent that, in fact, many industries are highly concentrated (such as automotive, petroleum, metallurgy in the USA) strict observation of changes in a number of companies that are born and die mask the fact that most companies remain stable. Moreover, this relative stability is only apparent because nothing excludes that changes have taken place within these large blocks which seem stable at the macro level.

This touches on one of the limits of the analogy with the models of natural selection since, as Perrow (1986) noted, although it continually regenerates, a living body cannot change its internal operating like structural changes of social organizations. Like Weick (1979) we believe that companies, especially large ones, have the ability to modify their environment by managerial decisions, or even create it. Industrial strategies such as those leading to cases of massive concentration in the automotive field, both in the USA and in Europe, structure the economic environment more than they suffer from it. The environment is not the only social science to make choices as suggested by the model of population ecology. The company has the power to create, to bend external constraints making them highly dependent on endogenous factors. This is probably why the selections made by the negative nature result in corporate bankruptcies almost exclusively of small structures, rarely of large institutions. The obvious lack of analysis on the internal skills of the organization that suffers from the model of population ecology can be filled if we place ‘routine’ at the heart of the business organizations as suggested by evolutionists. More recently the work of Feldman and Pentland (2003) shows that the routine is an intense place of learning and can lead to the creation of new organizational routines if the interface leaves a margin of freedom to the players in the organization. Organizational inertia would then find additional justification to better inform the death of certain organizations or adaptation and maintenance of others, than the time required for the establishment of routines. Nelson and Winter (1982), for whom the environment is not only outside the company, show that if the complex processes of innovation appear first in the external forms with strong organizational inertia, they are also sources of longer-term adaptation. Rereading the literature on ecological populations in the light of these internal mechanisms, Burgelman (1991) mentions that inertia is relative since it is the manifestation of induced strategic processes within the organization, the signs of an intense activity paradoxically perceived as immobile by outside observers.

A contextualization of forgetting through the creation of routines: empirical evidence

The Renault case was studied extensively at the end of the 1980s by the research group of the French École des mines, the CRG (Aggeri, 1992). The Valeo case was studied by the author.

The Valeo case: forgetting core knowledge in the search for breakthroughs

Starting position: Valeo is a large firm, active in the automobile industry. It provides parts and subsystems to the main auto companies. At the beginning a firm without many distinctive abilities, Valeo has become a technological and market leader in many parts systems. We have in particular looked at the lighting systems for which the company is a world leader. How to maintain its competitive advantage to bring bold solutions to car constructors?

Organizational shock: acquiring a structure favouring breakthrough innovations. Valeo's top management was concerned about how to maintain the firm's technological advantage in auto lights. They were convinced that the firm had pushed the improvements within the existing lighting system paradigms to their limits. Their engineers did not seem to see the point and were resistant to moving away from the existing R&D framework. To increase the chances of change, the executive group decided to create a 'meta-level of analysis and thinking', intended to question the existing lighting paradigm. To such a meta-level, specialists from other non-traditional technical traditions were associated. In so doing, management succeeded in creating the conditions that promote integration rather than specialization alone.

Change process: by creation of meta-levels of analysis forcing a multidisciplinary design phase. For example, instead of thinking about lighting, the meta-level suggested that lighting groups should think in terms of driving comfort. In so doing, they could modify not only the nature (colour and focus) of lighting and its intensity, which were the traditional concerns of lighting engineers, but also other factors related to visibility by using infrared cameras. As a result, improvements could affect the whole glass system (lateral, back and in particular front), without even changing the lighting system itself. Also, mobility of the lighting system, in relation to the drivers' intent, in particular in taking turns, could improve comfort and visibility considerably. As a result, many new possibilities were discovered and implemented.

The research process in still going on, but it is expected that new transformations in the work process will take place. Thus, managerial intervention to change the problem definition has led to more integration among technologies available, internally and externally, and has transformed the thinking about lighting. In return, one can expect all the product divisions of the firm to start thinking outside of the box, by encouraging cooperation and interpenetration with other technical specialties, inside or outside the firm. This way of thinking could be institutionalized and transform the firm's technical and managerial systems.

The various areas of expertise will probably be forced to change considerably, becoming much more subtle, with a different type of training for engineers and technicians, and with the requirement to develop new know-how, to respond to the need for competitive advantage. Such know-how will be dominated by the new philosophy of innovation, with an emphasis on integration and on the contributions of new sciences and technologies. Thus, the lighting experts could become integrators using lighting technologies, but in connection with several others.

The push for integration, which was expressed through the integration meta-level and a wider definition of the lighting problem, will most probably be followed by an institutionalization of the new approach to product development, centred on a finer understanding of consumer expectations. Such an institutionalization will then push for new developments to respond to newly defined needs.

In terms of forgetting, we can see that here again the intervention of higher level executives has contributed to making some older learning more 'rigid', facilitating their obsolescence and forgetting, in favour of new and stimulating learning focused on how cross-fertilisation among technologies, and with external actors, can bring in new knowledge and lead to new products. In such a case, forgetting is not about lighting knowledge and techniques, but their contextualized use. These cannot be considered without replacing them in a larger set. Their position has been uprooted to be replaced by a new learning package.

The Renault case: the institutional pre-eminence of the top management

Starting position: the competition of Japanese automakers was very threatening in terms of cost-time-range of products on the market. Renault is a famous French firm, known for its innovative capabilities. It has been a technological and product pioneer in many areas of automobile design and parts. It is also known for an emphasis on sophisticated technological breakthroughs, with little attention to their implementation. Thus the Renault car has been the typical advanced car which is poorly built.

Organizational shock: the challenge is to learn to operate in continuous improvement (logic of 'kaizen'). To correct such a deeply rooted and institutionalized behaviour, Renault's executives set up study teams to make proposals for improvement that take into account the nature of the firm, its history and the challenges that it has been able to solve. The project focused more specifically on institutionalized behaviour. The teams concluded that, given the expected difficulties of implementing a rigorous quality management policy, and the skills available to suppliers, it would be effective and economical to externalize many production systems, which were then internally produced.

However, unconvinced, the top managers' position was based on the belief that to transform the firm's production systems, it was necessary to internally master quality management techniques. So it was resistant to the idea of depending totally on external actors. Externalizing was seen as appropriate only if the firm's mastery of quality management knowledge and techniques was sufficient, to be able to dialogue effectively with suppliers. In so doing, the top management intervened decisively to institutionalize the idea that only those skills that were mastered internally could be externalized.

The R&D managers were frustrated at first to see their study discarded, and forced to go back to what they perceived to be a well-known but hard to resolve issue. Nevertheless, they accepted giving Total Quality Management (TQM) another try. They improved significantly the firm's ability to work systematically on quality. In so doing, the collective learning on the older systems was very important. Internal expertise in building and designing cars was strengthened, and under competitive pressures a new generalized knowledge about the use of quality as a managerial method was developed (Pascale and Athos, 1981).

This specialized knowledge led to a new understanding of the determinants of quality in car manufacturing in general and in Renault products in particular. A finer understanding of how integration can improve the final product quality was derived. Such a learning finally justified the externalizations that were made in the following decade.

Hence, Renault top management started with a decision which modified norms, cognitive orientations and standard operating procedures. These critical changes increased the firm's focus on quality and brought new needed learning about each of the areas of technical expertise, and on the product total quality. Only then did the new institutional transformation with progressive and effective externalization of some subsystems become possible.

Change process: by the establishment of operating modes consistent with TQM and simultaneous engineering. To understand the forgetting effect, it is necessary to propose sociological interpretations of the French way of industrial manufacturing. Crozier (1963) suggested that French manufacturing has been traditionally dominated by specialized engineers. These were focused on technology and did not take into account management. Renault, like many other large French state-owned firms, was typical. Each function was remarkably apt at technical sophistication, but the whole was clumsy. The engineers' study of the manufacturing problem ended with a predictable solution. 'Renault knows how to design, but does not know how to manufacture. So let's subcontract the more difficult parts of the process', said an engineer involved in the evaluation's study. The top management took a different perspective, a broader view, taking the risk of confronting the engineers, which were seen as the key to the firm's competitive advantage. In so doing, they made the forgetting of specialized practices easier, making them obsolete, and strengthening the broader, more strategic perspective. It was not that technical knowledge was not necessary, but its use was to favour the operations of the firm as a whole. The new perspective led to new learning, more functional and more relevant. Externalizing some activities became a part of the new learning, a part of the whole rather than a specialized decision. Again, the decision premises shifted, making some learnings less relevant than others and making it easy to forget.

Essentially reducing the phenomenon of innovation in the creation of routines, the evolutionary approach provides a partial picture since the more 'political' dimensions, principles such as authority, or the method of allocation of resources, will significantly influence the change process of the company. Some consider that this ballet is played between the system that produces knowledge in the company, and the political process at work, the essence of organizational structures (Hatchuel, 1990). It would therefore also play, beyond the technical system, the managerial infrastructure would therefore also play an essential role in the organizational capabilities for adaptation by encouraging forgetting: by interrupting routines installed in structures social functioning of the company by legitimizing the taking of initiative and exploration. It is this second source of forgetting that we will now develop as a perspective of organizational change.

Informal structures such as motor creativity in organizational practices of the company

The informal in the literature

In their study of the regulation of complex systems, Crozier and Thoenig (1976) concluded on the fact that during an adjustment process in a turbulent environment, it is better to blame the change on some individuals or groups in the organization capable of rapid adaptation rather than disrupting the overall structure. For the authors, the guiding principle of such a movement in the organization is based on the emergence of an intuitive understanding that can only carry out an optimal evolution of the structure vis-à-vis environmental insults. Such a finding reinforces the idea that a coexistence of a bureaucratic organizational model with more decentralized linking mechanisms and informal learning is possible without the domination on one form over the other. For Crozier and Freidberg (1977), these are indeed the actors who carry the 'strategic capability' that play a central role in organizational adaptation more than the structure itself, contrary to submissions by contingency scholars. The historical feud between the proponents of the bureaucratic form based on the principle of unity of command and those of the completely decentralized organic form (Burns and Stalker, 1961) found here an argument that limits the scope of such a theoretical debate. There is no ideal form of organization, but differentiated forms which collide with bureaucratic principles and more organic solutions. Lawrence and Lorsch (1967) already made this link clear by showing how environmental contingencies determine the organizational structure of the company.

Analysing the way in which high-risk industries respond to shocks of great importance, such as the accident at the nuclear plant at Three Mile Island in the United States, Perrow (1984) puts forward a principle of organizational learning called 'interactive complexity'. In large post-industrial sectors that rely on new technologies demanding continuous production with high levels of security, Perrow observes that technological control requires high levels of abstraction modes of representation of complex equipment from the agents. Yet it is this type of knowledge that is necessary when a system undergoes a technical event that brings out its regulated operating mode. In these situations, non-routine releases the organization's liaison mechanisms and evolves into a hierarchical system that will regulate more organically. From this perspective, Weick (1979) sees technology as an abstract event made by invisible equipments to the extent that operators must master all the sequences and interdependences of the process. These authors show that the formal organization is no longer able to control the level of complexity between agents, and the solution that emerges is to delegate responsibilities, decentralize decisions by giving more autonomy to human resources. This question of creativity in business struggles with contradictions such as investing in new projects or current ones, just as differentiating or producing low-cost is at the heart of developments in the field of organizational ambidexterity. The pioneering work of Duncan (1976) and Tushman and O'Reilly (1996) emphasizes the need to build dual organizations capable of articulating at different mechanistic or organic structures. This thinking is fuelled by recent developments in Birkinshaw and Gibson (2004) who oppose this type of structural ambidexterity, contextual form where members of the organization are able to link the two together in aligning existing procedures while being able to adapt by exploration.

For March (1988), foolishness suspending rational imperatives, whether in personal choices or organization, is a means of exploring alternative pathways, and thereby generating experience. From the behavioural point of view, this step aside vis-à-vis the official rules of conduct does not replace the principles of rational action, but it ends up enriching solutions that purely intelligent processes of decision would not detect. The mode of organization that results from such a vision is not a frozen structure but a system that grows by leaps, finishing, on one hand, to complete these routines Zuscovitch cherished in his work on the meso-economic technical progress (1984), and by disturbing experience of intuition on the other. The organization no longer appears as a fixed structure, but more as a differentiated matrix sending signals to actors concerned in order to initiate rationality and creative foolishness. This call to action is essential in the original phases of innovation. It depends in part, as shown by Zmud (1982) in an example of the dissemination of software, on the parallel coexistence of mechanical and organic structures within the same organization, or as argued by Birkinshaw and Gibson (2004), by studying a large number of 'business units', the introduction of managerial forms at any time encouraging the contextual ambidexterity.

A contextualisation of forgetting through informal structure: empirical evidence

The SEW USOCOM and ES cases were studied through different research projects by the author.

The SEW USOCOME case: the sleepiness of design patterns of work stations

Starting position: maintaining a global leadership in the field of medium size electric motors. Despite a context of growing uncertainty and higher technological complexity, the Taylorian model of industry is still the reference for organizations in different continents, even if several adjustments have been made. If, since the first applications in the USA, it has never been replaced, it has been the target of several arrangements always searching to distort temporary (and in certain segments only) hierarchical bureaucratic tools. The study of industrial organizations provides us with a number of examples of such an oversight of hierarchical links by amending the chart flow through horizontal mechanisms. We illustrate here this aspect of forgetting through the SEW USOCOME case. If the corrections applied to the Taylorian model are oriented in that direction, contrary to what happened in the technical area, the forgetting described here requires unlearning some bureaucratic rules. These rules being unable to provide an answer to a non-standard situation, the case shows that adaptation will be found by giving some actors the possibility to elaborate solutions on the basis of pure intuition or tacit knowledge on the basis of experience.

Organizational shock: redistribution of responsibilities and roles in the redistributed design process. SEW USOCOME is a French subsidiary company of a German international industrial group called SEW

EURODRIVE. World leader in automatic motion systems (mid-size electrical engines; electrical reversing frequenzors; servo-conductor systems), the whole company represents 1.5 billion euros of sales and 11,000 employees located in a large number of countries all over the world. The French subsidiary located in the north of Alsace is an assembly plant of electrical engines and associated servo-conductor modules. A large number of technological references are produced (more than 10,000) to give an appropriate answer to industrial needs. The scope of the products generates a strong complexity of production managed by employees of the company. The unit in Alsace is the main production location in France with an amount of 1300 employees for a total of 1850 in the country as a whole.

Change process: skip from a Taylorist bureaucracy to a participatory organization with job enrichment. In 1996, the French company, strongly involved in Total Quality Management (TQM), decided to create a pilot experiment concerning the design of the 'on line process' of setting. Without manufacturing activities, costs of assembly are the main sources of financial loads. Due to the adoption of the continuous learning programme, which conforms with TQM, associated with a strong belief that 'without appropriation by actors of the company there are no solid results' (the Quality Manager), the company decided to break away from former practices in the designing process. While maintaining the leadership of management in the long term, the managers asked operators to design themselves their own work places by going beyond the traditional office methods. The result quickly became noteworthy in terms of productivity improvements with rates close to 40 per cent. Two rules inspired the decision process. The first came from the appropriation of change by operators based on empirical experimentations (observation of facts by people on the shop floor). The importance of experimentation as a source of creativity and radical innovations in companies was stressed recently in the field of dynamic capacities (Teece *et al*., 1997), by scholars such as Thomke (1998), Von Hippel (2005), Schreyögg and Kliesch-Eberl (2007). The authors explained that the rupture in the innovation process for higher creativity was coming more from the capability to chain up experimentations based on trial and error in the organization than from a centralized R&D investment. The second is related to the autonomisation process of operational employees in terms of means and actions in accordance with an organic structure. At the end of this first experimentation, the middle management of SEW USOCOME understood the double advantage of this approach. Indeed, the mobilization of employees was improved in the enterprise as a whole and not only in the experimental area.

This approach was generalized in the following decade to the whole company. It has become today a permanent and transversal structure called WIEPRO-Consulting (a German expression). This is an original strictly internal structure of consultancy. The announced objective is the development of knowledge and the optimization of the processes at stake in the organization. This internal consulting structure is in charge of operational actions of organization with internal clients. Devoted to production only at the beginning, it has been generalized to administrative processes such as the management of production, the purchasing activity or suppliers' accounting. The SEW USOCOME Group, which is starting a new activity in China, has transferred the same approach by soliciting the local employees. In all experimentations, the progress realized is communicated to all employees of the production unit by using an indicator board informing on the improvement (before and after indicators), as well as a personalized presentation by the operators themselves to other employees including the staff of managers. One of the last building blocks concerned a setting line of breaks for electrical engines. The productivity obtained by a new design of the linear process went up from 10 breaks per capita per hour to 16 without changing the conditions of work, thus improving productivity by more than 50 per cent.

In that case, forgetting concerns two elements, and takes the obvious form of a voluntary strategy of change. A first manifestation of forgetting for the organization concerns the rule of management which forgets the formal structure of the methods office during the implementation phase of the project. This office is normally in charge of all transformations on the production process. The head manager of this new transversal structure WIEPRO-Consulting (the former quality manager of France) argues that the management 'produces the actors' (in the theatrical sense) of change by empowering them, i.e. by giving them total responsibility for their actions. When the solution is completely realized, the management get their rights back. During that digression, the formal component devoted to the elaboration of solutions optimizing the process (the organization and method office) is fully at the disposal of operators for technical support. The second form of forgetting concerns operators themselves who are encouraged by the management to challenge their own routines and to create innovations by using their practical senses considered irreplaceable by the hierarchy. Indeed, who knows better the sources of productivity of their work stations? Forgetting the process seems to be perfectly managed by the enterprise through the creation of a transversal service (WIEPRO-Consulting) encouraging the operators to forsake certain practices to be replaced by personal 'home made' solutions.

The Électricité de Strasbourg (ES) case: an endogenous learning/unlearning phenomenon

Starting position: the deregulation of the electricity sector looking at European and competitiveness objectives. March (1991) emphasizes the danger of exploiting excessively standard practices and technologies at the expense of research for new technological opportunities. This decoy could create a learning trap leading to exploiting the routine at stake, preventing learning of exploration as demonstrated by Bénézech *et al.* (2001) in the context of quality certification of French SMEs. It just so happened in a local subsidiary of Électricité de France (called Électricité de Strasbourg) where adopted routines do not freeze learning but apply forms of unlearning at the heart of the routine. The display of quality documentation is devoted to routinization in order to make the behaviour of members of the organization predictable. Then, a following question consists in determining if a learning process could emerge from the simple utilization of a procedure, a best practice or instruction card. Électricité de Strasbourg (ES) carries and distributes electricity produced by EDF (the main shareholder with 75 per cent of the capital of ES) for the Alsace region. Manager and operator of the electric network, the company commercializes the electricity on 376 communes of the Bas-Rhin area, representing more than 446,000 clients, 1100 employees and 425 million euros of sales. An investigation of this company (based on a PhD project) was realized in a recent period using a longitudinal approach over three years.

Organizational shock: a reorganization process around value chains. The quality manager of ES mentions that a general policy is defined on the basis of a strategic vision. Then this policy is communicated to middle managers of macro-processes who receive specific goals. These people negotiate resources with functional managers and have afterwards total autonomy concerning the allocation of resources and the realization of the process; the only constraint is the realization of goals. Hence the breathing space and freedom of agents create an elaborate organizational interface allowing the creation of large distortions between the norm (ISO 9000), and procedures or derived work instructions, with the 'performative' aspects of the organizational realization.

Change process: a job enlargement and integration of the value chain. A major transformation in the organization seems to result from a new representation of the activity in terms of process coming from the framework of ISO 9000 quality norms, especially at the level of complaints from clients. Traditionally the hotline service receives demands from clients who require answers. Operators fill in an intervention card explaining the nature of the problem, and then communicate it to the appropriate intervention service. Due to quality management requirements, the setting-up of macro-processes such as 'commercialization to individuals' puts these two departments in the same process. Then, it becomes obvious that an optimization of this process as a whole is possible. The 30 employees of the hotline received a training giving them adequate knowledge in order to solve some specific problems by phone thus avoiding pointless interventions. The result has been a meaningful reduction of interventions on the ground, e.g. for operations such as reading the meter following simple moving in and out. The number of ground operators has decreased from 80 to 50 proving an economy of 30 persons full time re-affected to other missions in the company. This process of improvement in the interface with clients generates also for them a competitive advantage due to a reduction of the time to decision.

Through standardization of solutions to current problems based on coordination between efforts of commercial and technical operators, the company has generated a routine constantly renewable by repetitive practices. The quality manager certifies the existence of endogenous learning effects expressed by an increasing level of competences of the hotline operators as well as a reduction of the level of stress in their work for agents implied in this routine.

This case reveals that the adoption of routines makes it possible to save time, financial and human resources on the one hand, but also generates improvement of the quality of services on the other. Mechanisms of coordination and routines improve step by step through practice. A real learning effect is at work inside the routine itself. The eruption of forgetting grows up as a natural phenomenon, generated by repetitive executions of routines by operators. Learning is not yet exogenous coming from top-down stimulations but comes clearly from endogenous dimensions and bottom-up initiatives either intentional or non-intentional (see Felman and Pentland, 2003; Lambert and Ouédraogo, 2010).

Table 2.1 summarizes the main steps of this phenomenon of forgetting in each company studied.

Conclusion

The shock of change in organizations seems to refer to two distinct phenomena of forgetfulness. One located in the technical system is the establishment and development of routines in order to increase performance (March and Simon, 1958). The other looks more or less for the interruption of deep modes of administration inherited from a bureaucratic structure. This is, first, the individual's ability to unlearn the socio-organizational structure of the company that forgetting, as agent of change, comes into its own. It is, second, the potential role of strategic management at the organizational level by initiating structures that encourages the taking of initiative and informal communications within the company.

Forgetfulness is often seen as a failure of memory related to the context of its use. It thus represents a negative externality of memory. Forgetting to create spaces to make way for new knowledge. According to Feigenbaum and Simon (1964), this loss of information is closely linked to a physical phenomenon of learning because of the increasing number of associations to remember. In this interpretation the issue is not to forget but to remember. However, we argue that in certain situations such as where the individual and the company must generate creativity to survive the competition, forgetting becomes a major strategic issue rather than memory. After this phase, a new permanent system for the sake of stabilizing the operation of the organization is set up. Memorizing routines soon becomes a strategic issue when that is the strength of bureaucracies.

Table 2.1 Forgetting in some companies

Cases	*Valéo*	*Renault*	*SEW-USOCOM*	*Électricité de Strasbourg*
Starting position	To maintain its competitive advantage to bring a bold solution to car constructors	Competition of Japanese automakers very threatening in terms of cost-time-range of products on the market	Maintain global leadership in the field of electric motors, medium size	Deregulation of the electricity sector looking at European and competitiveness objectives
Organizational shock	Acquiring structure favouring breakthrough innovations	Learn to operate in continuous improvement (logic kaizen)	Redistribution of responsibilities and roles in the design process redistributed	Reorganization process around value chains
Change process	Creation of meta-levels of analysis forcing a multidisciplinary design phase	Establishment of operating modes consistent with TQM and simultaneous engineering	Skip a Taylorist bureaucracy to a participatory organization with job enrichment	Job enlargement and integration of the value chain
Manifestation of forgetting	Change of routine imposed by the direction	Change of routine compliance with the new values of the direction	Forget the formal structure (Methods Office) generally in charge of the workstation's design	Replacing a routine using endogenous learning

When in 1938, Barnard described the physiological function of the informal organization, he assigned an organizational function of cement to preserve the formal structure ensuring the integrity of the players who ran it. Now for this ideologist of management working before the Second World War, compared to formal management style, these groups of informal structure mean irrational impulses which nested pell-mell: misunderstandings, customs, habits and the full weight of the institution. Decades later, and in a certainly different context of technological evolution, these informal structures are to play the same role, this time not by protecting individuals from alienating aggressions of the rational management style, but by feeding it creative and original solutions which aim only at perpetuating the organization through change. Between these two visions, one period has passed, and with it perhaps the idea of the human factor as a single source of errors and missteps in the system.

Notes

1 Thank you to T. Hafsi who I worked with on a similar project for inducing me to reread my classics, especially the first work of H. Simon on the premises of the decision (Hafsi and Lambert, 2008). My thanks go especially to C. Lerch who helped, with intelligence and finesse, to improve the quality of that reflection.

References

Aggeri, F. (1992) 'Quels enjeux pour le calcul économique: décider ou apprendre?', *Cahier de Recherche*, 2, Centre de Gestion Scientifique, École des Mines de Paris.

Ancori, B. (ed.) *Apprendre, Se Souvenir, Décider: une nouvelle rationalité de l'organisation*, Paris: Edt. du CNRS.

Aldrich, H. (1979) *Organizations and Environments*, Englewood Cliffs, NJ: Prentice Hall.

Barnard, C.I. (1938) *The Functions of the Executive*, Cambridge, MA: Harvard University Press.

Bénézech, D., Lambert, G., Lanoux, B., Lerch, C. and Loos-Baroin, J. (2001) 'Completion of knowledge codification: an illustration through the ISO 9000 standards implementation process', *Research Policy*, **30**(9): 1395–1407.

Birkinshaw, J. and Gibson, C. (2004) 'Building ambidexterity into an organization', *Sloan Management Review*, **47**(2): 209–26.

Burgelman, R.A. (1991) 'Intraorganizational ecology of strategy making and organizational adaptation: theory and fields research', *Organization Science*, **2**(3): 239–62.

Burns, T. and Stalker, M. (1961) *The Management of Innovation*, London: Tavistock.

Carroll, G.R. (1984) 'Organizational ecology', *Annual Review of Sociology*, **10**: 71–93.

Cohendet, P. and Llerena, P. (1992) 'Nature de l'information, coût de transaction et organisation de l'entreprise', in B. Ancori (ed.) *Apprendre, Se Souvenir, Décider: une nouvelle rationalité de l'organisation*, Paris: Edt. du CNRS.

Crozier, M. (1963) *Le phénomène bureaucratique*, Paris: Edt. du Seuil.

Crozier, M. and Friedberg, E. (1977) *L'acteur et le système*, Paris: Edt. du Seuil.

Crozier, M. and Thoenig, J.-C. (1976) 'La regulation des systèmes organises complexes. Le cas du système de decision politico-administratif local en France', *Revue Française de Sociologie*, **16**(1).

Cyert, R.M. and March, J.G. (1992) *Behavioral Theory of the Firm*, 2nd edn, Cambridge, MA: Blackwell Publishers.

De Holan, M. and Phillips, N. (2004) 'Remembrance of things past? The dynamics of organizational forgetting', *Management Science*, **50**(11): 1–11.

Di Maggio, P. and Powell, W. (1983) 'The iron cage revisited: institutional isomorphism and collective rationality in organizational fields', *American Sociological Review*, **48**: 147–60.

Duncan, R.B. (1976) 'The ambidextrous organization: designing dual structures for innovation', in R. Kilman and L. Pondy (eds) *The Management of Organizational Design*, New York: North Holland: 167–88.

Dunne, T., Roberts, M.J. and Samuelson, L. (1988) 'Firm entry and post-entry performance in the US chemical industries', papers 7 -88-2, Pennsylvania State Department of Economics.

Feigenbaum, E.A. (1970) 'Information processing and memory', in D.A. Norman (ed.), *Model of Human Memory*, New York: Academy Press: 451–68.

Feigenbaum, E.A. and Simon, H.A. (1961) 'Forgetting in an association memory', *Reprints of the 1961 National Conference of the Association for Computing Machinery*, **16**, 2C2–2C5.

Feigenbaum, E.A. and Simon, H.A. (1964) 'An information-processing theory of some effects of similarity, familiarization, and meaningfulness in verbal learning', *Journal of Verbal Learning and Verbal Behavior*, **3**(5): 385–96.

Feldman, M.S. and Pentland, B.T. (2003) 'Reconceptualizing organizational routines as a source of flexibility and change', *Administrative Science Quarterly*, **48**(1): 94–118.

Freeman, J. (1990) 'Ecological analysis of semiconductor firm mortality', in J.V. Singh (ed) *Organizational Evolution: New Directions*, Newbury Park; London; New Delhi: Sage Publications.

Galbraith, J. (1973) *Designing Complex Organizations*, Reading, MA: Addison-Wesley.

Galbraith, J. (1975) *Organization Design*, Reading, MA: Addison-Wesley.

Hafsi, T. and Lambert, G. (2008) 'Oubli et changement stratégique: une exploration', Congrès Annuel de l'ASAC sur 'Managing the Responsible Enterprise', Halifax, Nova Scotia, Canada, 24–27 May.

Hannan, M.T. and Freeman, J. (1977) 'The population ecology of organizations', *American Journal of Sociology*, **82**(5): 929–64.

Hannan, M.T. and Freeman, J. (1984) 'Structural inertia and organizational change', *American Sociology Review*, **49**(2): 149–64.

Hatchuel, A. (1990) *Production de connaissances et processus politiques dans la vie des entreprises*, Paris: Ecole des Mines.

Hedberg, B. (1981) 'How organizations learn and unlearn', in P. Nystrom and W. Starbuck (eds) *Handbook of Organizational Design*, Vol. I, New York: Oxford University Press: 3–27.

Lambert, G. and Ouédraogo, N. (2010) 'Normes, routines organisationnelles et apprentissage d'entreprise', *Revue Française de Gestion*, **36**(201): 65–85.

Lawrence, P.R. and Lorsch, J.W. (1967) *Organization and Environment: Managing Differentiation and Integration*, Boston: Graduate School of Business Administration, Harvard University.

March, J.G. (ed.) (1988) *Decisions and Organizations*, Oxford: Basil Blackwell.

March, J.G. (1991) 'Exploration and exploitation in organizational learning', *Organization Science*, **2**(1).

March, J.G. and Simon, H. (1958) *Organizations*, New York: Wiley.

Nelson, R.R. and Winter, S.G. (1982) *An Evolutionary Theory of Economic Change*, Cambridge, MA: Harvard University Press.

Pascale, R. and Athos, A.G. (1981) *The Art of Japanese Management*, Harmondsworth: Penguin Books.

Perrow, C. (1984) *Normal Accidents: Living with High-Risk Technologies*, New York: Basic Books.

Perrow, C. (1986) *Complex Organizations: A Critical Essay*, 3rd edn, Glenview, IL: Scoot, Foresman.

Postman, L. and Underwood, B.J. (1973) 'Critical issues in interference theory', *Memory and Cognition*, **1**: 19–40.

Quinn, J.B. (1978) 'Strategic change: logical incrementalism', *Sloan Management Review*, **20**(1): 7–19.

Schreyögg, M. and Kliesch-Eberl, G. (2007) 'How dynamic can organizational capabilities be? Towards a dual process model of capability dynamisation', *Strategic Management Journal*, **28**: 913–33.

Simon, H. A. (1971) 'Designing organizations for an information-rich world', in M. Greenberger (ed.), *Computers, Communications, and the Public Interest*, Baltimore, MD: John Hopkins Press: 37–52.

Simon, H. A. (1997) *Administrative Behavior*, 4th edn, New York; Macmillan.

Stinchcombe, A.L. (1965) 'Social structure and organizations', in J.G. March (ed.) *Handbook of Organizations*, Chicago: Rand McNally.

Teece, D.J., Pisano, G. and Shuen, A. (1997) 'Dynamic capabilities and strategic management', *Strategic Management Journal*, **18**(7): 509–33.

Thomke, S.H. (1998) 'Managing experimentation in the design of new products', *Management Science*, **44**(6): 743–62.

Tucker, D.J., Singh, J.V. and Meinhard, A.G. (1990) 'Founding characteristics, imprinting, and organizational change', in J.V. Singh (ed.), *Organizational Evolution: New Directions*, Newbury Park; London; New Delhi: Sage Publications.

Tushman, M. and O'Reilly, C. (1996) 'Ambidextrous organizations: managing evolutionary and revolutionary change', *California Management Review*, **38**(4): 8–30.

Von Hippel, E. (2005) *Democratizing Innovation*, Cambridge, MA: MIT Press Books.

Weick, K. (1979) *The Social Psychology of Organizing*, 2nd edn, Reading, MA: Addison-Wesley.

Winter, S.G. (1990), 'Survival, selection, and inheritance in evolutionary theories of organization', in J.V. Singh (ed.), *Organizational Evolution: New Directions*, Newbury Park; London; New Delhi: Sage Publications: 269–97.

Zmud, R.W. (1982) 'Diffusion of modern software practices: influence of centralisation and formalisation', *Management Sciences*, 28: 1421–31.

Zuscovitch, E. (1984) 'Une approche Méso-économique du progrès technique: diffusion de l'innovation et apprentissage industriel', Thèse de Doctorat d'État Es-Sciences Economiques, Université Louis Pasteur, Strasbourg.

3 Product lines innovation

On the use of technical laws of evolution

Christophe Lerch, Eric Schenk and Denis Cavallucci

Introduction

The metaphor of life sciences was widely developed in industrial economics in order to understand the dynamics of sectors against a backdrop of innovation. Thus, approaches taking inspiration from the ecology of populations as an extension of Schumpeter's destructive creativeness concept, studied industrial dynamics by means of complex predator-prey systems (Goodwin 1967; Hofbauer and Sigmund 1988; Modis 1999).

In the fields of economics and management of innovation, the theory of dominant design (Abernathy and Utterback 1978) echoes the concept of life cycles: according to Abernathy and Utterback, each product class evolves over three successive phases. The first phase is the fluid phase, featuring a high rate of product innovations and a low rate of process innovations. This phase can be assimilated to the emergence of a family of products following a breakthrough innovation. In the second transitional phase, the ratio between process innovation and product innovation reverses. This is a consolidation phase for the products family. The third phase is the specific phase, during which both the rate of innovation in products and processes drops, signaling maturity of the products family. According to the authors, evolution of the ratio in products innovation is attributable to the emergence of a dominant design. For Christensen *et al.* (1998), a dominant design constitutes the requirement that determines the structure of the product family. This requirement can be a particular design trait or the combination of several design traits. A strong similarity appears therefore between the concept of dominant design and the notion of standard technology (Arthur 1989; David 1985). Consequently, dominant design reflects a stabilized system with a great degree of consistency between underlying technologies, their use and methods of production. This consistency is the result of a non-deterministic process of competition, leading to the possible emergence of a de facto standard (Arthur 1989).

The theory of dominant design can be viewed as a theory of selection which describes the emergence of dominant product types. It can be compared to the theory on natural selection found in biology. Indeed, the emergence of a dominant design constitutes both a means of leverage for industrial development and a major constraint for the evolution of product families. Dominant design is associated with increasing returns through economies of scale in production and network economies and it is a development factor influencing the products family. Since it is a de facto standard, a dominant design constitutes a restrictive framework in terms of product architecture that it is sometimes impossible to break free from. Indeed, dominant design can be the cause of a lock-in (Arthur 1989) that leads to a form of technological ossification.

Rather than adopting a market selection perspective of technological evolution, our point is to understand the role of engineers in product design. More precisely, we question their ability to predict the evolution of product lines likely to generate future dominant designs.

This leads us to propose an analysis of the innovation process based on an evolutionary approach to technical objects. In this way, we follow on from the work of Deforge (1985), of which Boldrini (2012) offers a recent view. According to Boldrini, the central issue revolves around to what degree the nature of the innovation is deterministic. This author rejects the proposition that evolution in technical objects responds to internal laws of products. In this sense, his writings are critical of a certain interpretation of the TRIZ laws of evolution[1]. Our work adopts a perspective complementary to that of Boldrini. As with him, the TRIZ laws of evolution are at the core of our reflections, in as much as they constitute a culmination of the process that aims at exploiting the theory of evolution of species, here in the domain of product design. While it is true that the laws of evolution for TRIZ resulting from observation of past occurrences feature primarily an empirical status, from our perspective they also constitute useful and effective principles for guiding research of product designers. Thus because of their implementation by designers, they can be validated through their very existence.

Our argumentation is presented according to the following plan. In the first section, we present a taxonomic approach to products as proposed by Deforge (1985), leading to the determination of a lineage concept and the emergence of an evolutionary approach for technical objects.

In the second section, we concentrate on managing lines of descent or families of products, emphasizing strategic issues linked to the generation of product diversity and rate of renewal among products. The conception of lineage, or lines, is at the core of the creation of new types of distinctive skills by adopting an intermediary exploration strategy, between incremental innovation and breakthrough innovation. However, this approach appears difficult to implement because of the unpredictable nature of the selection process.

Thus, in the third section, we show that the *laws of evolution* cited by Altshuller (1988) represent a means for designers to guide their research activities down selected paths. We have chosen changes in descent lines of computer keyboards in reference to the laws of evolution as an illustration.

Descent lines of products: from a naturalist view to the economic approach

From the evolution of species to the evolution of technical objects

The first edition of the major work by the botanist, physician and zoologist Carl von Linné (1710–1778) *Systema Naturae* was published in 1735. He produced a scientific classification of three realms of nature encompassing the animal, plant and mineral kingdoms, which was to serve as a model for several generations of naturalists. The 11-page first edition of *Systema Naturae* would be constantly expanded to incorporate the progressive discoveries of Carl von Linné and his colleagues, eventually comprising a 13th edition containing 3,000 pages. During this period, the animal kingdom was divided into six groups: quadrupeds, birds, amphibians, fish, insects and invertebrates. Linné's quest is based on the assumption that species are immutable: "Linné believed that he did nothing more than describe and classify divine creations and that he would be able to complete his inventory before he died" (Wikipédia *Systema Naturae*).

Lamarck's (1744–1829) theory of transformation opened the way to calling into question the immutability of species. Species evolved according to two principles: increasing complexity of organic beings when activated by factors within their metabolisms and the specialization of living beings into multiple species when influenced by varied environments to which they are forced to adapt. The evolutionary approach reached a decisive culmination with Darwin's (1809–1882) publication of *The Origin of Species*, which purported the principle of natural selection in which all the individuals within a species present at least slight differences and only those best adapted to their environments will participate in the following generation.

The study of technical objects: the taxonomic phase

At the end of the nineteenth century, with the works of Armengaud, Borgnis and Reuleaux (see the summary offered by Koenigs 1905), a research movement grew up that strove to apply the principle of taxonomies of the natural sciences inspired by Linné to objects, particularly to machines: "Mechanization is a veritable artificial zoology, in which the creator is man himself, so to round out its evolution, we must create an analogous system for mechanization, with a comparative anatomy and a psychology" (Koenigs 1905, quoted by Deforge 1985, p. 24). This constitution of a taxonomical basis for technical objects reached a culmination in the products descent line concept of Deforges, presented in his work *Technology and Genetics of*

Industrial Objects. For this author, the classification of objects into families and descent lines is done according to four criteria.

The first criterion returns to the *object's function*. Yet, as Deforge emphasizes, using this criterion raises several difficulties, including the degree of accuracy brought into play in determining functions. So, as stated by Deforge, "The makeup of an object and of its originality is a question of judgment" (p. 97). This difficulty of making the concept of function operational leads Deforge to develop a second criterion of classification: the *principle of the object*. For simple objects, the principle harks back to the primary physical-chemical phenomenon that was initially implemented. More complex objects do not fit into this approach because the principle is not visible from the exterior, but rather results from a subtle combination of technology and the shape of an object. Therefore, the *technological solution* and *shape* constitute Deforge's third and fourth criteria of classification. These situate the problem in its achieved solution, found essentially in the designers' *functional diagrams*. A comparison of two objects according to four general criteria generates 16 combinations ranging from full resemblance to total dissimilarity. Also according to Deforge, two objects that share a function are part of the same family. A descent line then is defined as all of the objects of a family that share a common principle. For instance in Table 3.1, products labeled A to H share a common function and therefore belong to the same family. Within this family, products A to D (respectively E to H) share a common principle and hence form a descent line.

The study of technical objects: emergence of the evolutionary approach

Even though it was elaborated under the assumption of fixity of species, the Linnaean taxonomy was probably a decisive step in the emergence of evolutionary theories. Indeed, without taxonomic criteria it would have been difficult to assess the evolution of species over time. Similarly, taxonomies of technical objects are probably responsible for the emergence of an evolutionary theory of technical objects.

The architect and engineer Lafitte (1972) observed through a Lamarckian approach that objects and machines have general differential features and are susceptible to evolution as a function of human requirements and economic constraints. The origin of this idea is with Leroi-Gourhan, who, in his ethnology of pre-historic objects, defends the existence of a general evolutionary *tendency* of technical objects: "A tendency exists for a piece of hand-held flint to develop a handle, for a load dragged between two poles to take on a wheel … a tendency also for the wheel to be complemented by a crank handle, then to acquire belts and down gearing" (Deforge 1985, p. 82). Another appropriate source to mention is Moles (1970), a former professor at Strasbourg University. He is known for his studies in information and communications sciences and is also considered along with Zwicky (1969) as

Table 3.1 Families and descent lines according to Deforge

Products	A	B	C	D	E	F	G	H	I	J	K	L	M	N	O	P
Function	**1**	**1**	**1**	**1**	**1**	**1**	**1**	**1**	0	0	0	0	0	0	0	0
Principle	**1**	**1**	**1**	**1**	**0**	**0**	**0**	**0**	1	1	1	1	0	0	0	0
Technology	1	1	0	0	1	1	0	0	1	1	0	0	1	1	0	0
Shape	1	0	1	0	1	0	1	0	1	0	1	0	1	0	1	0

one of the fathers of morphological analysis.[2] The tools developed by Moles such as attribute constellations, symbolic equations, similarity matrices, can be used to compare and classify objects and to clearly outline their evolutionary trajectories. In a study of the genesis of the evolution of technical objects, Simondon (1989) maintains that there exists a co-evolution of man and machines. He thus defends the existence of a fulfillment process by which technical objects acquire a form of autonomy and self-sufficiency. According to him, this "natural" evolution of descent lines is subject to economic and cultural constraints that disturb it.

At the same period as Simondon and Deforge, the Soviet engineer Altshuller, unaware of the work of these two technicians, carried out a monumental task in which he attempted to analyze 400,000 patents and 1,500 descent lines of products in order to understand the determinants influencing the evolution of technical objects. A part of this work is described by Salamatov (2002),[3] who formulated a hypothesis similar to that of Simondon, in as much as it defends negentropic evolution of objects, moving from the simple to the complex, and on to self-organization.

This vision recalls the one conveyed implicitly by Leroi-Gourhan in as much as it upholds the hypothesis postulating that there exists a relative independence between the conception process of designing new technical systems and the will of man:

> The technical arena evolves entirely independent of the participation of man. Even when men play a role in evolution, it is no more significant than the role of a bee in the wild that transports pollen from one flower to another ... man may remove an element from the technical evolution processes, he may transfer from one machine to another, he may for a time go against the tide of technical advances or move a considerable distance ahead of science, but it is impossible for him to change natural laws.
>
> (Salamatov 2002, p. 4)

Precursors of the evolutionary approach of technical objects of the French and Soviet schools have a rather deterministic approach where the engineers and product designers are somehow the "discoverers" of paths of technical evolution which are largely already drawn. But the management of product lines can also be viewed as a way to manage innovation with a strategic perspective.

Product families and lineage as instruments in the strategic process

When a company undergoes a rapid evolution brought on by technological progress, a change in consumer behavior or a different functional use for a product, it not only has to be able to offer a wide range of products but it also must be capable of generating a high rate of renewal of these same products. With this as a backdrop, managing product descent lines and

Table 3.2 Paths of evolution (from Uzumeri and Sanderson 1995)

	Low diversity	*High diversity*
High renewal rate	Commodity	Variety Intensive
Low renewal rate	Change Intensive	Dynamic

families – more so than managing isolated products – is a strategic process that seeks to adapt to the specific requirements of customer niches and to changes in requirements.

Product descent lines and the building of competitive advantages

In this area, the work of Sanderson and Uzumeri (1995) is particularly illuminating in as much as it characterizes product families according to two criteria: The number of different products in the family at an instant *t*, which may be either large or small, and the renewal or replacement ratio of products in the family that may be intense or moderate, with an intensive renewal rate of products implying short life cycles of the products.

These two criteria are used in determining four distinct evolution paths for both products and product families (see Table 3.2):

The *Dynamic* type family features intensive variety and a high renewal rate, which makes it the most challenging scenario to manage. Against this backdrop, the continuous flow of new product design is likely in some cases to increase diversity within the family, or to reduce it by generating obsolescence among older products.

The Sony Walkman case presented by Sanderson and Uzumeri (1995) foreshadows the case of a company that builds a competitive advantage through its capability of managing a product family of the dynamic category. Sony developed descent lines of products – Sanderson and Uzumeri use the terms *sub-families* or *generations* – with clear objectives in terms of design that targeted precise market segments. Sony generated a larger variety of products than its competitors. As an example, for the American market alone, the company developed more than 250 different models of the product in the 1980s. It should also be noted that these products were widely differentiated, with Sony offering products specific to the Japanese, European and American markets, varying to target such categories as customer lifestyles, sports enthusiasts, music lovers and children. Product longevity was better also, with two years compared to an average of 1.2 years for the competition.

In the case of Sony, the platform design process at the origin of the new descent lines was significantly different from the design process within the descent line. The conception of the new descent line generally involved major technological changes, in addition to changes in the production process.

Yet the new technologies retained links with the dominant design and consequently could not be assimilated with technological breakthroughs. For this reason, several product descent lines could coexist within the Walkman family. Thus the limited number of technical platforms made it possible to generate this wide variety of products through incremental changes occurring within the descent lines.

A modular strategy is at the core of generating diversity in the case of Sony. In reference to Sanchez and Mahoney (1996), the modular strategy is used primarily to increase the variety of products within the same product family at the lowest possible cost, by combining components that adapt to standardized interfaces structuring the architecture of products. This is important in managing *Variety Intensive* and *Dynamic* type families. However, the modular nature of the product also makes it possible to significantly accelerate the renewal rate of products within a family, or even renew the families themselves, a fundamental element for *Dynamic* families. On the one hand, the design process of a product is accelerated by aligning the design process of elements – assuming that departments responsible for designing components can function independently, which drives the renewal rate of products within a family. On the other hand, the process is accelerated by dissociating the learning phases of product architecture from the learning phase of component design, which promotes the renewal rate in families of products.

Nonetheless, the modular strategy alone cannot control the issue of strategic management of product descent lines or of product families. As pointed out by Le Masson *et al.* (2010), the issue of creating new products within an existing line is probably a different type of problem from that of creating an entirely new descent line.

Contribution of the C-K theory

In product development logic within an existing descent line, designing new products involves resolving a problem relatively similar to previously handled problems. The issue is relatively well defined, its complexity can be mastered with a sequential problem resolution approach and the potential for conflict between participants is limited. The problem resolution process focuses on finding a satisfactory solution within a set of partially known solutions, in accordance with the principles of procedural rationality. However, once on the path toward a breakthrough innovation with the objective of bringing out a new line or new family of products, the difficulty resides in determining exactly which problem to resolve. The underlying concept of the product and of the technology used to implement its functionalities may be called into question. It is in this context that Hatchuel (2002) questions the Simonian approach that includes design in the paradigm of decision theory. According to Hatchuel, the design function is not limited to a problem resolution process that seeks to select satisfactory solutions within a bounded perimeter. It is more a question of creating new solutions areas using an expansive rationality.

According to the C-K theory (Le Masson *et al.* 2010) innovative design is present as the creation of an expansive partition in domains of concepts (C) and knowledge (K). From our perspective, the C-K theory also makes it possible to draw a distinction between the new product family creation issue and that of creating a new descent line as per Deforge.

Designing in search of a new value: a strong expansion of the concept with a limited increase of knowledge (Δ C – ∂ K)

The concepts underlying products and their components of products are re-evaluated, but the technology used remains relatively standard. In other terms, validating proposals requires a reasonable investment and the acquisition of knowledge for resolving the problem remains limited. Designers control the process: They offer new solution concepts that must subsequently be validated. Economic value rests on the principle of exploitation of good design standards. As costs for exploring knowledge are limited, there is room financially for increasing the number of concepts tested. In determining new concepts we distance ourselves from functionality of existing products by generating new product families as per Deforge.

Generating a new product line: design by technological mutation and regeneration of business lines, resulting in significant increase in knowledge, but limited development of the concept (∂ C – Δ K)

It is also possible to conceive of a situation where modifications of design remain limited or controlled, while creating knowledge by searching for new technological solutions is likely to be long and costly. In this scenario, it is R&D that leads in exploring knowledge, while the role of innovation function is to focus the cognitive attention of researchers on the most promising concepts. Creating economic value is contingent on the principle of overcoming technical and physical contradictions. This implies an in-depth exploration process of a limited number of concepts. This time the technological breakthrough in the framework of an existing product concept will lead to the emergence of new descent lines in the meaning of Deforge.

These two approaches are especially interesting in that we believe that they can be used to develop a type of limited creativity or *controlled breakthrough* comparable to that observed by Reich *et al.* (2010), where the challenge is to apply the principle of "being creative while staying in the box" (Reich *et al.* 2010, p. 1).

Combining scientific research and conceptual innovation: significant development of both knowledge and concepts (Δ C – Δ K)

Le Masson *et al.* (2010) consider a third solution in which the design of a new product is done completely separate from existing products. This involves conceiving science based products. These products present a dual problematic: determine the product's functional aspects and, simultaneously, produce scientific knowledge concerning the main phenomena associated with the product. The authors mention the Swatch watch example:

> In the 1970s, the Swiss watch industry was confronted with the emergence of a new technique, quartz. Thanks to this technology, Japanese manufacturers had started to inundate world markets with watches which were not only cheap but also far more accurate than the best Swiss watches. Nicholas Hayek, the co-founder of Swatch, saved the Swiss watch-making industry by operating a dual expansion: first, in concepts, by turning watches into disposable fashion objects, and, second, in knowledge, since not only did the industry appropriate quartz technology but it also reinvented the architecture of watches by combining cases and supports to offer a wider variety of watch faces.
>
> (Le Masson *et al.* 2010, p. 289)

However, in this type of situation, the company generates a breakthrough offer that changes the rules of the marketplace, resulting in a general calling into question of the benchmark offer as well as of a majority of existing offers. This situation does not truly correspond to an issue involving the evolution of a portfolio in an existing product, but is rather a question of establishing the basis for a radically new strategic business area.

Toward the generation of controlled diversity: the laws of evolution of technical systems

Can the creation of radically new solution spaces (Le Masson *et al.* 2010) be based on some experience from the past? By definition, radical innovation implies a break with the past. However, there is theoretically an infinite number of ways to reformulate an original problem and to rethink a product. One way to restrict the set of possibilities is to look for general trends in the evolution of products.

The analysis of patents undertaken by Altshuller follows precisely this path, as it is reflected in his *laws of evolution of technical systems*, which subsequently became *laws of evolution*. After setting out these laws, we shall briefly address the issue of their status: Should laws of evolution be interpreted solely as a synthesis of observed statistics or as innate trends that govern the evolution of industrial objects? In the conclusion, we describe the advantages that these laws present to making the concept of descent lines operational from a perspective of innovation management.

The laws of evolution of technical systems

According to the work of Altshuller (1988, 1989, 1999), the laws of evolution are ranked into three groups: static, kinematic and dynamic laws. The objective of static laws (Laws 1, 2, 3) is to verify the structural and functional

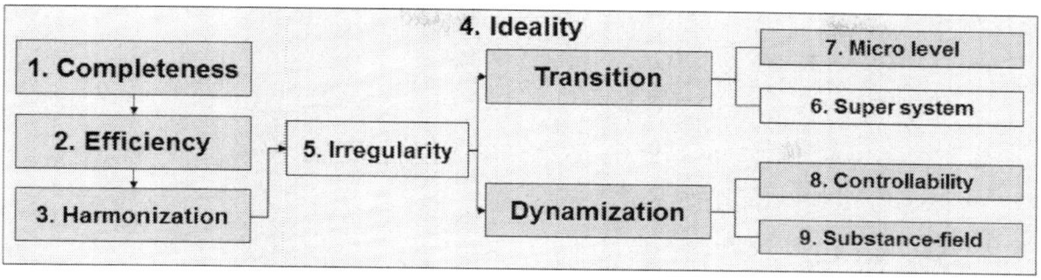

Figure 3.1 Overview of the laws of evolution

integrity of a system while leading designers to ask themselves what the object is missing to satisfy the requirements of the law. Laws 4, 5 and 6 represent kinematic laws where a system is seen in a wider time-space area and analysis focuses on the system's history. Lastly, laws 7, 8 and 9 are dynamic laws, i.e., possible projections into the future. Logically, a system can only follow one of the possible paths. Figure 3.1 gives an overview of the laws of evolution.

A summary of the laws of evolution is given in the remainder of this section. We will use a case of computer keyboards as an illustration.

Law 1: Completeness of parts

In order for a system to carry out its principal function, it is essential that it possesses four fundamental parts that properly fulfill their roles in the functioning of the system: an engine, a transmission element, a working organ and a control organ.

Computer keyboards:

- *engine: the upper surface of keys;*
- *transmission: the key switches and system cables;*
- *working organ: the interface with the mother board, which varies depending on models;*
- *control organ: the groups of springs inside micro-switches that receive messages sent by one's fingers and are used to toggle between two states, 0 for no letter and 1 for a letter.*

If all the parts are available, the control function seems somewhat diminished in its capacity for helping carry out the main function. The individual progress of the keys, springs, cables and interface can thus be perceived as development of the keyboard with regard to law 1.

Law 2: Energy conductivity

One condition necessary for the functioning of a system is unimpeded and effective circulation of energy through its parts. An important indicator of whether or not law 2 is being adhered to is an accounting of loss ratios when input energy is compared to output energy.

> **Computer keyboards:** *Total energy expenditure (muscular, optical ...) required to enter text into a computer should be as low as possible. Certain developments of standard keyboards will make it possible to minimize effort required for pressing a key – think of the first examples of keyboards of manual typewriters, where keys required substantial pressing effort for proper functioning. The development of tactile keyboards advances the concept further in this sense. The Dvorak keyboard may be associated with a reduction of energy. This is because the amount of energy expended by finger movements for entering standard text is lower with Dvorak keyboards than with a QWERTY keyboard.*[4]

Law 3: Rhythm coordination of parts

Wherever the functioning rhythm of one part is inadequate compared to another part there is an inevitable loss of efficiency that is detrimental to system performance. As such, it is essential to establish harmony between the parts of a system to attain optimal functioning.

> **Computer keyboards:** *Microsoft's natural keyboard harmonizes the distribution of letters and the natural position of the arms and hands to form a V. Other concepts go even further with a system of depressing keys by two inverted half-spheres, a little as if the sophistication of concept had been pushed to the point of reaching the tip of each finger.*

Law 4: Increasing ideality

The development of every technical system seeks to reach the highest level of perfection, the ideal. In practice, systems evolve toward the ideal when their functional performance improves with simultaneous decreases in costs. Taken to its extreme, ideality should be imagined as the achievement of the primary function in the physical absence of matter.

> **Computer keyboards:** *With handwriting and voice recognition systems, the keyboard no longer physically exists, but its function is carried out.*

Law 5: Uneven development of system parts

The parts of a system develop in an uneven fashion. The more uneven the evolution of parts, the more the system becomes complex, which is due primarily to one or another of its parts.

> **Computer keyboards:** *The weak link of the system is the contact zone with fingers. Consequently, any brainstorming project bringing this area to the level of the other keyboard parts would be considered progress.*

Law 6: Transition to a super system

A system that has used up all of its development possibilities can vanish, its principal function emerging later within a super system. In this case it becomes a component of the super system.

> **Computer keyboards:** *The screen of touch pads or the table upon which the keyboard was formerly placed, touch tables, actually become keyboards.*

Law 7: Transition from macro to micro level

This law is a reflection of the evolutionary trend of technical systems toward miniaturizing system components. The ultimate phase of development could possibly be interpreted as an evolution toward the nano-world.

> **Computer keyboards:** *The reconfigurable liquid crystal keyboard valid to infinity, for placing keys of a size of one's own choosing and in the location one wants them placed. We can also cite the complex laser optical projection keyboards, or more recently, holographic projection.*

Law 8: Increasing dynamism and controllability

The development of technical systems continuously seeks the highest level of dynamism and controllability, which is reflected by a progressively greater degree of segmentation within a like physical state without impacting the controllability of the system.

> **Computer keyboards:** *Keyboards that can be folded in two or three, or flexible keyboards.*

Law 9: Increasing use of substance-fields

In order to avoid making systems excessively complex, a substantial contribution in terms of functionalities can justify increasing the number of components in the system. This evolution will only take place if the functionality – new cost ratio that is engendered – is positive.

> ***Computer keyboards:*** *Adding a solar cell or a wireless auto-power supply, or adding a biometric sensor for security and/or customization.*

Use of laws of system evolution

In a deterministic approach of evolution, such as that of Salamatov (2002), these laws are often interpreted as innate elements guiding the evolution of the entire system, independent of human actions. According to this perspective, the designer is nothing more than a catalyst of evolution, a tool for bringing the laws to light. As regularities observed on the basis of statistical studies done by Altshuller, the laws may be considered in another way. Thus, we maintain along with Boldrini (2012) and Choulier (2000) that the laws of evolution primarily describe tendencies resulting from the observation of past evolution of objects. Nonetheless, we also believe that these tendencies make up structuring principles that guide designer activities in their research work. Indeed, knowledge of the laws of evolution provides the most logical paths for future evolution, with regard to the situations observed by Altshuller. For engineers, the laws of evolution also constitute a "counter-analysis" system with relation to the standard stage-gate model process involving analysis of the need, etc. Lastly, the laws of evolution lead to designing directions that initially appear appropriate, which helps reduce the problem of combinations of possibilities that arise when seeking to use the descent lines concept as a prospecting tool.

A radar graph may be useful to provide a graphic representation of the laws and to diagnose the direction in which it is expedient to direct designers' attention and efforts. Figure 3.2 shows the maturity index for each branch of a law using the computer keyboard example.

Conclusion

In our economies, there is a consensus concerning the importance of innovation in building competitive advantages. For creative and innovative firms aiming to systematize innovation, a product line based management (product creation within existing lines and creation of family lineage or new product) is an interesting way to consider the strategic management of innovation. Moreover, we wished to tackle the idea that observations of the past evolution of the products can be used to build interesting principles or laws to guide the exploration of breakthrough solutions of tomorrow.

Deforge's work, which describes the evolution of industrial products in the form of descent lines of products, is an important basis for reflection to us, as modern innovation management theories can serve as complements to it. Still, it is rapidly evident that it is impossible to implement this approach in a strategic way in as much as the number of trajectories for evolution grows

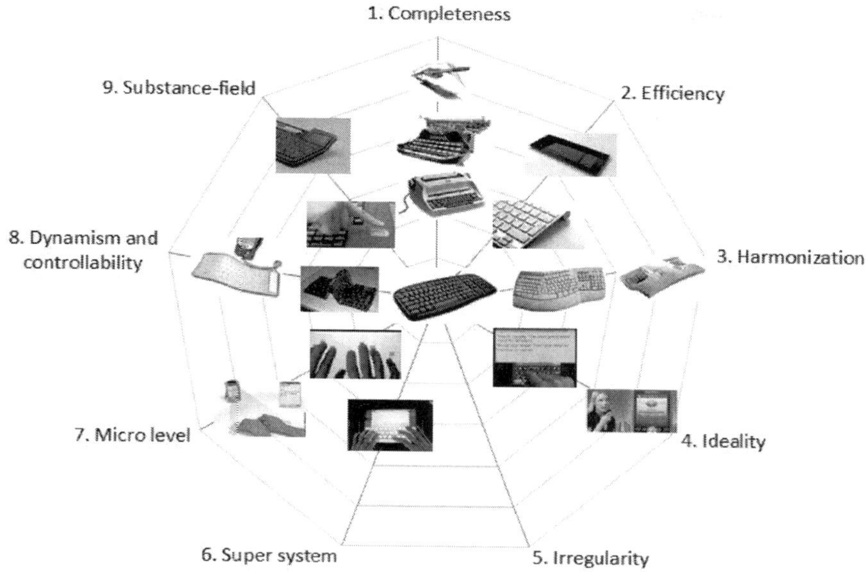

Figure 3.2 Laws of evolution for computer keyboards

exponentially. Using Altshuller's laws of evolution of technical systems appears illuminating when faced with this issue. While the laws primarily reflect an empirical observation, they may also serve as a benchmark for designers. The nine laws of evolution therefore represent product evolution strategies that at first glance are relevant.

Our example of the evolutionary paths of computer keyboards illustrated how they can be analyzed with regard to the laws of evolution. However, this illustration is of a retrospective type. It is a question of anticipating what the keyboard could evolve into when examining it from the perspective of the item's potential. This task requires an extremely precise knowledge of the field and cannot be achieved without the participation of experts (Bootz and Schenk forthcoming).

Notes

1 TRIZ is the Russian acronym for the theory of inventive problem solving. This theory was developed by the Soviet inventor Genrich Altshuller, beginning in 1946.
2 A recent analysis of the link between morphological analysis and innovation can be found in Aloui (2008).
3 A student of Altshuller, whose work has been partly translated by researchers from INSA Strasbourg.
4 David (1985) offers an economic analysis of the emergence of the keyboard QWERTY standard and its persistence in view of Dvorak keyboards.

References

Abernathy, W.J. and Utterback, J.M. 1978, "Patterns of Innovation in Industry," *Technology Review*, **80**: 40–7.

Aloui, A. 2008, "La question de la définition organique de systèmes techniques en conception: cadre théorique et éléments de méthode," Thèse en Sciences de gestion, Univ. Jean Moulin, Lyon 3.

Altshuller, G. 1988, *Creativity as an Exact Science*, New York, Gordon & Breach.

Altshuller, G. 1989, *And Suddenly The Inventor Appeared*, Moscow, Detskaya Literatura.

Altshuller, G. 1999, *The Innovation Algorithm. TRIZ, Systematic Innovation and Technical Creativity*, Worcester, MA, Technical Innovation Center.

Arthur, W. 1989, "Competing Technologies, Increasing Returns, and Lock-In by Historical Events", *Economic Journal*, **99**: 116–31.

Boldrini, J.-C. 2012, "Le rythme implacable de l'innovation est-il gouverné par des lois d'évolution internes aux objets?" *Revue Française De Gestion Industrielle*, **31**: é107–31.

Bootz, J.P. and Schenk, E. forthcoming, "L'expert en entreprise: proposition d'un modèle définitionnel et enjeux de gestion," *Management & Avenir.*

Choulier, D. 2000, "Synthèse sur les lois d'évolution. Comparaison entre les vues de Simondon, Deforge et Altshuller," *Rapport Interne N°05, Mécatronique 3M, Thématique Conception Innovante et Distribuée*, UTBM.

Christensen, C.M., Suarez, F.F. and Utterback, J.M. 1998, "Strategies for Survival in Fast-Changing Industries," *Management Science*, **44**(12): 207–20.

David, P.A. 1985, "Clio and the Economics of QWERTY," *American Economic Review*, **75**(2): 332–7.

Deforge, Y. 1985, *Technologie et génétique de l'objet industriel*, Paris, Coll. Université De Compiègne.

Goodwin, R.M. 1967, "A Growth Cycle," in C.H. Feinstein (ed.), *Socialism, Capitalism and Economic Growth*, Cambridge, Cambridge University Press.

Hatchuel, A. 2002, "Towards Design Theory and Expandable Rationality: The Unfinished Program of Herbert Simon," *Journal of Management and Governance*, **5**: 260–73.

Hofbauer, J. and Sigmund, K. 1988, *The Theory of Evolution and Dynamical Systems*, Cambridge, Cambridge University Press.

Koenigs, G. 1905, *Introduction a une théorie nouvelle des mécanismes*, Paris, A. Hermann.

Lafitte, J. 1972, *Réflexions sur la science des machines*, Paris, Vrin.

Le Masson, P., Weil, B. and Hatchuel, A. 2010, *Strategic Management of Design and Innovation*, Cambridge, Cambridge University Press.

Modis, T. 1999, "Technological Forecasting at the Stock Market," *Technological Forecasting and Social Change*, **62**: 173–202.

Moles A. 1970, *Créativité et méthodes d'innovation*, Paris, Editions Fayard.

Reich, Y., Hatchuel, A., Shai, O. and Subrahmanian, E. 2010, "A Theoretical Analysis of Creativity Methods in Engineering Design: Casting ASIT within C-K Theory," *Journal of Engineering Design*: 1–22.

Salamatov, Y. 2002, *Les Lois D'évolution Des Systèmes Techniques*, mimeo, Traduit par J. Stien, INSA De Strasbourg.

Sanchez, R. and Mahoney, J.T. 1996, "Modularity, Flexibility and Knowledge Management in Product and Organization Design," *Strategic Management Journal*, **17**: 63–76.

Sanderson, S. and Uzumeri, M. 1995, "Managing Product Families: The Case of the Sony Walkman," *Research Policy*, **24**: 761–82.

Simondon, G. 1989, *Du mode d'existence des objets techniques*, Paris, Aubier.

Uzumeri, M. and Sanderson, S. 1995, "A Framework for Model and Product Family Competition," *Research Policy*, **24**: 583–607.

Zwicky, F. 1969, *Discovery, Invention, Research: Through the Morphological Approach*, Toronto, The Macmillan Company.

4 Tools and implications of managing knowledge and creativity

Claude Guittard, Claude Paraponaris, Jean-Louis Ermine and Pascal Lièvre

Introduction

Interests in knowledge management (KM) come from various disciplines: economy, social, engineering. These interests are not often linked. Within the firms, KM practices vary a lot depending on national cultures and on activity sectors. Some of the missions of the scientific community are to identify these interests, to analyze the potentialities, and to propose an evolution of concepts and practices. However, one can identify a potential symmetry between the diversity of practices and the partitioning of scientific approaches.

In an integrated way, we will present in this chapter a state of the art of a set of research developed in the French speaking world during the past four years. Although they come from different perspectives, some French speaking researchers have decided to exchange and compare their problematic and methodologies. Since 2008, an annual congress (GeCSO, Gestion des Connaissances: Société et Organisation), has created an opportunity to consolidate exchanges. In this chapter, we show the common bases of the different fields, as well as their complementarities.

The GeCSO congress started with the objective of establishing a dialogue between the different approaches dealing with dynamics of knowledge in the society and in the firms. The originality of the congress is, from one side, the rich interaction between different disciplines and, on the other side, the importance of the relation with practice, as nearly all the presented researches are based on practical experiences.

This dialogue has been going on for four years[1] without favoring a specific discipline. The constant increase of the number of participants leads to the conclusion that a community is emerging in the French speaking world: it gathers researchers coming from psychology, computer science, economy, management, sociology, communication, learning sciences, and cognitive science.

The different events have gathered researchers from about 30 institutions and from six countries. They have also gathered case studies presented by firms in special sessions.[2] These kinds of special sessions give the opportunity to compare experiences in modeling and managing knowledge and concepts, methods and researches in a more academic framework.

We first present the methodology used to structure this synthesis, and then we will develop each research track including the different studies.

Methodology

In order to synthesize the different approaches in KM, we have studied all the 62 papers published in the first three conferences. To have a first representation of the diversity, we analyzed the keywords that were used: among 173 keywords used by the authors, 8 keywords represent 27 percent of the total amount. They are, in decreasing order:

- knowledge management
- knowledge
- community
- modeling
- sharing
- innovation
- "savoir" (in French, another concept for knowledge)
- engineering.

Several dimensions emerged from the articles: the practical dimension, the epistemological point of view, and the intention to define a theory of action.

The practical dimension can be seen in cases studies in technical or economical environments that are rapidly changing: it may concern questions related to knowledge identification, modeling or sharing. The technological component is important, and the problematic is either to use concepts and methods to achieve a real life specific objective, or to study how knowledge creation and sharing operate, or to build a specific tool. In this dimension, there is a real convergence of the disciplines: social sciences, cognitive sciences, or engineering. Most of the papers describe the functioning or the success of knowledge systems.

The epistemological point of view is characterized by the deep precision given to key concepts and terms in analyzing and modeling. It seems mandatory to give definitions linked to knowledge. The authors stress the importance of giving their own approach on the different cognitive aspects of knowledge, the subjectivity of the notion, and the different modes of expression.

Regarding the theory of action, studying the knowledge dynamics often means adopting an anthropological approach to analyze the link between the subject and his/her environment. Some research deals with the link to procedures or techniques; other research is interested by the link to power or economical efficiency. Then, knowledge is linked to individuals together with the groups they constitute, the decisions they take, within socialization processes that are always specific.

This state of the art is not a statistical study. It is based on theories and methods used, explicitly or not, in the different researches described in the articles. This first synthesis is not limited to the different authors, their disciplines, and their results. It is oriented towards the underlying concepts of the articles. Concurrently with the progress of analyzing the papers, approaches and concepts were discovered, as numerous mark-ups, to explain KM processes.

The identification of the different fields was conducted through a progressive reading, and by referencing the presented concepts and theories. As a matter of fact, for each article, in a systematic way, we have searched for the theories, concepts, and methods (and their authors) that are at the heart of the prerequisites of the authors. Each of the 62 articles was part of one to three different fields. We used a saturation methodology to stabilize each and to operate differentiations.

The different researches are grouped into six main fields:

- knowledge generation from activities;
- epistemological foundations of knowledge as situated action;
- knowledge transfer;
- knowledge capitalization;
- knowledge modeling, with and around the MASK method;
- epistemic communities, and communities of practice;
- creativity and knowledge management.

Knowledge generation from activities

Activities, as any complex system (organizational system, or informational device), are "knowledge intensive." From that observation, which is mostly shared by management sciences, and knowledge engineering, one needs to have pertinent analysis to identify, in the activities, the knowledge creation and dissemination processes, and then build methodologies to facilitate these processes.

Creativity in the activities, especially in design, is also linked to the dissemination of strategic knowledge within the concerned organization. Then, we are within the organization, and we deal with its routines. We highlight the spaces to create and share knowledge in an abductive approach.

KM in design activities

The question is to adapt different timescales in the organization, one for activities or projects, and one for knowledge capitalization.

An example is given in the Myriade project, with a design model based on contradictions (microsatellites design), in CNES (French National Center for Space Studies) studied in Belleval *et al.* (2010). This program implies for the actors an evolution of their theories of action in the sense of Argyris and Schön (1978). The point is to transform a representation based on "Mission Success First" (that targets reliability and robustness of systems) into a representation "Smaller, Faster, Better, Cheaper" inspired by the NASA policy implemented by Dan Goldin, in the early 1990s. This evolution is central to the design process.

In that case, the design problem is ill-defined and contradictions may appear during the process. The need is to overcome that problem, and not to stay in the "problem solving" model initiated by Simon and Newell (1972). Different theories of design are then mobilized, highlighting the interdependency between three spaces (Pimmler and Epinger, 1994):

- the space of product (what to do?)
- the space of the design project (how to do?)
- the space of actors (who does what?)

Interaction between problem solving activities and the design process organization has an impact on the number of iterations in the design process, and on the performance (cost, quality, time). Accepting perturbations occurring between these three spaces, we have a new definition of innovative design. Its process is non-structured, because its objective is built following the pace of the activities. The problem to solve is contradictory: a set of technical, organizational, or cognitive contradictions has to be cleared (Belleval *et al.*, 2010).

System engineering is a good candidate for such an approach, especially the classical "V life-cycle" of Forsberg and Mooz (1991). This model allows the co-evolution of the problem together with the solution.

Requirement analysis leads to defining precisely the technical specification that can be discovered as contradictory. These contradictions have to be solved during the architecture design. For that, one can use an existing knowledge base, or give up the project because it is not feasible, or use the contradictions to create new combinations (Altshuler, 1984), by expanding the solutions space (Le Masson *et al.*, 2006).

In each case, one considers the interactions between the technical characterization of the product, the organization of the design project, and the organization of the company, especially the actors. These three dimensions are deployed in different V cycles to structure design building blocks. The design space is then an open space, with opportunity to solve problems or contradictions by expanding the cognitive, technical, and organizational dimensions.

This study shows that a dead end at only one contradiction can cause the project failure (an incremental innovation instead of a disruptive innovation).

Towards a generalized approach

Mastering of interactions between those three dimensions is then strategic for orienting the design project. This leads to an attempt to generalize what is knowledge generation from activities. This is the proposal of the "Bordeaux School" (Moradi *et al.*, 2008, Brunel *et al.*, 2010) which is a systemic approach that defines an "ingenition" cycle. Those researchers are inspired by the engineering method, and consider that different complementary theories are not linked: design, knowledge, and learning. After several actions in industrial domains, they conclude that it is false to think that "good" specialists know how to manage their knowledge and the way they learn. Their goal is to manage the produced knowledge in the design project by taking into account the whole knowledge value chain in a specific organization. For that purpose, a filter "object–semantics–social and structure" is given to design step by step a product together with the appropriate data, then information and then knowledge. All along that value chain, knowledge, competencies, and actors are shared. The objective is to build knowledge bases available when designing and developing a product.

The integrating knowledge process has five phases:

- modeling the context;
- designing the product usage by the client;
- knowledge generation during the extended life cycle of the product;
- comparative scale of internal and external resources;
- value chain–value sequence–knowledge generation.

This process facilitates integration of resources and cognitive capabilities of the firm.

Epistemological foundations of knowledge as "knowing in action" and impact on knowledge management

Specific research is concerned with knowledge management from a point of view of epistemology of practice. It concerns the definition of knowledge as a situated practice, as "knowing in action" in the sense initiated in the 1950s by Polanyi.

We can distinguish three types of research:

- The research initiated by Lièvre *et al.* (Rix and Lièvre, 2008; Lièvre and Rix-Lièvre, 2010), in the field of "organization studies" intends to give a more theoretical and precise meaning to the notion of tacit knowledge. They prefer to talk about "knowledge in action," in a sense given by Piaget, and consider that "tacitness" is a characteristic of knowledge in action, and prefer to use "implicit" for that non-explicit aspect. This concerns only individual knowledge. They propose to restrict the tacit dimension of knowledge to a characteristic of collective knowledge, as in law one refers to something that is not yet elicited, but already known by different people. This research is to use the Polanyi intuition, the Piaget concepts about "succeed and understand" in the 1970s, and the methodological works of Pierre Vermersch, in a psycho-phenomenological perspective, in the 1990s, to clarify the notions of tacit and explicit knowledge that are central to knowledge management. The basic idea is to consider that practices and then attached knowledge are in a pre-reflective state, then non-conscious, that operates a fundamental distinction between acting and consciousness of acting. When I read, I read. I don't need to be conscious of reading to read. But I can consciously analyze the way I read to progress. The actor is not able to refer spontaneously to his knowledge, because he rationalizes. He makes his personal experience coherent with his counterpart and his identity. One needs then a specific way to access this "knowledge in action": elicitation interview (Vermersch, 2003), self-confrontation interview (Theureau, 2006), subjective re-situ interview (Rix and Lièvre, 2008). It is impossible to access that type of knowledge with a classical simple sociological interview after the action. These authors propose to study the implications of that theory on knowledge management, for instance for knowledge codification.

- The research of Hakim Hachour and Safia Abouad (2008) in information and communication sciences (Paragaphe laboratory, Université Paris 6), in the sociophenomenological framework initiated by Schutz (1953, 1976), distinguishes three types of knowledge: fundamental knowledge, routine knowledge, and specific knowledge. Schutz (1953, 1976; Schutz and Luckman, 1973) defines three types of knowledge (Table 4.1) and is interested in the cognitive styles for achieving a goal. That process is structured in different steps; the articulations between steps are precisely studied. This approach ("course-of-action patterns") is developed in Theureau (2006). Tacit or fundamental knowledge is seen as "obvious." This is the result of purely subjective processes. Specific knowledge is separated from action, it is objective. Routine knowledge is attached to practical goals of the action; it is linked to the potential realization included in the daily environment (signification in action).

From these principles, and from analyzing the organization of a socialized knowledge system, those researchers are designing knowledge management systems as a new organization "memory-centered," in a sense given by Lemoigne, which implies three fundamental levels of modeling: the syntactic level (technical), the semantic level (referential), the pragmatic level (implication). These three levels are dependant from a primordial meta-level: ethics.

Table 4.1 Characteristics of the three types of knowledge (Schutz, 1953, 1976, Schutz and Luckman, 1973)

Fundamental knowledge	Routine knowledge	Specific knowledge
Internalized elements (capacities)	Elements with growing specificities	Externalized elements (data)
Attached to the situation	Dependent on the situation	Independent from the situation
Subjective	Inter-subjective	Objective
Tacit	Observable – describable	Explicit
Obvious	Incidental	Questionable

- The research of Brassac (2008) distinguishes two types of studying cognitive processes:
    - the "internal" perspective, where cognition is an information processing system operating on cerebral entities and mental states corresponding to real life states that the actor tries to understand, memorize, tell, and manipulate;
    - the "external" perspective that relates brain, body, and external world. Cognition is then an activity that doesn't represent the characteristics of the world, but makes them present.

Most of the papers analyzed in the present state of the art are from the second perspective, concerning essentially the managers, the routines, and the communities. We are then far from Simon's model of "rationality" (Simon, 1972; Simon and Newell, 1978) and we are more interested in the practices that participate in knowledge elaboration.

Knowledge transfer

Knowledge transfer practices are various and ancient. One can classify them in two parts: the organizational training systems and the specific knowledge transfer systems.

Knowledge transfer with training systems

Wannenmacher (2010) presents a large synthesis of training systems: tutoring, pairing, mentoring, etc.

- The tutoring function is a process of professional socialization and a process of transmission of professional practices. The tutor guides the newcomer and helps him in his learning problems (Boru and Leborgne, 1992).

- The pairing function facilitates the newcomer's socialization. He is associated with an experienced worker for all of his learning period (Meignant, 2001).
- The mentoring function disseminates know-how, competencies within specific working communities.

These three functions favor interactions. One cannot disseminate know-how without simple actions such as observation, impregnation, imitation, practical experience, advising by experienced workers, demonstration, trial-and–error, exercises. But to be effective, some conditions for learning must be fulfilled.

Knowledge transfer between "Knower" (K) and "Learner" (L)

Research of specific conditions to realize critical knowledge transfer within an organization leads to a better understanding of the knowledge transfer process. Brassac (2000, 2008) has developed a methodology for supporting knowledge transfer, based on an approach of social psychology of cognitive processes. The most relevant situations are the retirements of a technician or engineer.

Knowledge acquisition is a process where at least two actors, "Knower" (K) and "Learner" (L), are co-responsible. This process takes place in a set of social interactions, discursive exchanges, body language, and machines manipulations. The actors must maintain these manipulations. Under those conditions, there is no real knowledge transfer, but a construction of significations that are to be used afterwards by L, after acquisition.

To facilitate that acquisition, Brassac has developed the following methodology:

- speech collection is not enough;
- you should not disconnect the speech of K and the apprehension of this speech by A;
- you should not abstract the expression of the expertise of the concrete place of effectuation;
- you should not prevent both actors K and A from representing diagrammatically working elements.

This is the opportunity for K and L to have a direct relationship: for knowledge, hesitations, oversights, documents, or limits dependencies. This preserves the possibility of asking, raises incoherencies or conflicts, and proposes new perspectives. For that reason, it is necessary to have three persons: K, L, and a knowledge engineer (or the researcher). One uses overviews in a concrete working context, such as films, interviews, or comments on films.

Transfer possibilities are possible because knowledge is distributed among K and artefacts (documents, equipment ...). A good strategy is to have a constant feed-back with the different ways of collecting knowledge (diversity, redundancy). This is the way to reveal discrepancies between performative statements and experiences, and give recommendations and performative reformulations.

To summarize, two practical recommendations are formulated:

- retention: organization of the documentation, expertise retention, organization of resources after corpus analysis;
- capitalization: save operational protocols, anticipate the next transfer, and design a back-up for L's activity.

Industrial knowledge capitalization and management

Lessons learned for technical objects

In an engineering context, working on technical objects, knowledge management is useful for the actors to give sense to information on objects, and browse among the different objects they manipulate in their activities. An example is given by TEEXMA (Rachedi *et al.*, 2010) used as a support for lessons learned. In that tool, a situation can be attached as a fact to a technical object, and can be documented with a short piece of knowledge, that can be appropriated by the concerned actors, in an objective of continuous improvement (Figure 4.1).

Capture professional gestures

Another given example of knowledge capitalization technique is given by video-based knowledge capture. This technique is developed in a sensible domain: nuclear energy (EDF, the French power supplier), to capture the know-how incorporated in technical gestures called "rare." A very deep research has been made about diversity and quality of video-taping and about elicitation modes by actors during recordings (Le Bellu *et al.*, 2009).

The EDF approach is based on a subjective view of the operator's activity, and on a protocol for collecting subjective cognitive elements concerning technical gestures in context. Two cameras are used, one outside the operator, one carried by the operator, nearby his gestures. The choice of "first-person" filming fosters the identification process to the operator by the observer. It also increases the quality of elicitation by the operator.

The questions are: are the operator's routines unconscious (proportionally to his expertise)? How to make him elicit his routines? What are the limits?

The filming technique is very specific. The operator is mentally prepared, and is asked to think of his gesture, to have a reflexive posture regarding his gesture (tell successively goals and results) in order to give the best possible description, in a format adequate to have a cognitive modeling and then a structure for training documents.

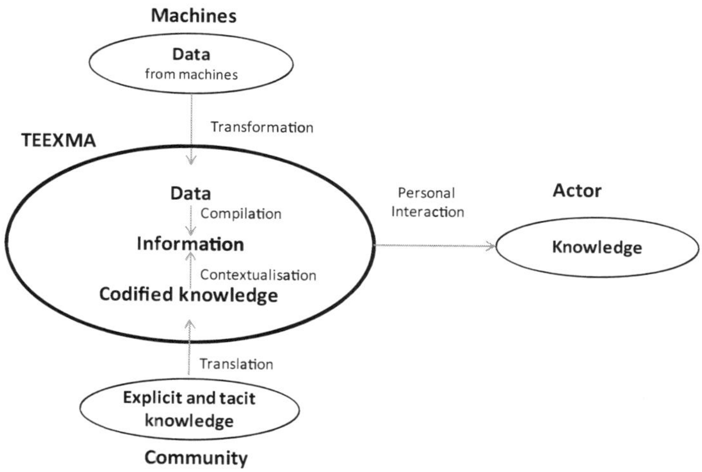

Figure 4.1 The TEEXMA tool for technical knowledge management (Rachedi *et al.*, 2010)

Creative system

Saulais and Ermine (2011) propose a "creative system" within the company, which implies all the knowledge actors by exploiting the databases related to the company's intangible assets. This creative system led to the creation of intellectual property works which will be accumulated in the intellectual corpus and which will generate new value within the intellectual capital. As the intellectual corpus acts as a sub-system of the knowledge corpus, the model proposed is based on the AIK model proposed in Ermine (2003). The authors put emphasis on the creator (mainly as involved in his own creation) as a necessary condition to stimulate his inventive activity. Furthermore, as the intellectual corpus of the organization is intimately linked to the personal intellectual corpus of all the experts (creators, inventors), encouraging the creation of an intellectual property work demands investments in the cognitive profile of the experts and, more globally, in knowledge. Managing the inventive activity means managing human beings (experts) and knowledge.

Knowledge modeling around and with MASK

A lot of research refers to the MASK method (Method for Analyzing and Structuring Knowledge), developed by Jean-Louis Ermine (1996, 2008), quite frequently used for industrial use and others.[3]

In the last version, the MASK method has two parts: the first one (MASK II) is a method for identifying an organization's strategic knowledge portfolio

Table 4.2 Three types of approach for cognitive resources mapping

Functional classification	It tags the competencies. It is easy to make, and is privileged by the management. It is based on the organization chart of the company, and then depends a lot on that organization.
Knowledge domains classification	A knowledge domain is related to the activity of a group of actors. This approach is adapted when the complexity level is high. The models may be, for instance, trees or cognitive maps (Ermine, 2007).
Processes classification	It links knowledge to activity. It is rather complex to operate. It comprises four steps: sensible processes and collective consensual issues; critical activities linked to those processes; knowledge needed to solve the main problems; measurement of the criticality of this knowledge.

(knowledge mapping), the second one (MASK I) is a method for knowledge preservation, based on modeling techniques for tacit knowledge.

MASK II: knowledge identification with knowledge maps

Knowledge mappingKnowledge mapping is a modeling method to classify knowledge and access to the right knowledge at the right moment. This is a management process to identify the portfolio of strategic knowledge, and for the design of a KM plan of the organization. Different approaches can be used with different issues for KM (Table 4.2).

The knowledge map may be used as a data repository, and processed by data mining techniques to progressively elaborate knowledge (Menaouer and Baghdad, 2009).

The connection between the knowledge map and the firm's units is essential. The knowledge systems can be maintained through two logics (Figure 4.2). One is from knowledge to action; the other is from the operational level to the strategic level.

Critical knowledge maps

Knowledge mapping can be extended, in some sensible domains (agriculture, nuclear energy, personal cares, extreme environments …), to minimize the "knowledge crash" risk (i.e., loss of knowledge that can be fatal). According to Grundstein (2002), crucial knowledge is necessary for companies to solve their critical problems. Each knowledge domain is tagged by a criticality measure that is based both on vulnerability of knowledge (rareness, accessibility, cost and delays for acquisition), and its strategic importance.

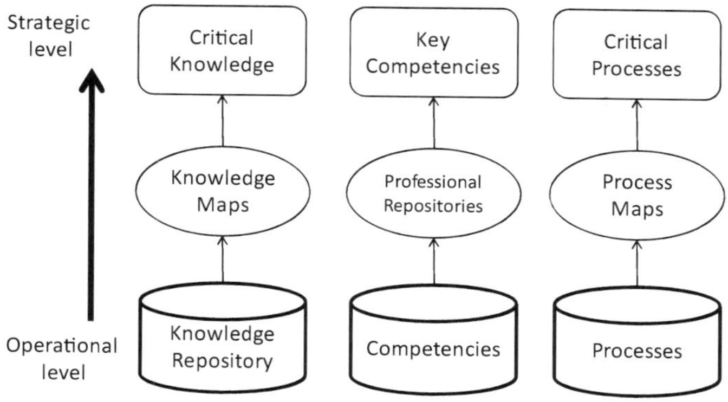

Figure 4.2 The three approaches: knowledge/competencies/processes (Aubertin, 2007)

This analysis is made by semi-directive interviews with institutions and practitioners, in three parts. We give an example for a system in organic farming (Soulignac *et al.*, 2010).

The first part gives the general contexts and infrastructures and identifies where the concerned actors are. The second part develops the informational activities. The third gives a model of information flows between the institutions and other actors. Each type of information is related to the support (technological dimension), the origin (organizational dimension), and the content (informational dimension) (the SOC model: Support, Origin, Content). Relational matrices are then established to identify flows (information-prescription): weak knowledge flows and strong knowledge flows.

After this phase of identification of the knowledge system involved in organic farming, the map of the different knowledge domains is built, and for each knowledge domain, the criticality is calculated, and domains are classified (Figure 4.3). Criticality is proportional to a combination of four parameters: usefulness, rareness, difficulty of acquisition, and difficulty of use. Criticality evaluation is qualitative; it is calculated with a grading system.

Reducing criticality requires knowledge development and acquisition by a significant number of actors. This must be designed according to the double nature of knowledge.

MASK I: knowledge elicitation with modeling techniques

MASK I is a knowledge engineering method. Its objective is to collect and capitalize tacit knowledge of experts with interview and modeling techniques (Ermine, 2008). The objective is to deliver a structured synthesis of the knowledge on a given domain in a form named (as a metaphor) a "knowledge book."

Knowledge themes
Weed
Phosphated fertilization
Nitrogenous fertilization
Climat, ground
Crop rotation
Market
Sulphurated fertilization
Harvest, storage
Potassium fertilization
Varieties
Slugs
Insects
Airborne diseases
Ground diseases

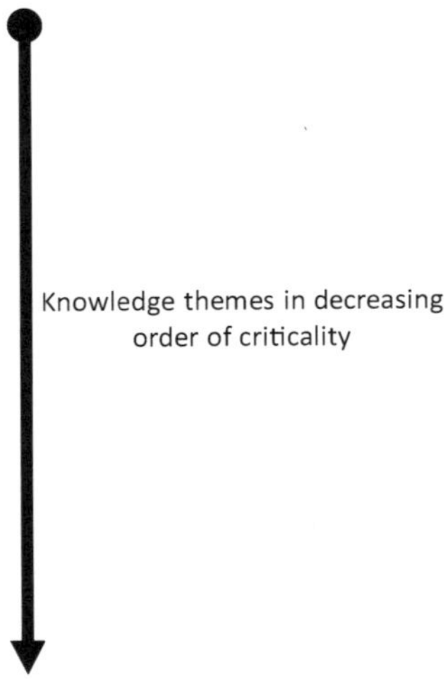

Figure 4.3 Identifying the critical knowledge in organic farming (Soulignac *et al.*, 2010)

This method is a tool for an expert, in interaction with a knowledge engineer, to analyze and formalize his knowledge to create a knowledge base. It is based on a conceptual framework given by the "knowledge macroscope" (Figure 4.4). It is a modeling method based on six different models for six different fundamental types of knowledge (Table 4.3).

In practice, the MASK I method is implemented in three steps:

- Knowledge elicitation: that process consists in making visible an added-value part of a personal or collective tacit knowledge ("pushing away the tacit frontier"). Elicitation is made by interviews with knowledge holders, co-construction of different knowledge models, peer reviews, and assessments.
- Knowledge book, or knowledge repository design, structured by the explicit knowledge collected in the first step.
- Implementation of IT-based knowledge systems (knowledge servers, learning systems, collaborative systems for communities).

The limits of this elicitation process are regularly discussed in the literature (Lièvre and Bonnet, 2010). The transfer of real life situated experience of experts to modeling of this experience with the MASK I models is related to very important questions. How to skip from the expert's practices, which are "unlogical" in the sense given by Bourdieu (1994) to a formal

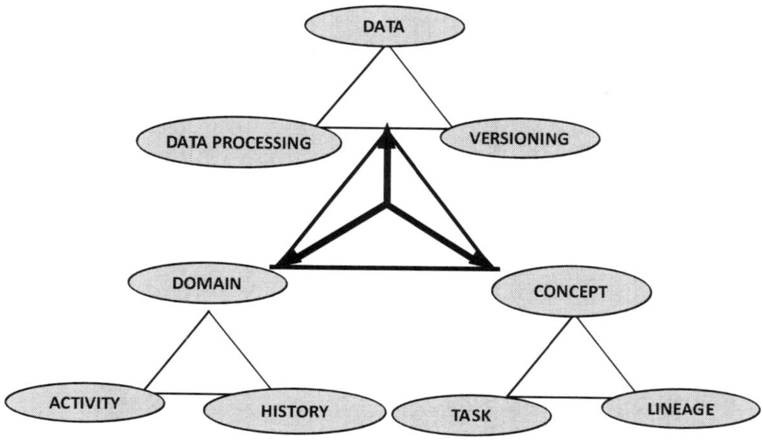

Figure 4.4 Knowledge macroscope in MASK I (Ermine, 1996)

Table 4.3 The different models in MASK I

Domain: model general phenomena that must be mastered in the domain
Activity: model the processes of the domain. Identify and decompose the activities, the flows between activities, the actors, the resources. Identify the knowledge, know-how …
Concept: model the conceptual structure of an expert (semantic network)
Task: model know-how as problem solving methods
History: model the historical context explaining the evolution of the domain
Lineage: model genealogic trees of solutions, ideas, products … with the attached pros and cons

logic framework given by the MASK I models? The basic idea is to start from investigating "knowledge in action" (Piaget, 1974a, 1974b), with a methodology as the elicitation interview (Vermersch, 2003), and then build a link between the level of "knowledge in action" and the formal level of MASK.

Epistemic communities and communities of practice

Specific research on communities is conducted at the Strasbourg school, within the BETA lab (Bootz, Burger-Helmchen, Créplet, Cohendet, Dupouët, Guittard, Kern, Munier, Pénin, Schenk). Its interest is in knowledge dynamics, and its relationship to authority.

Different approaches for communities

Research on communities of practice and on epistemic communities has now a well-known genealogy. The first identification of communities was made in case studies.

Orr (1990) is recognized as the founder of this research relative to learning, socially designed by the employees. More generally, the literature on community is presented with four main streams, based essentially on North American sociology:

- learning by practice (Brown and Duguid, 1991);
- the situated characteristic of knowledge (Greeno and Moore, 1993);
- learning as a social construct (Lave, 1988; Lave and Wenger, 1991);
- learning as a social practice where significations linked to action are negotiated (Wenger, 1998).

The relationship between communities and knowledge is that of local and socialization. Communities are analyzed as places where, permanently, local models, shared representations, and jargons are built, hence enabling synergies of individual varieties (Amin and Cohendet, 2004; Cohendet and Llerena, 1999). A community is founded on self-made norms that are de facto guides for action. Not to respect the norms may jeopardize the community (Munier, 2009).

Most organizations are made up of several communities. How then can functional organization and autonomous communities be coordinated?

The study of community implies the study of problematic interactions. Interactions with hierarchical structures ask the question of knowledge dissemination and combination. In that case, communities are not much concerned. Interactions between autonomous communities ask the question of knowledge transfer, with the background of organizational pressure.

The antinomy hierarchy–community

The rules for the hierarchy and the communities are antagonist. The community can have pressure from the hierarchy in its organization. But the community is creative through its own norms, and more generally through the interactions of its members. The hierarchy is interested in the community to develop its activities, to innovate, and to control. From a synchronic point of view, the relationship firm > community can be apprehended as a pressure (Munier, 2009). A community is viable if it respects the social norms elaborated by its members, it acts regarding a repository of norms. Hence, the community can stand the pressure, if and only if, the required norms belong to the repository of possible norms.

If a firm is always objective-driven, a community evolves regarding its past, its action is structured by constraints on its resources and orientation on its rules. The firm, as a regulator, must manage a double bind: respect

Table 4.4 The two types of community

Communities of practice (CoPs)	Epistemic communities (EC)
Their objective is to improve their activity by leading a reflection on the practices, building, and sharing of a common repository of resources. Socialization process made up of tacit knowledge, socially localized in shared repository of common resources (routines, artifacts, vocabulary, and styles).	Small working groups composed of actors working on subset of mutually recognized problems linked to a type of knowledge, which, at least, accept a procedural authority, recognized by everybody in the same way and seen as essential to the success of their cognitive activity (Cowan et al., 2000). They use codebooks: dictionaries, grammar rules, to facilitate communication and synergies. They create knowledge with the use of externalization.

Table 4.5 Entrepreneur and manager roles (Bootz and Kern, 2009; Créplet *et al.*, 2001)

Entrepreneur role	Manager role
Oriented towards dissemination and management of new knowledge, tacit activity, difficult to identify. Intention + attention. Intention: wish to implement a given strategy and rally the company. Attention: observation and assimilation of weak signals from all communities. The entrepreneur tries to maintain knowledge diversity together with dialog.	Oriented towards internal/external links. Guarantee an environment favorable to community creation, and manage the knowledge by codification. Preoccupied by interfaces management to keep control. Use of frontier objects to guarantee communication.

the norms of the community, and capture and orientate this community. We have then a double governance structure, which justifies the interest of studying the complementarities between "entrepreneur" and "manager" from a management point of view.

From a classical point of view, the community is considered as a risk of losing control. Advanced studies prove now that is not really true. The strange characteristics of a knowledge community must be seen as a managerial innovation, and not as an organizational risk (Bootz and Kern, 2010).

From an evolutionist point of view, several ways for promoting communities are explored (Créplet and Dupouët, 2009). The authors recommend not directly managing the communities, in order to create an efficient environment for communities development, and not edicting contradictory injunctions of rules or norms. The evolution approach is also the framework used by Saulais and Ermine (2011) to propose useful tools for knowledge

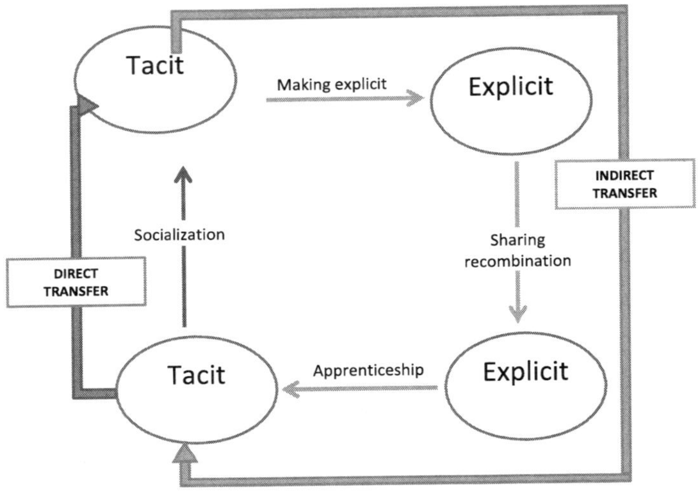

Figure 4.5 Indirect and direct knowledge transfer, according to Nonaka (Ermine, 2007)

management. These authors propose models to understand the creative system within a company and knowledge evolution system

Conclusions

As a synthesis of this transverse analysis, we can give two converging main streams:

> Knowledge cannot be conceived as objects relevant to physical or economical manipulations.

A clear distinction is made between information and knowledge:

- knowledge is structured in personal schemes and in dialogist processes "person/environment";
- information is the code used by knowledge to disseminate. It may be the support for knowledge transfer.

Two knowledge management strategies

Two KM strategies can be adopted by the organizations. The first one is a defensive strategy to prevent risk (knowledge loss, knowledge crash). Knowledge codification is a main tool in that strategy (knowledge books, procedures, documentation …). The second one is an offensive strategy, where knowledge is a resource giving the opportunity to develop new products and improve the processes in the firm (creativity, innovation). There are many tools for that strategy, which can be seen as supports and combinations of two types of knowledge transfer (Figure 4.5).

Exploring those two strategies will be, with no doubt, the next challenge to integrate more and more the different scientific domains of knowledge management.

Notes

1 Groupe ESC Troyes (2008), École de Management de Bordeaux (2009), BETA and École de Management de Strasbourg (2010), Centre de Recherche Clermontois en Gestion et Management, Groupe ESC, Clermont – Ferrand (2011). The 2012 congress is at École Supérieure de Gestion, UQAM, Montréal.
2 The presentations in special sessions are not included in this analysis.
3 See for instance: http://aries.serge.free.fr/.

References

Altshuller, G.S. (1984) *Creativity as an Exact Science: The Theory of the Solution of Inventive Problems*, New York: Gordon and Breach Science Publishing.

Amin, A. and Cohendet, P. (2004) *The Architecture of Knowledge*, Oxford University Press, Oxford.

Argyris, C. and Schön, D. (1978) *Organizational Learning*, Addison-Wesley, Reading, MA.

Aubertin, G. (2007) "Cartographier les connaissances critiques: une démarche stratégique pour l'entreprise," *Management des connaissances en entreprise*, Lavoisier-Hermes Science, Paris: 125–44.

Belleval, C., Deniaud, I., and Lerch, C. (2010) "Modèle de conception à base de réseau de contradictions. Le cas de la conception des microsatellites au CNES," *Actes du 3ème Colloque Francophone de Gestion des Connaissances*, Université de Strasbourg.

Bootz, J.P and Kern, F. (2009) *Les communautés en pratique: leviers de changements pour l'entrepreneur et le manager*, Lavoisier-Editions Hermès, Paris.

Bootz, J.P and Kern, F. (2010) "Les communautés de connaissance: leviers de changement organisationnel pour le manager et l'entrepreneur," *Actes du 3ème Colloque GeCSO Gestion des Connaissances*, Université de Strasbourg.

Boru, J.J. and Leborgne, C. (1992) "Introduire et développer la fonction tutorale dans les entreprises," *Actualité de la Formation Permanente*, **119**: 24.

Bourdieu, P. (1994) *Raisons pratiques*, Editions du Seuil, Paris.

Brassac, C. (2000) "Intercompréhension et Communiaction®," in A.-C. Berthoud and L. Mondada (eds.), *Modèles du discours en confrontation*, Peter Lang, Berne: 219–28.

Brassac, C. (2008) "L'acquisition de savoirs comme activité située et distribuée," *Actes du 1er Colloque GeCSO*, ESC Troyes.

Brown, J.S. and Duguid, P. (1991) "Organizational learning and communities-of-practice: Toward a unified view of working, learning and innovation," *Organization Science*, **2**(1): 40–57.

Brunel, S., Zolghadri, M. and Moradi, M. (2010) "Global approach for knowledge management in design," *Information Sciences for Decision Making*, **40**: 703.

Cohendet, P. and Llerena, P. (1999) "La conception de la firme comme processeur de connaissances," *Revue d'Economie Industrielle*, **88**, 2eme trim. 99: 211–36.

Cowan, R., David, P., and Foray, D. (2000) "The explicit economics of knowledge codification," *Industrial and Corporate Change*, **9**(2): 211–53.

Créplet, F. and Dupouët, O. (2009) "Communautés d'apprentissage et pratiques prospectives: de l'aide à la décision au management des connaissances," in J.P. Bootz and F. Kern (eds.) *Les communautés en pratique: leviers de changements pour l'entrepreneur et le manager*, Lavoisier-Editions Hermès, Paris.

Créplet, F., Dupouët, O., Kern, F., and Munier, F. (2001) "Dualité cognitive et organisationnelle de l'entreprise: le rôle différencié du manager et de l'entrepreneur," *Revue d'Économie Industrielle*, **95**: 9–22.

Ermine, J.L. (1996) *Les systèmes de connaissances*, Editions Hermès, Paris.

Ermine, J.L. (2003) *Gestion et Ingénierie des connaissances, modèles et méthodes*, Editions Hermès, Paris.

Ermine, J.L. (2007) "Introduction au Knowledge Management," *Management des connaissances en entreprise*, Lavoisier-Hermès Science, Paris: 23–45.

Ermine, J.L. (2008) *Management et ingénierie des connaissances, modèles et méthodes*, Lavoisier-Hermes Science, Paris.

Forsberg, K. and Mooz, H. (1991) "The relationship of system engineering to the project cycle," *INCOSE*: 57–65.

Greeno, J.G. and Moore, J.L. (1993) "Situativity and symbols: Response to Vera and Simon," *Cognitive Science*, **17**: 49–59.

Grundstein, M. (2002) *Gameth: un cadre directeur pour repérer les connaissances cruciales pour l'entreprise*, Lamsade Université Paris-Dauphine.

Hachour, H. and Abouad, S. (2008) "Une contribution de la sémiotique des dystèmes d'action à la gestion de la connaissance," *ISDM*, **33**.

Lave, J. (1988) *Cognition in Practice*, Cambridge University Press, Cambridge.

Lave, J. and Wenger, E. (1991) *Situated Learning: Legitimate Peripheral Participation*, Cambridge University Press, Cambridge.

Le Bellu, S., Lahlou, S., and Le Blanc, B. (2009) "Comment capter le savoir incorporé dans un geste technique à l'aide d'outils numériques?" *Actes du 3ème Colloque GeCSO Gestion des Connaissances*, BEM, Bordeaux.

Le Masson, P., Weil, B., and Hatchuel, A. (2006) *Les processus d'innovation: conception innovante et croissance des entreprises*, Editions Hermès, Paris.

Lièvre, P. and Bonnet, E. (2010) "Retour sur la mise en œuvre de la méthode MASK à la conduite du blé biologique: la question du mode d'investigation des connaissances tacites," *Actes du 3ème Colloque GeCSO Gestion des Connaissances*, Université de Strasbourg.

Lièvre, P. and Rix-Lièvre, G. (2010) "La dimension 'tacite' des connaissances expérientielles: une mise en perspective théorique et méthodologique," *Actes du 3ème Colloque GeCSO Gestion des Connaissances*, Université de Strasbourg.

Menaouer, B. and Baghdad, A. (2009) "Vers une cartographie des connaissances guidée par fouille de données," *Actes du 3ème Colloque GeCSO Gestion des Connaissances*, BEM, Bordeaux.

Moradi, M., Brunel, S., Zolghadri, M., and Vallespir, B. (2008) "Design for learning and teaching: A knowledge-based approach to design products," in T. Yamaguchi (ed.) *Practical Aspects of Knowledge Management*, Springer, Berlin: 244–55.

Munier, F. (2009) "Viabilité des communautés de pratique pilotée: un modèle dynamique," *Actes du 2ème Colloque GeCSO Gestion des Connaissances*, BEM, Bordeaux.

Orr, J. (1990) "Sharing knowledge, celebrating identity: War stories and community memory in a service culture," in D.S. Middleton and D. Edwards (eds.) *Collective Remembering*, Sage, Beverly Hills, CA: 169–89.

Piaget, J. (1974a) *La prise de conscience*, PUF, Paris.

Piaget, J. (1974b) *Réussir et comprendre*, PUF, Paris.

Pimmler, T.U. and Eppinger, S.D. (1994) "Integration analysis of product decompositions," *Proceedings of the ASME Design Theory and Methodology Conference*, **68**: 343–51.

Polanyi, M. (1962) *Personal Knowledge: Towards a post-critical philosophy*, Harper Torchbooks, New York.

Rachedi, S., Boujut, J.-F., and Poussier, G. (2010) "Etude de cas d'une offre industrielle de management des connaissances techniques," *Actes du 3ème Colloque GeCSO Gestion des Connaissances*, Université de Strasbourg.

Rix, G. and Lièvre, P. (2008) "Towards a codification of practical knowledge," *Knowledge Management Research & Practice*, **6**: 225–32.

Schön, D. (1983) *The Reflective Practitioner: How professionals think in action*, Basic Books, New York.

Schön, D.A. (1996) *The Reflective Practitioner: How professionals think in action*, Ashgate Publishing Limited, Aldershot.

Schutz, A. (1953) "Common-sense and scientific interpretation of human action," in *The Problem of Social Reality*, Collected Papers I, Martinus Nijhoff, The Hague, 1962: 3–47.

Schutz, A. (1976) *The Phenomenology of the Social World* [1932], HEB, London.

Schutz, A. and Luckmann, T. (1973, 1989) *The Structures of the Life-World* (Volume I and II), Northwestern University Press, Evanston, IL.

Simon, H. (1978) "Rationality as process and product of thought," *American Economic Review*, **68**(2): 1–16.

Simon, H. and Newell, A. (1972) *Human Problem Solving*, Prentice Hall, Englewood Cliffs, NJ.

Saulais, P. and Ermine, J.L. (2011) "Creativity and knowledge management," *Actes du colloque GeCSO*, Clermont-Ferrand.

Soulignac, V., Ermine, J.L., Paris, J.L., Devise, O., and Chanet, J.P. (2010) "Gestion informatisée des connaissances pour une agriculture durable," *Actes du 3ème Colloque GeCSO Gestion des Connaissances*, Université de Strasbourg.

Theureau, J. (2006) *Le cours d'action: méthode développée*, Octares, Toulouse.

Vermersch, P. (2003) *L'entretien d'explicitation*, ESF, Paris.

Wannenmacher, D. (2010) "Le partage de connaissances tacites: l'importance de l'interaction in situ," *Actes du 3ème Colloque GeCSO Gestion des Connaissances*, Université de Strasbourg.

Wenger, E. (1998) *Communities of Practice: Learning, meaning and identity*, Cambridge University Press, New York.

5 Innovation, territories and creativity

Some reflections about usual and less usual instruments for innovation-driven regional policies

Emmanuel Muller, Jean-Alain Héraud and Andrea Zenker

Introduction: addressing the issue of regional innovation policy instruments from a different perspective

The recent decades have witnessed an increasing prominence of the region as a relevant level for innovation policy. Theoretical discussions particularly underlined this issue in the 1990s (see, e.g., Amin and Thrift 1994; Cooke 1997; Malmberg and Maskell 1997; Morgan 1997; Storper 1997). It started being universally accepted (and it is not a paradox) that, within the general framework of increasing globalization, regions are key sites for knowledge creation and innovation. Therefore, instruments such as territory-based clusters linking private and public (in particular academic) actors developed worldwide around the turn of the millennium. They complemented overall national instruments (sector-oriented policies, tax-credit systems, etc.) with more refined and focused devices. In a way, it was the discovering or re-discovering of the idea of the innovation "ecosystem."

Such policies fully recognize the fundamental nature of innovation as an interactive learning process. As a consequence they are multi-actor, multi-sector and multi-level.[1] What we want to underline in this chapter is the even more complex network of knowledge combination in action behind any innovative activity, especially if we consider the whole process, from the very up-stream creative segments to the diffusion-adoption phases. Creativity in earlier stages is also a network phenomenon, but not necessarily among institutional actors such as firms or public labs.

Cohendet *et al.* (2010), while presenting "the anatomy of the creative city," explain the complex process of creation linking three levels: The underground where relatively loose communities explore possible new concepts in various fields of knowledge or cultural spheres; the upperground formed of explicit organizations (firms and institutions), focusing on the exploitation of the new knowledge and therefore possibly performing the relevant innovations;

and the middleground (specific actors, platforms, events, places or spaces ...) which is essential for catalyzing the transmission of novel opportunities between underground and upperground. The innovation policy on a territory like a creative city must understand this complex creative scheme and decide where to operate: Stimulating underground actors? Supporting upperground structures? Implementing places for middleground functions like fairs, festivals, ideas labs ...?

Considering this category of creative processes, the sorts of policies that must be implemented in order to stimulate innovative behaviors are completely different as we can see. This is particularly true when addressing the issue of collective creativity as it will be detailed later on in the chapter.

In this respect the main issue of the chapter is to provide a frame for the comparison of different instruments that may support innovation-led regional development through a reinforcement of collective creativity. To this end, three different instruments will be examined with the help of this frame as examples of possible options (we do not aim at any form of exhaustiveness in terms of instruments). Beside the previously evoked cluster policies we will consider two further examples that cast light on creative mechanisms. The second example operates at a meso (but local) level and involves actors that are not necessarily "legitimate" from an institutional point of view to participate in the creativity/innovation process: the *ideas labs*. The third example is at microeconomic level; it points to the role of specific knowledge actors within firms, i.e., *knowledge angels*.

The structure of the chapter is as follows. The next section displays the proposed dimensions setting the frame for the intended comparison whereas the third section considers comparatively the innovation policy instruments selected for the sake of the analysis. The fourth section broadens the reflection in stressing the need for contextualization of regional policy instruments and in making the link with current policy trends. Finally, the conclusion will give us the opportunity to provide some elements nurturing what could become a counter-manifesto for creativity-led regional innovation policies and instruments.

Proposing five dimensions for the analysis of selected innovation policy instruments

The starting point of the investigation is the following. If we take for granted that collective creativity constitutes an avenue for regional innovation-led development, then we need to think about how the relevance of different policy instruments can be appreciated in this respect.

Nevertheless, and as a prerequisite, it seems worthwhile first better understanding the complex relations between innovation and collective creativity. According to Parjanen (2012) who investigates collective creativity in the context of practice-based innovation, innovation processes are triggered by problem(s) utilizing scientific and practical knowledge production and creation in cross-disciplinary innovation networks. Within these networks diversity or distances between innovation actors are essential. Innovation potential may be found in exploiting different kinds of distances (such as cognitive, functional and organizational). In other words, efficient regional innovation policy instruments with regards to collective creativity should allow and/or foster "strange" if not "bizarre" or at least "unexpected" encounters between ideas, people and organizations.

The aim here is to provide a very basic frame which makes it possible to put side by side (and intrinsically to compare) different policy instruments considering the deeply changing nature of innovation processes and systems and its implication for regional policies and instruments.

Moreover, the "changing nature of innovation processes and systems" (cf. Smits and Kuhlmann 2004, pp. 5–8) – which leads also to the concept of transformative innovation (in the sense given by Steward 2008) – can be characterized with the help of three trends, i.e., i) the end of the linear model; ii) the rise of the systems approach; iii) inherent uncertainty and need for learning. According to Daimer *et al.* (2012, p. 220):

> Increasingly, phenomena like social innovation, service innovation, low-tech innovation, frugal innovation, relational innovation and value innovation are being recognised as highly relevant innovation arenas that are challenging the standard definition … This goes hand in hand with a perception of innovation as a local phenomenon that is emerging across the globe wherever it is required and not just exclusively in few innovation hubs.

Consequently, and considering previous works dealing with the examination of innovation policy instruments, in particular Wieczorek and Hekkert (2012), we propose the following five-dimension framework summarized in Table 5.1.

Considering three regional innovation policy instruments

Following the logic of an exploratory exercise, the review we propose is based on three types of policy instruments. It must be clearly stated that we do not claim any form of exhaustiveness. The choice made was only led by a willingness of exemplarity as well as by our previous knowledge concerning the selected instruments particularly due to numerous comparative analyses performed on French and German cases.[2] We will also stress here the ontological limitation of such an archetypical approach, being perfectly aware that the reality of the instruments is much more complex than their simple description or assessment. This does not only concern the design of policy instruments but also their implementation.

Table 5.1 The proposed framework

Dimension 1: "Participation" i.e., to which extent does the concerned instrument stimulate the participation of various actors (firms, academia, public agencies, (creative) individuals, etc.)?
Dimension 2: "Interaction" i.e., to which extent does the concerned instrument stimulate the occurrence of interaction among heterogeneous actors?
Dimension 3: "Equilibrium" i.e., to which extent does the concerned instrument prevent ties that are either too strong or too weak and/or prevent institutions involved from being too weak or too stringent?
Dimension 4: "Resources" i.e., to which extent does the concerned instrument require a high level of resources (financial means, physical infrastructure, etc.) to be mobilized and/or require huge investments by public authorities due to complex and cost-intensive management?
Dimension 5: "Knowledge" i.e., to which extent does the concerned instrument create space for actors' capability development and favor knowledge creation and circulation between them?

The first type of instrument we consider for this analysis can be seen nowadays as an archetypal instrument of both innovation and regional policy: it is cluster policies. This is a worldwide phenomenon, though presenting slightly different applications following national traditions. It is worthwhile noting that before becoming the name of a policy, the word "cluster" was used for describing spontaneous concentrations of specialized activities. For A. Marshall, the firms "tend to cluster" in specific geographical environments to benefit from what we now call "knowledge externalities," as well as from other spatial externalities like infrastructures. According to Porter (1998, p. 78) clusters are:

> geographic concentrations of interconnected companies and institutions in a particular field. Clusters encompass an array of linked industries and other entities important to competition. They include, for example, suppliers of specialized inputs. Clusters also often extend downstream to channels and customers and laterally to manufacturers of complementary products and to companies in industries related by skills, technologies or common input.

The most famous innovative clusters often referred to in the policy literature and in the administrative/political documents resulted more from self-organized historical processes than deliberate policies: this is true for the mythical Silicon Valley as well as for the more recent concentration of nanotechnologies in the Grenoble area. Public (multi-level) policies and institutions contribute to the development of such "ecosystems" of innovation, but the most efficient ones were not designed and planned by any local or national authority.[3] Generally, public action just accompanies existing or potential clusters. This is one important ambiguity of the concept of "cluster policy" and a noticeable difficulty when one comes to policy evaluation. Nevertheless, cluster policies can be broadly sketched as requiring huge investments, very long term vision and a complex governance. Moreover, most (successful) clusters appear to be controlled by big players and quite often strongly technology oriented.

The second type of instrument we consider in the present review may be designed with different but nevertheless similar names such as "ideas labs" but also "living labs" or "innovation labs." Even if some literature can be found about this type of initiative (cf. for instance Niitamo *et al.* 2006 and Kusiak 2007), no generally accepted definition can be provided as in the case of clusters. Based on empirical observations performed in the frame of an applied research project comparing different cases in France, Germany, Spain and Canada,[4] a tentative definition can be proposed. In fact, ideas labs (and similar organizations) can be described as improbable places that allow a not very likely mix (i.e., between ideas, people, institutions, needs and so on) to take place. In other word, ideas labs provide (or at least attempt to) a form of institutionalization (or a kind of "production framework") supporting or accelerating serendipity phenomena. The paradoxical approach often adopted can be probably best described with the help of oxymora such as "happy accidents" or "to be expected surprises." Most often, "opposites" (such as artists and scientists, users and producers, young and old people, young business creators and retired professionals, etc.) are expected to meet and match so that creative spillovers result from this confrontation. Turning to the support of ideas labs and similar organizations as instruments for innovation-led regional policies some basic common features can be identified. First of all, setting up, managing and supporting ideas labs imply a complex governance but the level of investments required is less huge than in the case of cluster policies. In the same line, a less strong technology orientation is to be expected and some quicker (if any) results can be anticipated. Finally, the relationships between the different actors involved may be intrinsically more strongly equilibrated and more interactive than in the case of clusters but less motivating or attractive in terms of communication (in particular when considering the image effect attached to certain clusters).

The last type of instrument relates to knowledge angels-supporting policies. The core assumption of the research string dealing with knowledge angels consists in the existence of specific individuals who perform tasks – based on their creative abilities – that significantly increase the creative capacities of the firms they are embedded in (Muller 2008). Individuals designed as knowledge angels may be detected within KIBS (knowledge-intensive business services).[5] The underlying assumptions of investigations devoted to

Table 5.2 A synoptic comparison of the three types of instruments

Instruments dimensions	*Clusters policies*	*Ideas labs*	*Knowledge-angels supporting policies*
Participation	+++	+	++
Interaction	++	+++	++
Equilibrium	+	+++	++
Resources	+++	++	+
Knowledge	+	+++	+++

knowledge angels can be briefly summarized as follows. Knowledge angels are (or may be) specific individuals, who: (i) typically act as consultants (but not necessarily exclusively); (ii) may have the talent to "sense" things before they happen, or make them "happen" (from the subjective point of view of an external observer); (iii) make a difference in the way knowledge is created, organized and flowing within the firm and between the firm and its partners. In other words, knowledge angels are "suspected" of being able to generate their own markets (and/or to create their own jobs and working environments) to a certain extent. It is assumed that these key actors have the talent and creativity to evaluate externally available knowledge and to match it with the needs of the companies they are working for. Until now, investigations performed about knowledge angels were restricted (for methodological reasons) to KIBS, nevertheless and at least from a theoretical point of view, nothing excludes the possibility of the existence of knowledge angels in further types of (manufacturing or service) firms as well as in public administration or academic institutions. So far (and to the knowledge of the authors) no existing instruments that are explicitly displayed as knowledge angels-supporting policies can be detected, nevertheless this could become the case in a near future. More generally any kind of instruments supporting creativity and innovation in KIBS could be considered as a form of knowledge angels-oriented policy. For the sake of the demonstration, we assume that such kinds of instruments constitute to a certain extent an alternative way to the two types of instruments presented before. In other words, supporting at local level the development of knowledge angels would mean supporting some form of "mobile and agile one guy ideas lab" within mini-clusters constituted by the concerned KIBS and related clients (cf. Muller *et al.*, 2012a, 2012b). Compared to instruments devoted to clusters or ideas labs, this category of policy actions would offer an interesting mix of possibilities as shown in the following synoptic review (cf. Table 5.2) of the three types of possible instruments along the five dimensions detailed in the previous section.

Regional innovation policy instruments and contextualization

Referring to a black box of knowledge creation processes, Karlsen *et al.* (2012, p. 124) put forward that: "The key to opening the black box is to understand the conditions in which bridging develops into knowledge co-generation." This may be particularly true when considering the issue of contextualization of regional innovation policy instruments. In general and from the viewpoint of economic development, innovation is not so much an ultimate goal as such, but rather a means to achieve and enhance competitiveness and welfare. Existing experience of innovation-related policies has shown that there is a variety of means and measures developed to foster innovation. However, the spectrum of possible instruments is limited and is therefore implemented in diverse forms and modes. In particular, it is important to keep in mind the two ideal modes of innovation that were displayed by Jensen *et al.* (2007), i.e.:

i the STI mode (science, technology, innovation) corresponding to the production and use of codified scientific and technical knowledge; and
ii the DUI mode (doing, using, interacting) that, again, is based on an experience-based mode of learning.

Concerning the effectiveness of policy measures, it seems reasonable to state that successful ones are those that are best adapted to the very specific background conditions in the given place or territory in which they are implemented. This calls for a strong socio-economic embeddedness of policy measures and – consequently – for a best possible adaptation to the specific territory, thus for a strong contextualization of policies. This is not a new debate as such; discussions on endogenous potentials, regional innovation strategies and so forth date back to the 1970s but witnessed new impetus in the context of European measures and recent discussions on place-based approaches or the new orientation of cohesion policy on smart specialization arguments. A common denominator of all these reflections and approaches is the focus on the regional level occurring in parallel with the conception of the systemic view on innovation. The latter focuses – among others – on the strong social component of innovation, thus also points to the importance of interactions that specifically rely on modes of behavior, mentalities, cultures, etc., besides of course the existing "hard facts" of innovation-related actors in the territory. On the European level, the focus of the Europe 2020 growth strategy is on smart, sustainable and inclusive development. In this general frame and in the context of EU cohesion policy, "smart specialization" receives crucial importance. This argues in favor of a context-specific approach, including regional stakeholders and economic actors instead of a top-down definition of (technological) priorities. In this line, regions are invited to identify individual development paths based on their strengths, resources and potentials (see Barca 2009 or Foray *et al.* 2011).

This identification process and the following elaboration of strategic measures is in the ideal case strongly based on collective creativity of the local actors. In the frame of participatory processes, coupled with analytical evidence, the creative potential of local actors can be used for the development of future-oriented regional strategies. In this respect, promising insights are given by authors such as Chantelot (2010) displaying a TAO (for Tolerance, Amenities and Opportunities) model for policy.

Conclusion: towards a counter-manifesto for creativity-led regional innovation policies and instruments

The emergence of the concept of creativity as a recent core element for innovation and regional development policies is the central element of this chapter. It leads us now to discussing whether "creation processes" or "collective creativity" will be more frequent or effective when supported by specific policy instruments. In other words, if the aim is to support the "creative economy" at local level (i.e., from the point of view of public authorities in charge of agglomerations, metropolitan areas, cities, regions, etc.) what can be the possible tools? And what are the ideas, principles and hopes (and even maybe wishful thinking) behind? The analysis attempts to contribute to an enhanced understanding of synthetic mechanisms at work at the level of creative territories (based mainly on experience gained in comparing French and German examples). Considering in particular the impact of local scale and proximity, the work is conceived as a review of different types of instruments such as policies and incentive mechanisms affecting institutions and based on various combinations of actors such as clusters, ideas labs or knowledge angels.

Taking for granted that creative destruction applies resources, knowledge and skills to existing or future problems in new ways, the nature of regional influences on creative destruction is according to Gilbert (2012) threefold, i.e.: i) structural; ii) social; and iii) political. As a consequence, creating new forms of demand, supply or supporting resources leads to a possible rethinking of local innovation-supporting policy underlying principles and aims. Referring to previous works by Muller *et al.* (2012a) and embracing the knowledge gained from observations about clusters, ideas labs and knowledge angels that are detailed in the previous sections, different ideas can be examined for future policies and policy instruments devoted to the support of "creation processes" or "collective creativity" at local level.

The elements contained in Table 5.3 are presented under the form of short injunctions, which due to their voluntarily somehow provocative character, aim at becoming a "counter-manifesto" for local innovation-supporting policy.

For instance, we assume that cluster policies may constitute important "triggers" for innovation in a broad range of fields. Nevertheless, cluster policies should not be considered as the "one and only" means of fostering

Table 5.3 Elements of a counter-manifesto for rethinking local innovation-supporting policy

There is a need to be clustered	Virtually interconnected islands of knowledge (of wisdom)
Proximity of big science infrastructures	Ability to access (close or remote) knowledge
Clear specialization	Eclectics
Bigger is better	Smaller is faster
R&D-driven is smarter	Creativity-led is funnier
Planning is everything	Expect the unexpected (or just nothing)
Solution-oriented	Problem-driven

innovative actions. It is the reason why ideas labs or knowledge angels (as possible examples) should also be taken into consideration as possible alternatives, between probably many others, some of them being still to be conceived, developed or implemented. In this respect, the concluding elements displayed in this counter-manifesto, even if provocative in their formulation, constitute a plea for open, flexible and experimental supportive measures that may also embrace unconventional ways.

Notes

1 A strand of literature deals with multi-level governance relations, looking at the roles different tiers of governments play in decision making and policy implementation (vertical links), as well as the roles and interests of public and private actors involved (horizontal links) (Marks 1993; Hooghe 1996; Rhodes 1997; Hooghe and Marks 2001; Bache and Flinders 2004).
2 Cf. in particular the activities resulting from the cooperation between BETA (University of Strasbourg and CNRS, France) and Fraunhofer ISI (Germany) and carried out within the frame of the evoREG initiative. Detailed information is available at www.evoREG.eu.
3 Cf. notably working documents that can be downloaded at: www.evoreg.eu/docs/files/shno/Note_biblio_EM_impact_et_evaluation250111.pdf and www.evoreg.eu/docs/files/shno/Note_clusters_All_Can_evoREG_13.pdf.
4 Cf. notably working documents that can be downloaded at: http://www.evoreg.eu/docs/files/redo/laboratoire_SXB.pdf.
5 Cf. also Muller (2008), Muller *et al.* (2009, 2010).

References

Amin, A. and Thrift, N. (eds.) (1994) *Globalization, Institutions and Regional Development in Europe*, Oxford: Oxford University Press.

Bache, I. and Flinders, M. (eds.) (2004) *Multilevel Governance*, Oxford: Oxford University Press.

Barca, F. (2009) "A place-based approach to meeting European Union challenges and expectations. An Agenda for a Reformed Cohesion Policy." Independent

Report prepared at the request of Danuta Hübner, Commissioner for Regional Policy; J. Bachtler, European Policies Research Centre, University of Strathclyde: Place-Based Policy and Regional Development in Europe, http://strathprints.strath.ac.uk/15693/ (last accessed September 6, 2012).

Chantelot, S. (2010) "La géographie de la classe créative: une application aux aires urbaines françaises," CJRS (Online)/ RCSR (en ligne) ISSN: 1925–2218, 33(3): 89–108.

Cohendet, P., Grandadam, D. and Simon, L. (2010) "The anatomy of the creative city," *Industry and Innovation*, **17**(1): 91–111.

Cooke, P. (1997) "Regions in a global market: the experiences of Wales and Baden-Württemberg," *Review of International Political Economy*, **4**(2): 349–81.

Daimer, S., Hufnagl, M. and Warnke, P. (2012) "Challenge-oriented policy making and innovation systems theory: reconsidering systemic instruments," in K. Koschatzky (ed.), *Innovation System Revisited: Experiences from 40 years of Fraunhofer ISI research*, Stuttgart: Fraunhofer Verlag: 217–34.

Foray, D., David, P.A. and Hall, B.H. (2011) "Smart specialization: from academic idea to political instrument, the surprising career of a concept and the difficulties involved in its implementation," MTEI Working Paper 2011–001, and http://ec.europa.eu/europe2020/index_en.htm (last accessed September 6, 2012).

Gilbert, B. (2012) "Creative destruction: identifying its geographic origins," *Research Policy*, **41**: 734–42.

Hooghe, L. (ed.) (1996) *Cohesion Policy and European Integration: Building multi-level governance*, Oxford: Clarendon Press.

Hooghe, L. and Marks, G. (2001) *Multi-level Governance and European Integration*, Boulder: Rowman & Littlefield.

Jensen, M.B., Johnson, B., Lorenz, E. and Lundvall, B.-Å. (2007) "Forms of knowledge and modes of innovation," *Research Policy*, **36**: 680–93.

Karlsen, J., Larrea, M., Wilson, J.R. and Aranguren, M.J. (2012) "Bridging the gap between academic research and regional development in the Basque country," *European Journal of Education*, **47**(1): 122–38.

Kusiak, A. (2007) "Innovation: the living laboratory perspective," *Computer-Aided Design & Applications*, **4**(6): 863–76.

Malmberg, A. and Maskell, P. (1997) "Towards an explanation of regional specialization and industry agglomeration," *European Planning Studies*, **5**(1): 25–41.

Marks, G. (1993) "Structural policy and multilevel governance in the EC," in A. Cafruny and G. Rosentahl (eds.), *The State of the European Community*, Vol. II: The Maastricht Debates and Beyond. Boulder: Rienner: 391–410.

Morgan, K. (1997) "The learning region: institutions, innovation and regional renewal," *Regional Studies*, **31**(5): 491–503.

Muller, E. (2008) "Tracking knowledge angels: a research proposal," *International Journal of Services Technology and Management*, **10**: 343–8.

Muller, E., Zenker, A. and Héraud, J.-A. (2009) "Entering the KIBS' black box: there must be an angel! (or is there something like a knowledge angel?)," *Working Papers Firms and Region*, Karlsruhe: Fraunhofer ISI.

Muller, E., Zenker, A. and Héraud, J.-A. (2010) "Inside KIBS' black box: exploring the existence and function of knowledge angels," in D. Doloreux, M. Freel and R. Shearmur (eds.), *Knowledge-Intensive Business Services: Geography and innovation*, Farnham: Ashgate, 211–32.

Muller, E., Zenker, A. and. Baier, E. (2012a) "Knowledge angels or how creative people foster innovation in the service industry – emerging concepts and international observations," in K. Koschatzky (ed.), *Innovation System Revisited: Experiences from 40 years of Fraunhofer ISI research*, Stuttgart: Fraunhofer Verlag: 153–69.

Muller, E., Zenker, A. and Ramos, J.-C. (2012b) "Knowledge angels, creative behaviors and emerging innovation modes: observations from Alsace, Baden-Württemberg and Catalonia," in E. Di Maria Grandinetti and B. Di Bernardo (eds.), *Exploring Knowledge-Intensive Business Services: Knowledge management strategies*, New York: Palgrave: 35–55.

Niitamo, V.-P., Kulkki, S., Eriksson, M. and Hribernik, K.A. (2006) *State-of-the-art and Good Practice in the Field of Living Labs*, Proceedings of the 12th International Conference on Concurrent Enterprising: Innovative Products and Services through Collaborative Networks, Milan, Italy: 349–57.

Parjanen, S. (2012) *Creating Possibilities for Collective Creativity, Brokerage Functions in Practice-Based Innovation*. PhD thesis, Lappeenranta University of Technology UDC 001.895:159.928:65.011.8:658.589.

Porter, M.E. (1998) "Clusters and the new economics of competition," *Harvard Business Review*, **76**(6): 77–90.

Rhodes, R. (1997) *Understanding Governance*, Buckingham & Philadelphia: Open University Press.

Smits, R. and Kuhlmann, S. (2004) "The rise of systemic instruments in innovation policy," *International Journal of Foresight and Innovation Policy*: **1**, 4–32.

Steward, F. (2008) *Breaking Boundaries: Transformative innovation for the global good*, NESTA Provocation 07, online: www.nesta.org.uk/library/documents/Prov%20-%20BtB%20v8.pdf (last accessed September 6, 2012).

Storper, M. (1997) *The Regional World: Territorial development in a global economy*, New York: Guilford.

Wieczorek, A.J. and Hekkert, M.P. (2012) "Systemic instruments for systemic innovation problems: a framework for policy makers and innovation scholars," *Science and Public Policy* **39**: 74–87.

6 Governing communities of experts

Jean-Philippe Bootz, Olivier Dupouët, Francis Kern and Eric Schenk

Introduction

For many authors (e.g. Amabile, 1988), competence in a given field is viewed as an important (though obviously not the only) ingredient of creativity. Experts are particularly interesting in this context. Since these individuals possess rare forms of knowledge and a high creative potential, the management of experts becomes a central issue for companies seeking to create sustainable competitive advantages. This chapter raises the question of the governing of experts within organisations.

According to Bootz and Schenk (2009, forthcoming), it is meaningful to characterise the expert according to two criteria. First, from the cognitive perspective, experts are those who are skilled in the analysis and solving of various non-trivial problems, through the use of their ability to combine and rapidly mobilise theoretical and empirical knowledge. Then, from the sociological point of view, experts enjoy the legitimacy and recognition which allows them to be consulted in situations beyond those of their original social spheres.

In so far as these actors are part of the human and intellectual capital of companies, any active effort involving the management of knowledge must take such experts into account. Indeed, for the company it is a matter of both mobilising expert competences to confront critical situations, and of ensuring that the experts are suitably trained.

In this regard, two approaches can be identified. The first one relates to human resource management (HRM) practices. The firm can implement tools allowing the careers of its experts to be controlled and managed. Such an approach allows expertise to be treated uniformly, and makes it possible to orient the training and mobilisation of their knowledge towards the strategic ambitions of the firm. However, when such an approach is established within the boundaries of formalised procedures, it is too rigid and is necessarily generalised to a level which may not be well adapted to the specificities and individual career paths of each expert (Oiry, 2005).

The second approach relies on communities of practice, self-organised entities created by actors who share the same concerns (Wenger, 1998).

Within such entities, the expert can develop and share his/her knowledge in a flexible and progressive manner, while adapting the progression of this knowledge and its propagation to a local environment. However, within such a framework, it may be difficult to recognise and enlist expertise beyond the boundaries of the community, since the latter generally remains invisible to the management.

Here, we propose that relatively new objects, namely governed communities of practice, could allow the difficulties encountered with these two approaches to be overcome, while at the same time preserving their positive characteristics. Governed communities of practice are organised in accordance with the formal structure of the company to ensure the coherence with its strategy (McDermott and Archibald, 2010). The management nevertheless ensure that they are given sufficient autonomy to avoid stifling their self-organised qualities. This balance is reached through a combination of various forms of control, some of which are determined by the hierarchy, and others by the community. Such structures thus allow the flexible development of expertise, while at the same time ensuring that this expertise is useful and can be enlisted by the company.

The chapter begins with a definition of the concept of expert, which we define in terms of its cognitive and social dimensions. We then present the two traditional forms of governance of expertise within organisations, through human resource policies and self-organised communities of practice. Then, in the final section, we present the concept of a governed community of practice, in order to demonstrate how this new managerial entity can lead to the improved governance of expertise within the company, through a hybridisation of formal and informal relationships.

Experts within organisations

Although the terms expert and expertise have become part of everyday language, there is some confusion surrounding the limitations and extent of their meaning. On the basis of previous studies (Bootz and Schenk, 2009, forthcoming), we have adopted the model of an expert organised according to both cognitive and social elements.

According to a commonly accepted definition, an expert is a subject who is assumed to 'know and express the truth in the particular field in which he is considered to be competent' (Ancori, 2009). This vision, which predominates in studies of Political Science or Sociology, positions the expert between the scientist and the politician (Weber, 1919). The expert indeed possesses legitimacy and knowledge, and is consulted for his opinion and asked to provide support for a political decision. In the corporate world, or in academic studies of Artificial Intelligence (Chi *et al.*, 1982) and Management Science (Hatchuel and Weil, 1992; Cowan, 2001), the term *expert* is often used to qualify an individual having a high level of competence or knowledge in a given field (Salas *et al.*, 2010).

Different disciplines mobilise the notion of expert, placing emphasis on two essential dimensions of this concept. On the one hand, the expert has cognitive characteristics which differentiate him from the specialist. On the other hand, the expert is characterised by his social dimension. Indeed, the role of an expert is only really established when a person (or an establishment) consulted for his (its) expertise agrees to respond to this request. This consultation-acceptance process is directly related to the notion of legitimacy. We thus consider that an expert is an individual (or an establishment) consulted in a specific decisional or action-related context, and whose advice or judgement has a legitimate character.

The cognitive dimension of an expert

As it may be assumed that the expert possesses specific knowledge and competences, it is quite natural to look into the cognitive processes which characterise this expertise. In the comparison between the novice and the specialist, Ericsson *et al.* (1993) emphasise that the expert attains very high levels of performance through his experience, based on many years of practice in a given field.

The framework in which the expert intervenes is positioned at the intersection between the scientist and the specialist. He is involved in the solving of complex situations, which are also contextual or even multi-contextual (Salas *et al.*, 2010). As opposed to the specialist, he does not intervene in a precise and well-defined field, but rather in various, ever-changing contexts. In addition, as opposed to the specialist, the expert does not use formal rules or procedures, but evaluates the situation on the basis of a small number of relevant indicators, which he distinguishes in 'record time'. The expert thus makes use of intuitive reasoning and individual experience (Abernathy and Hamm, 1995; Klein, 2003). According to Baylor (2001), the relationship between the level of expertise and access to intuitive understanding takes the form of a U-shaped curve. In the case of a novice, the use of intuitive ideas is a consequence of his immaturity. With experience and learning, this immature intuition gives way to the use of well-integrated codified approaches and procedures. Finally, in the case of an 'expert' individual, the use of intuition again becomes the predominant approach. As emphasised by Salas *et al.* (2010), the expert develops mixed strategies, which call for the use of intuition on the one hand, and deliberate choices on the other hand. The expert's competence is thus based on knowledge, both tacit (experience-based know-how, cognitive know-how, social know-how), and explicit (theoretical knowledge), which is mobilised in difficult situations.

According to Bootz and Schenk (2009), the expert is not a homogeneous and static figure, but rather a person who can be consulted in various contexts, ranging from concrete action to the contribution of decisional support (Table 6.1).

Table 6.1 The three faces of the expert

Context of the consultation	*Type of expert*
Problem solving	Expert-specialist
Decisional support	Expert-expert
Evaluation support	Expert-scientist

The social dimension

One cannot proclaim oneself to be an expert. A potential expert will be consulted only if he is credited with legitimacy and confidence by an individual or organisation in a situation for which expertise is required. In most cases, the expert consultation is based on a priori recognition, relying on strong signals which can be observed by the applicant, or on weak signals, such as the expert's reputation and image.

From the managerial point of view, it is necessary to address the question of the legitimacy enjoyed by experts. This legitimacy can be based on a formal institutional structure (a professional order, an accreditation system), or on a self-organised social phenomenon. The legitimacy associated with an institution is codified, and may be understood when the observer is external to the institution awarding the expert status. On the other hand, the informal or self-organised legitimacy of the expert is based on the implied dimensions of the expertise, and is difficult to understand from outside the peer group in which the expertise has been established. It is thus reasonable to query the extent to which such an expertise can reach beyond the boundaries of the peer group.

The expert's legitimacy can be thought of as the confidence with which the potential expert is viewed by the expertise applicant (Bootz and Schenk, 2009, forthcoming). Following Nooteboom (2002), we distinguish between confidence in the expert's competence and confidence in his intentions. The need for confidence in the expert's intentions helps explain the requirement for transparency, with respect to possible conflicts of interest under which the expert could be placed. Moreover, the requirement for confidence in the expert's competence brings a paradox to light: how can competences which one does not possess oneself be evaluated? The cognitive distance (Nooteboom, 2002) separating the expert from the expertise applicant is a hindrance to the building of confidence in his competence, with this confidence nevertheless being a necessary element in the relationship between the applicant and the expert. In addition to theoretical knowledge, the cognitive distance between expert and non-expert can be accounted for by the experienced-based and cognitive know-how of the expert, which make up his implicit resources (Polanyi, 1966; Von Hippel, 1998). According to the SECI (Socialisation, Externalisation, Combination, Internalisation) model

(Nonaka and Takeushi, 1995), the mitigation of this cognitive distance can be achieved by two different means. Socialisation implies the exchange of implicit knowledge within a peer group recognising the expert, whereas codification allows validation and expertise criteria to be externalised with respect to the group. In the case where the expertise applicant and the expert are positioned within the same social group, trust can be built through interactions in accordance with the specific modalities of the social group (Cohendet *et al.*, 2004). Conversely, when the applicant is positioned outside the group promoting the emergence of expertise, the building of confidence in an expert is based on strong signals given by the expert or his peer group.

Towards the governing of experts

In a knowledge society, the governing of experts represents a major management asset. As emphasised by Bootz and Schenk (forthcoming), the governance of experts can be approached by conventional HR methods, or via communities of knowledge.

Experts and human resource management

Starting from a vision of the expert as an individual having specific competences, a legitimacy, and whose activities are transversal to the processes of the company, the HRM of experts could be organised along three lines.

Evaluation of competences

The identification of levels of competence and potential represents the first step in the governance of experts. In so far as expert competences have a strongly implicit nature, evaluations based on the profession-qualification pair have little relevance (Oiry, 2005). In practice, companies are thus often obliged to be content with general evaluation criteria such as '(the expert) is capable of managing atypical situations encountered in the manufacturing process', or with the implementation of expensive methods such as *assessment centres*.

Development of competences

The first difficulty is that of identifying high potential individuals, in whom the organisation's efforts should be vested. In view of the strong implicit dimension of expert competences, the means which can be envisaged for the development of these competences are tutoring and internal mobility schemes. Two difficulties should however be mentioned: tutoring requires the expert's active participation, which is difficult to control. In addition, mobility does not guarantee the accumulation of experience.

Management of the social dimension

For the reasons mentioned above, it is essential to take the expert's legitimacy factors into account. The expert's legitimacy is built up progressively through the use of strong signals (labels) or weak signals (reputation). In a management approach, the control of strong signals is a major issue, which no doubt partially explains the increasing use of external certification bodies. For example, organisations such as AFAQ-AFNOR offer competence certification services, which contribute to the recognition of expert competences.

The classical governance of an expert encounters a certain number of difficulties. First, the highly implicit and context-related nature of expert competences cannot be expressed by quantitative measurements on a scale of 'levels of competence'. In the best of situations, these measures allow the expert's 'professional knowledge' to be extracted, while leaving out the other dimensions which are nevertheless the basis of his cognitive dimension. In addition, management based on standard tools (mobility, training, tutoring …) is lengthy and expensive for the organisation. Even in the case where expert competences are present, the individuals having these competences do not necessarily have the legitimacy allowing them to make use of the expert function. In this context, a possible solution would be to transfer part of the governance of expertise to the individuals and intermediate social structures of the company, in particular the communities of practice. This operation can be implemented at a lower cost, in view of the fundamentally self-organised nature of these communities and the potential they represent in terms of expertise.

Governance of experts by communities of practice

The cognitive activity of the company develops mainly within communities of practice (CoPs). In fact, these communities produce individual and collective competences, and as such are efficient settings for the construction of not only cognitive, but also social expertise.

Cognitive and social dimension of experts within CoPs

According to Brown and Duguid (1991), CoPs develop a social learning construction and represent a select setting for the construction of experts' competences. In practice, when individuals observe the disparity between the prescribed process of tasks and organisational practices, they try, within a CoP, to develop understanding and actions anchored in their daily work. By taking part in a CoP, the individuals thus develop their practice autonomously, according to the elements they consider to be important. CoPs are settings in which the sharing of knowledge and learning takes place spontaneously, in a self-organised manner, without taking the constraints and methods defined by the organisation into account. Conventional contractual arrangements are thus replaced by the commitment and voluntary participation of the individuals (Wenger, 1998).

According to Wenger (1998), the mutual commitment characterising CoPs implies a mutual assistance relationship between the participants, such that the overriding competence is that which involves being able to aid, and to seek aid. McDermott (2001) agrees with this author on this point, and emphasises the fact that there is normally a climate of confidence between members of a CoP, provided they regularly share their experiences, objectives and ideas. CoPs thus represent a social environment which promotes the development of relationships of trust between individuals, and the legitimisation of rapidly recognised experts within the community.

Experts beyond their original community

CoPs allow the cognitive dimension as well as the social dimension of the expertise to be supported, and thus represent an 'expertise enabling context'.[1]

CoPs are social structures which are often invisible from the outside, and as a consequence the experts present within CoPs are not necessarily identified outside the social structure. Using the typology presented by Bootz and Schenk (forthcoming), one could thus consider CoPs to be above all expert-specialist incubators. The members of the CoP are indeed centred on the action and resolution of concrete problems, and their social sphere of contribution is clearly defined. However, a CoP is not made up of homogeneous degrees of competence (Cohendet *et al.*, 2006). In fact, it includes newcomers (at its periphery), confirmed members (at an intermediate level) and finally peers (at its centre). The latter enjoy legitimacy and a reputation which allows them to be approached by other members of the CoP, in order to resolve non-trivial problems, and are therefore internally considered to be experts. Experts from CoPs can nevertheless be distinguished from those who emerge from classical hierarchical structures, with respect to their competence construction process (*learning-by-doing* vs training) and the type of recognition they enjoy (hierarchical and codified vs community-related and implicit).

Since CoPs are favourable to the emergence of experts, the self-organised dimension of CoPs could lead one to believe that external managerial intervention is not useful, or could be counter-productive. However, a hands-off policy runs considerable risks. From the organisational point of view, biases can appear in the case of isolated CoPs. The strong anchoring of their identity can lead to stagnation in a routine, or rigidity. This can even stimulate discrimination with respect to other communities, or incompatibilities with respect to the hierarchical functioning of organisations (Cohendet *et al.*, 2004). When CoPs are not taken into account by the manager, these communities – thanks to their self-organised properties –nevertheless develop, but remain invisible and their expert potential will not be harnessed. There is thus a risk of a form of hyper-specialisation, at the base of the social dimension of the expertise, developing within these CoPs when they are not consulted from the outside. One solution thus involves acting directly on communities by trying to govern them. In this context, the governance of experts is achieved via governed communities of practice (GCoPs in the following), and involves the observation of the characteristics of these GCoPs (in particular, ways of reconciling self-organisation and control) and the 'profiles' of the coordinators.

On spontaneous communities and governed communities

Communities of practice have for a long time been of considerable interest to members who, in these social structures, detect the means to respond to major issues: production of good practices, improvement of activities, development of competences, construction of common identities. However, as a consequence of their self-organised nature, they have for a long time been considered by the management to be unidentified structures. However, more recently there has been an increase in the number of companies seeking to actively foster and develop such communities. As examples, Siemens (Probst and Borzillo, 2007), British Petroleum (Cohendet *et al.*, 2006, 2010), IBM (Gongla and Rizzuto, 2001), the European Council (Créplet and Dupouët, 2009), GDF-Suez (Blanchot-Courtois and Ferrary, 2009; Gosselin *et al.*, 2010; Dupouët and Barlatier, 2011), or even Schlumberger (McDermott and Archibald, 2010) have entered into a knowledge management approach centred on their communities of practice. The governance of experts via GCoPs implies the mobilisation of a sponsor and a coordinator, in order to successfully manage the tensions between control and self-organisation. It then implies the use of differentiated approaches, according to the nature of these GCoPs. Recent empirical studies emphasised a clear distinction between a community of exploration and operation.

Coordinated communities and the key role of the sponsor and the coordinator

For organisations entering into such an initiative, the difficulty then lies in the governance of the communities. It is indeed necessary that they produce elements which are in line with the company's concerns, without stifling their self-organised dimension, which is one of their main advantages. Two mechanisms are proposed in the literature to allow this delicate arbitration to be implemented: the fixing of objectives and deliverables, as well as governance based on the roles of coordinator and sponsor (Probst and Borzillo, 2007; Borzillo *et al.*, 2008; McDermott and Archibald, 2010).

According to McDermott and Archibald (2010), the most efficient GCoPs are those which are attached to themes of interest to the entire organisation. Thus, if the communities remain free in terms of the definition of the way in which they operate, the hierarchy continues to take care to maintain the coherence of these flexible structures with respect to the global strategy of the organisation. This organisation is controlled in a more operational manner by the introduction of clear objectives and associated performance indicators (Probst and Borzillo, 2008).

The second governance mechanism involves establishing links between the communities of practice and the formal structure. These links are secured by two actors, the sponsor and the coordinator. These two roles were initially identified by Wenger *et al.* (2002) and their importance was confirmed by ensuing studies (Probst and Borzillo, 2007, 2008; McDermott and Archibald, 2010).

It is likely that, each taken at his own level, the sponsor and the coordinator must arbitrate and find a balance between the need to control and maintain the autonomy of the communities of practice for which they are responsible.

The first role of the sponsor is to guarantee that the community for which he is responsible has access to the resources and time required for its operation. In this respect, the sponsor ensures that the CoP has a certain degree of autonomy with respect to the remainder of the company, and encourages a non-hierarchical operational mode of collaboration between peers and the distribution and sharing of knowledge. Moreover, he must look after the convergence of the community's themes and the company's concerns. The sponsor, who has the power to maintain or remove the CoPs for which he is responsible, thus plays the role of guarantor with respect to the effective usefulness of a CoP for the organisation

The coordinator plays the role of facilitator within the community. One of his main tasks is to ensure that knowledge and information circulate correctly, that the members of the community are able to communicate correctly, and more generally that collective dynamics are maintained. In this regard, the coordinator promotes the community's internal standards, which contribute to the construction of a certain degree of its autonomy with respect to the rest of the organisation. He must however at the same time supervise the work breakdown, and ensure that the community's deadlines are respected and that its activities are promoted along the initially defined lines (Wenger *et al.*, 2002). The nomination of coordinators and their professionalisation make it possible to ensure that the CoPs' productions remain connected to the strategic and operational objectives of the company.

Leveraging communities vs probing communities

In practice, governed communities involve various realities. There are in fact significant elements of diversification, such as the environment, the orientation, the level of maturity, the degree of institutional formalism, the size or the governance structure (CEFRIO, 2005). Recent empirical studies (Probst and Borzillo, 2007; 2008; Borzillo *et al.*, 2008; McDermott and Archibald, 2010; Dupouët and Barlatier, 2011) partly structure these diversification criteria and allow two types of GCoP to be globally discriminated:[2] *leveraging* communities and *probing* communities (Bootz,

2012). After having described their main characteristics, we determine their impact in terms of the governance of experts.

Probing communities of exploration

Probing communities[3] (Probst and Borzillo, 2007, 2008; Borzillo *et al.*, 2008) are centred on the production of innovative ideas. Exploration oriented communities look for knowledge which is distant from the current corpus of the organisation. They provide a certain degree of flexibility by proposing variations, which are different from the knowledge mastered by the organisation. In this sense, they correspond to the definition of an exploration GCoP (as defined by March, 1991), since the knowledge produced is intended for the production of substantially new knowledge having a strategic impact for the organisation.

These GCoPs are found mainly within structures for which innovation is a key competitiveness factor. For example, in IBM the e-business consultants meet in this type of community, in order to exchange innovative solutions (Probst and Borzillo, 2007; Borzillo *et al.*, 2008). At GDF-Suez, the role of the 'Scientific Committee' community is to propose lines of research detached from the traditional R&D culture of the organisation, in order to promote innovation (Dupouët and Barlatier, 2011).

They also benefit from the active support of the sponsor, who awards considerable resources aimed in particular at providing the members with the time and infrastructures necessary for their participation. This type of community is also supported by active leadership, in order to ensure not only a suitable internal environment for the sharing of knowledge, but also the reporting to the sponsor of the community's activities. CoPs focused on exploration communicate their productions to agents, able to make decisions and implement these changes within the formal organisation. The flow of knowledge arising from exploration is thus vertically oriented, towards the company's decision-making instances, where they can be evaluated and taken into account in the development of a strategy (Schulz, 2001). The use of CoPs increases the managers' ability to permanently maintain an exploration activity (Mom *et al.*, 2007, 2009). From this point of view, CoPs contribute to flexibility, by supplying the decision makers with a source of potential opportunities, which they may or may not decide to implement. In this sense, the role of CoPs can be thought of as being complementary to that attributed by Nonaka and Takeuchi (1995) to the intermediate management, for the transmission of new ideas from operational levels to the decision-making spheres.

Leveraging communities of exploitation

Leveraging communities (Probst and Borzillo, 2007, 2008; Borzillo *et al.*, 2008) are comprised of members who share technical and operational practices in order to optimise their daily activities. These communities are clearly oriented towards operations (Dupouët and Barlatier, 2011) in the sense that they are centred on the improvement of existing methods (Probst and Borzillo, 2007).

As an example, at Daimler Chrysler more than 70 communities of this type allow engineers and technicians from different production platforms to implement and share good practices related to production (Probst and Borzillo, 2007). Within GDF-Suez, the aim of the 'analysis and laboratory' CoP is to ensure the sharing of knowledge from the laboratories working with water and waste, related to measurement techniques, standards and safety, or the choice of machines (Dupouët and Barlatier, 2011). Finally, in the prospective team of EDF R&D (Bootz, 2009), the members built up a community which led to the development of an original tool, '*Aramis*', based on the calculation of Bayesian networks, which continues to be improved through the exchange of experience and collaboration with external structures (research laboratories, universities, consultants).

The objectives are determined by the community's coordinator, and are centred on the improvement of the technical dimensions related to practices, such as processes or methods. In this context, he is in charge of the controlling and validation of each member's practice, in order to preserve the operational excellence of the community. In this role, it is important that he enjoys a strong degree of legitimacy. This attribute allows him to evaluate the competences of the various members of the community, and to organise them in order to accomplish all the activities. He must therefore not only promote spaces for socialisation (face-to-face, telephone conferences, video-conferences), but also guide and direct them through the use of his high level of technical competence.

The flow of knowledge from CoPs dedicated to operations is mainly horizontal, aimed directly at persons able to put them immediately into practice (Schulz, 2001). The transfer of this knowledge is achieved *via* interpersonal relationships, which penetrate the internal boundaries of the organisation (Hansen *et al.*, 1999; van Wijk *et al.*, 2008; Kang and Snell, 2009). Frequent encounters and strong relationships allow in-depth and contextualised knowledge to circulate, which is a major advantage when it comes to the transmission of know-how and process-related knowledge. In addition, since they welcome individuals from all functions within the organisation, operation-oriented CoPs contribute to the development of knowledge common to the company's different entities (Gilbert, 2006; Raisch, 2008; Taylor and Helfat, 2009), and thereby contribute to the coherence of the company (Raisch *et al.*, 2009).

Usefulness, dosage and location of control in GCoPs

When companies are committed to the management of their intellectual capital, through the use of governed communities of practice, they try to preserve, or even promote the main qualities of spontaneous communities. In this sense, the company must pay particular attention to the implementation of appropriate controls, allowing the activities of the different communities to be kept in line with the global strategy of the company, and allowing the promotion the autonomy required for the knowledge sharing process, the development of competences, and the continual improvement which is the very basis of these communities.

Two essential forms of control exist within companies (Ouchi, 1979; 1980). The first of these is bureaucracy, in other words the entire set of formal and hierarchical controls, applied by a legitimate authority through the application of explicit rules. The second is that which Ouchi (1979) refers to as the clan, in other words control through socialisation, which is possible when the individual objectives to be coordinated are at least partially congruent. The 'clan' form corresponds to the development of internal standards in communities of practice.

Moreover, by analysing the communities' activities, three essential aspects need to be controlled: the inputs (human, material and financial resources, made available to the community), the processes (i.e. the way in which the community operates and the production modes it implements) and the outputs (the finished products – documents, benchmarks, good practice guides, etc. – which it makes available to the rest of the organisation). The control applied to the communities' activities will have an influence on the way in which they operate and on their cognitive productions (Cardinal, 2001; Turner and Makhija, 2006). Consequently, the activities of exploration and exploitation each require a specific combination of control mechanisms.

The management of governed communities of practice requires the alignment of these forms of control, which on the one hand guarantee a balance between alignment and autonomy, and on the other hand ensure adequate achievement of the specific cognitive task of the community under consideration. It is thus necessary to decide for each of these aspects (inputs, processes, outputs) which form of control (bureaucracy or clan) should be applied (Cardinal, 2001; Long *et al.*, 2002; Turner and Makhija, 2006).

In the case of operation-oriented communities, we indicated that knowledge circulates horizontally, towards peers. The knowledge recipients are thus the same persons as those who produce the knowledge, and are the best positioned actors for its evaluation; the outputs will thus be better controlled in the 'clan' mode. In order to develop expertise which is useful to the company, and to ensure that knowledge and good practice circulate correctly, it is in the management's best interests to exercise control over the processes, in particular through formalisation, without necessarily controlling their contents. For this, bureaucratic control can be used by the community coordinator, who will nevertheless simultaneously endeavour to maintain a certain degree of flexibility and freedom for the members of the community.

In the case of exploration communities, the management's attention and a bureaucratic type of control are applied to the outputs and the cognitive productions of the community. In fact, since by definition it is difficult to implement standards to control an emerging and uncertain process, and it is necessary to maintain a certain degree of autonomy, the internal functioning of the community is better managed through the development of internal social standards. On the other hand, the organisation will need to decide whether or not to integrate new suggestions and ideas produced by the community into its portfolio of activities. To make such a decision the company needs to have formal evaluation mechanisms in order to decide whether these suggestions should or should not be integrated, both into the processes and into the company's products. In parallel, this requires the community to have formalisation and reporting procedures which are suitable for the organisation.

Concerning the practical implementation of these controls, it may be necessary to distinguish between the case of communities created *ex nihilo* following a management decision, and the case of pre-existing communities integrated into the company's knowledge management techniques. In the first case, via the sponsor and the coordinator, the management must ensure that it promotes the emergence of internal standards. Bureaucratic control must therefore be sufficiently flexible and evolutionary to encourage this evolution towards the community's autonomy. In the second case, the implementation of bureaucratic control must not break down the social construct developed within the community, for example by placing excessive emphasis on processes in the case of operational communities.

Governance of CoPs and governance of experts

The governance of experts via probing communities

The governance of probing communities requires the ability to identify and select innovative subjects, and to then federate internal expertise from various fields related to these subjects. Within the exploration GCoP, expertise is in general not specifically related to trades, but to the ability to search for weak signals, the seeds of change or relevant information through processes such as a technology or an environmental watch, for example. Through their involvement in the cognitive process of knowledge creation, expert-scientist profiles will be needed.

In this framework, the governance of experts requires the identification, in the different layers of the company, and in particular in the CoPs, of pockets of knowledge relevant both to the enrichment of a creative thinking process, and to its development through collective thinking. This implies knowledge of intra-organisational networks and the ability to organise them to foster creative tension, which is a source of innovation. In this context, the coordinator is not necessarily an expert in the relevant field, in the sense that he plays less the role of a judge or a reference, than that of a coordinator. Indeed, he must be able to coordinate both different experts and communities, and to organise all of these with respect to the hierarchy, via the sponsor.

In the case of a pre-existing community, the coordinator's role is mainly that of legitimating the activities and promoting their development, by obtaining suitable resources via the sponsor. In this context, the coordinator's acuity allows him to identify the innovative potential of a community, which had remained previously invisible. He must therefore be particularly vigilant with respect to the control exercised by the sponsor on the community, by ensuring that the operational modes and internal standards built up by the members do not fall apart.

In the case where the GCoP is engaged, the absence of passion *stricto sensu*, combined with interventionism from the management, implies that the main danger threatening this type of community is that of it switching over to a traditional hierarchical function such as that of a project group (Gosselin *et al.*, 2010). The coordinator must therefore promote the progressive appearance of functioning based on a system of standards, which are shared and co-constructed, and through its adequacy with respect to the climate of induced trust. According to Gosselin *et al.* (2010), this tension can also be managed dynamically, through the sponsor's and the coordinator's progressive withdrawal, in favour of the community members. In the initial phase, the sponsor's role is central, since the resources he can award support the community's creation and reinforce its legitimacy with respect to the hierarchy. This intense commitment must then give way to progressive withdrawal, allowing the community and the experts to gradually acquire their required autonomy. An identical movement can be observed in the coordinator, who must perceive the correct moment for his progressive withdrawal, to the benefit of the members of the community, and must manage the process of transferring some of his responsibilities. Eventually, the members must acquire a greater degree of autonomy in terms of contribution, validation and formalisation, while at the same time taking advantage of the coordinator's stimulus actions. According to Gosselin *et al.* (2010), the success of the coordination thus requires its 'original artificiality' to be overcome, by promoting the progressive appearance of a form of non-coordination.

Finally, in so far as the coordinator of strategic communities tries to explore innovative subjects, which are often inconsistent with the organisation's dominant culture, the profile of the expert coordinator corresponds to that of an intrapreneur. According to Woodman *et al.* (1993), the latter in fact tries to develop the organisational creativity, which promotes innovation and the development of new activities. For this, he explores new opportunities outside the dominant activity, and Alter (2000) qualified such a person as being a 'positive deviant'. He is a 'hybrid being', between the figure of manager and that of the entrepreneur (Basso, 2004). Much as in the case of an entrepreneur, the motivating factors of the intrapreneur are both intrinsic (quest for responsibility and autonomy, will to achieve a project) and extrinsic (quest for social recognition or even enrichment) (Hornsby *et al.*, 1993, 2002).

Nevertheless, these elements are not satisfied in the same institutional environment. The intrapreneur is integrated into an organisation, and must therefore be sufficiently skilful to succeed in 'integrating his objectives into those of his company, and using the rules of the system to make his project succeed' (Blanchot-Courtois and Ferrary, 2009, p. 97). On account of his atypical profile, the intrapreneur often runs the risk of being marginalised. He is thus particularly sensitive to signs of social worth such as the enhancement of his autonomy and responsibilities, promotion, public recognition or resources approved for his projects (Morris and Kuratko, 2002). In this context, probing GCoPs represent a powerful promotional lever of intrapreneurial culture, by offering these 'positive deviants' a special place to express their ideas, and the form of institutionalisation which they need. The downside is that the intrapreneur profiles are undoubtedly less common and more difficult to 'control' than more conventional managers. The setting up of an intrapreneurial culture, necessary for the governance of these communities of exploration, and consequently of the experts, then requires an adaptation of the HR tools and processes (recruitment, training, mobility …). This institutionalisation can be achieved through the creation of an intrapreneur community of practice, as pointed out by Blanchot-Courtois and Ferrary (2009) in the case of GDF-Suez.

The governance of experts via leveraging communities

Leveraging communities are composed mainly of expert-specialists who mutually transfer practices of the operational type in order to optimise their daily activities and improve the technical mastery of their practices.

Within leveraging communities, the tension between control and self-organisation is a priori less pronounced than in the case of probing communities. Since they are centred on the optimisation of daily activities, the control carried out by the management via the sponsor is indeed more flexible and is not based on the setting up and monitoring of quantifiable objectives. In this context, the coordinator is mainly concerned with his ability to support the operational excellence of the community. To this intent, he controls the processes through the assessment and validation of the practices and competences of its members. He must therefore enjoy technical legitimacy. However, his main role is that of the coordinator, in order to maintain the community's unity and secure the practice sharing processes between experts. He must therefore ensure the promotion of spaces for socialisation, by relying more strongly on physical face-to-face encounters, rather than on the extensive use of IT tools. He will nevertheless resort to this in order to promote the best practice formalisation and distribution processes (control of the outputs). In fact, although the formalisation and distribution of best practices is for the company's management an essential driver of its desire to coordinate this type of community, these processes are in practice sometimes bypassed as a result of the wish to control the information used by these communities (Dupouët and Barlatier, 2011).

In order to support the formalisation and distribution of good practices, the coordinator must promote KM tools in phase with the operational modes of the members and the community culture, in order to avoid risks and resistance (Cohendet *et al.*, 2004; CEFRIO, 2005; Daunais *et al.*, 2010; Harvey, 2010).

With a view to intercommunity distribution, these tools must also correspond to the characteristics of the 'boundary objects' (Wenger, 1998), in other words modularity, polyvalence, abstraction and standardisation (Chanal, 2000). These tools must therefore be sufficiently modular to be adapted to the local constraints, and sufficiently robust to maintain an identity between the communities (Star and Greismar, 1989).

The circulation of good practices between different communities will also be reinforced in the case where the coordinator succeeds in playing the role of an interface actor, either as a 'translator' (Brown and Duguid, 1998), or as a 'knowledge broker' (Brown and Duguid, 1998; Wenger, 1998).

Globally, these processes of formalisation and good practice distribution will have the quality of reinforcing the social dimension of the expert-specialists situated in the communities, by developing recognition for their competences and a legitimacy extending beyond the boundaries of their original structure.

A very large majority of leveraging communities are composed of pre-existing communities. However, this type of community can be created *ex nihilo* in the case where the hierarchy detects that some key activities are not being continuously improved, and require the creation of expertise. In the first case, the coordinator will have already been designated by the members of the community, based on his leadership qualities and technical legitimacy. This person will need to be particularly attentive to ensure that the formalisation and control processes he implements do not challenge the community's pre-existing modes of operation.

If the community is initiated by the hierarchy, it is possible to envisage a coordinator who is not necessarily a technical expert, but whose competences are more strongly oriented towards guidance, in order to promote the experience and knowledge sharing processes, and thus to participate in the construction of a collective identity and an internal social standard. Nevertheless, the internal competence control and validation activity still requires sufficient mastery of the technical aspects of the practices. It is also possible in the long run, once the community's unity has been established, that another internal leader emerges and requires a handover in terms of coordination.

Conclusion

In a knowledge-based economy, the management of knowledge becomes a key strategic issue. Human resources should not be thought of as a collection of operators or performers, but rather as a group of actors constantly mobilising various forms of knowledge and competence, and who are found in situations which foster learning. Among these actors, we emphasise the key role played by the experts. The cognitive and social dimensions through which we perceive the experts lead us to propose them to be governed, not with standard HR tools, but through GCoPs. The governance of these experts is of such importance for the strategic management of the company that these communities of practice need to be governed in order to control the tension existing between their self-organised nature and their necessary integration into the company's organisational structure.

This managerial function allows the respective roles of the sponsor and moderator of these communities to be identified. However, the theory of ambidexterity also allows us to distinguish between probing communities oriented towards exploration and leveraging communities oriented towards exploitation. Whereas in leveraging communities the sponsor/animator tandem is a central theme for the management, in probing communities the function of knowledge entrepreneur becomes paramount, and is materialised in the person of the intrapreneur, a facilitator who is focused on the creation and circulation of knowledge.

In the management of knowledge, ambidexterity and the functional duality of entrepreneur / manager allow the governance process to be explained, the manner in which the experts are mobilised in different types of community to be specified and the tension between their self-organisation dynamics and modes of integration within the company to be removed. The solving of this tension is at the heart of the company's future and strategic perspectives. The management of knowledge can thus renew the management of human resources by highlighting the management of experts, through the governance of communities.

Notes

1 Here we refer to Von Krogh *et al.* (2000) who speak of a 'knowledge enabling context'.

2 In their typology, Probst and Borzillo demonstrate the existence of a third type of community: 'social and productive space' communities. In so far as these communities operate according to a self-organisation model close to that of spontaneous communities, they cannot be considered to be governed communities according to our definition. We did not integrate these into our analysis.

3 Term used by Borzillo *et al.* (2008).

References

Abernathy, C.M. and Hamm, R.M. (1995) *Surgical Intuition: What it is and how to get it*, Philadelphia: Hanley & Belfus.

Alter, N. (2000) *L'innovation ordinaire*, Paris: Presses Universitaires de France.

Amabile, T.M. (1988) 'A model of creativity and innovation in organizations', *Research in Organizational Behaviour*, **10**: 123–67.

Ancori, B. (2009) 'Expertise et citoyenneté: les Grecs anciens et nous', *Revue d'anthropologie des connaissances*, **3**: 485–529.

Basso, O. (2004) *L'intrapreneuriat*, Paris: Economica.

Baylor, A. (2001) 'A U-shaped model for the development of intuition by level of expertise', *New Ideas in Psychology*, **19**: 237–44.

Blanchot-Courtois, V. and Ferrary, M. (2009) 'Valoriser la R&D par des communautés de pratique d'intrapreneurs', *Revue française de gestion*, **5**(195): 93–110.

Bootz, J.P. (2009) 'Les communautés d'apprentissage: structuration de la littérature, illustrations et perspectives', *Gestion 2000*, juillet-août: 175–93.

Bootz, J.P. (2012) 'L'évolution du manager: un pilote de communauté de pratique entre l'expert et l'intrapreneur', *12ème rencontres Prospective des Métiers (AGRH)*, Paris.

Bootz, J.P. and Kern, F. (2009) *Les communautés en pratique: leviers de changements pour l'entrepreneur et le manager*, Traité Management et Gestion des STIC, Paris: Hermès Science Publications, Lavoisier.

Bootz, J.P. and Schenk, E. (2009) 'Comment gérer les experts au sein et en dehors des communautés?', in J.P. Bootz and F. Kern (eds), *Les communautés en pratique: leviers de changements pour l'entrepreneur et le manager*, Paris: Lavoisier.

Bootz, J.P. and Schenk, E. (forthcoming) 'L'expert en entreprise: proposition d'un modèle définitionnel et enjeux de gestion', *Management et avenir*, forthcoming.

Borzillo, S., Probst, G. and Raisch, S. (2008) 'The governance paradox: balancing autonomy and control in managing communities of practice', *Academy of Management Best Papers Proceedings*.

Brown, J.S. and Duguid, P. (1991) 'Organizational learning and communities of practice: toward a unified view of working, learning and innovation', *Organization Science*, **2**(1): 40–57.

Brown, J.S. and Duguid, P. (1998) 'Organizing knowledge', *California Management Review*, **40**(3): 90–111.

Cardinal, L.B. (2001) 'Technological innovation in the pharmaceutical industry: managing research and development using input, behavior, and output controls', *Organization Science*, **12**: 19–36.

CEFRIO (Centre francophone d'informatisation des organisations) (2005), *Travailler, apprendre et collaborer en réseau. Guide de mise en place et d'animation de communautés de pratique intentionnelles*, 116 pages.

Chanal, V. (2000) 'Communautés de pratique et management par projet: A propos de l'ouvrage de Wenger (1998) *Communities of Practice: Learning, Meaning and Identity*', *M@n@gement*, **3**(1): 1–30.

Chi, M., Glaser, R. and Rees, E. (1982) 'Expertise in problem solving', in R.J. Sternberg (ed.), *Advances in the Psychology of Human Intelligence*, I. Hillsdale, NJ: Lawrence Erlbaum, 7–75.

Cohendet, P., Créplet, F., Diani, M., Dupouët, O. and Schenk, E. (2004) 'Matching communities and hierarchies within the firm', *Journal of Management and Governance*, **8**, 27–48.

Cohendet, P., Créplet, F. and Dupouët, O. (2006) *La gestion des connaissances: firmes et communautés de savoir*, Paris: Economica.

Cohendet, P., Roberts, J. and Simon, L. (2010) 'Créer, implanter et gérer des communautés de pratique', *Gestion*, **35**(4): 31–5.

Cowan, R. (2001) 'Expert systems: aspects of and limitations to the codifiability of knowledge', *Research Policy*, **30**(9): 1355–72.

Créplet, F. and Dupouët, O. (2009) 'Communautés de savoir, mécanismes de gouvernance et apprentissage organisationnel', in J.P. Bootz and F. Kern (eds), *Les communautés en pratique: leviers de changements pour l'entrepreneur et le manager*, Paris: Lavoisier.

Daunais, S., Zibara, L., Beaulieu, M., Cohendet, P., Moreau, B. and Simon, L. (2010) 'Le pilotage des communautés de pratique au sein de l'industrie des télécommunications: une comparaison des sociétés Bell et Orange', *Gestion*, **35**(4): 47–55.

Dupouët, O. and Barlatier, P.-J. (2011) 'Le rôle des communautés de pratique dans le développement de l'ambidextrie contextuelle: le cas GDF SUEZ', *Management International*, **15**(4): 95–108.

Ericsson, K.A., Krampe, R.T. and Tesch-Roemer, C. (1993) 'The role of deliberate practice in the acquisition of expert performance', *Psychological Review*, **100**: 363–406.

Gilbert, G.C. (2006) 'Change in the presence of residual fit: can competing frames co-exist? *Organization Science*, **17**(1): 150–67.

Gongla, P. and Rizzuto, C.R. (2001) 'Evolving communities of practice: IBM Global Services experience', *IBM Systems Journal*, **1**(4): 842–62.

Gosselin, F., Barlatier, P.-J., Cohendet, P., Dunlavey, P., Dupouët, O. and Lampron, F. (2010) 'Le partage des rôles et des responsabilités à l'égard du pilotage des communautés de pratique', *Gestion*, **35**(4): 36–46.

Hansen, M.T., Nohria, N. and Tierney, T. (1999) 'What's your strategy for managing knowledge?', *Harvard Business Review*, **77**(2): 106–16.

Harvey, J.-F. (2010) 'Comment favoriser le partage des connaissances? Le cas des communautés de pratique pilotées', *Gestion*, **35**(4): 73–80.

Hatchuel, A. and Weil, B. (1992) *L'expert et le système. Quatre histoires de systèmes-experts*, Paris: Economica.

Hornsby, J.S., Naffziger, D.W., Kuratko, D.F., and Montagno, R.V. (1993) An interactive model of the corporate entrepreneurship process', *Entrepreneurship Theory and Practice*, **17**(2): 29–37.

Hornsby, J.S., Kuratko, D.F. and Zahra, S.A. (2002) 'Middle managers' perception of the internal environment for corporate entrepreneurship: assessing a measurement scale', *Journal of Business Venturing*, **17**: 253–73.

Kang, S. and Snell, S.A. (2009) 'Intellectual capital architectures and ambidextrous learning: a framework for human resource management', *Journal of Management Studies*, **46**(1): 65–92.

Klein, G. (2003) *The Power of Intuition*, New York: Doubleday.

Long, C.P., Burton, R.M. and Cardinal, L.B. (2002) 'Three controls are better than one: a computational model of complex control systems', *Computational & Mathematical Organization Theory*, **8**: 197–220.

March, J.G. (1991) 'Exploration and exploitation in organizational learning', *Organization Science*, **2**(1): 71–87.

McDermott, R. (2001) 'How to design live community events', *Knowledge Management Review*, **4**(4): 103–17.

McDermott, R. and Archibald, D. (2010) 'Harnessing your staff's informal network', *Harvard Business Review*, March, 1–7.

Mom, T.J., van den Bosch, F.A. and Volberda, H.W. (2007) 'Investigating managers' exploration and exploitation activities: the influence of top-down, bottom-up, and horizontal knowledge inflows', *Journal of Management Studies*, **44**(6): 910–31.

Mom, T.J., van den Bosch, F.A. and Volberda, H.W. (2009) 'Understanding variation in managers' ambidexterity: investigating direct and interaction effects of formal structural and personal coordination mechanisms', *Organization Science*, **20**(4): 812–28.

Morris, M.H. and Kuratko, D.F. (2002) *Corporate Entrepreneurship*, Orlando, FL: Harcourt Publishers.

Nonaka, I. and Takeuchi, H. (1995) *The Knowledge Creating Company*, New York: Oxford University Press.

Nooteboom, B. (2002) *Trust: Forms, foundations, functions, failures and figures*, Cheltenham: Edward Elgar.

Oiry, E. (2005) 'Qualification et competence: deux sœurs jumelles', *Revue française de gestion*, **31**: 13–34.

Ouchi, W.G. (1979) 'A conceptual framework for the design of organizational control mechanisms', *Management Science*, **25**: 833–48.

Ouchi, W.G. (1980) 'Markets, bureaucracies, and clans', *Administrative Science Quarterly*, **25**: 129–41.

Polanyi, M. (1966) *The Tacit Dimension*, London: Routledge.

Probst, G. and Borzillo, S. (2007) 'Piloter les communautés de pratique avec succès', *Revue Française de Gestion*, **170**: 135–53.

Probst, G. and Borzillo, S. (2008) 'Why communities of practice succeed and why they fail', *European Management Journal*, **26**: 335–47.

Raisch, S. (2008) 'Balanced structures: designing organizations for profitable growth', *Long Range Planning*, **41**(5): 483–508.

Raisch, S. and Birkinshaw, J. (2008) 'Organizational ambidexterity: antecedents, outcomes, and moderators', *Journal of Management*, **34**(3): 375–409.

Raisch, S., Birkinshaw, J., Probst, G. and Tushman, M.L. (2009) 'Organizational ambidexterity: balancing exploitation and exploration for sustained performance', *Organization Science*, **20**(4): 685–95.

Salas, A., Rosen, M.A. and DiazGranados, D. (2010) 'Expertise-based intuition and decision making in organizations', *Journal of Management*, **36**: 941–73.

Schulz, M. (2001) 'The uncertain relevance of newness: organizational learning and knowledge flows', *Academy of Management Journal*, **44**: 661–81.

Star, S.L. and Griesemer, J.R. (1989) 'Institutional ecology, translations and boundary objects: amateurs and professionals in Berkeley's Museum of Vertebrate Zoology, 1907–1939', *Social Studies of Science*, **19**: 387–420.

Taylor, A. and Helfat, C.E. (2009) 'Organizational linkages for surviving technological change: complementary assets, middle management, and ambidexterity', *Organization Science*, **20**(4): 718–39.

Turner, K.L. and Makhija, M.V. (2006) 'The role of organizational controls in managing knowledge', *Academy of Management Review*, **31**(1): 197–217.

Van Wijk, R., Jansen, J. and Lyles, M. (2008) 'Inter- and intra-organizational knowledge transfer: a meta-analytic review and assessment of its antecedents and consequences', *Journal of Management Studies*, **45**(4): 830–53.

Von Hippel, E. (1998) 'Economics of product development by users: the impact of "sticky" local information', *Management Science*, **44**(5): 629–44.

Von Krogh, G., Ichijo, K. and Nonaka, I.O. (2000) *Enabling Knowledge Creation: How to unlock the mystery of tacit knowledge and release the power of innovation*, Oxford: Oxford University Press.

Weber, M. (1919) *Le savant et le politique*, Paris: Union Générale d'Éditions [1963].

Wenger, E. (1998) *Communities of Practice: Learning, meaning and identity*, Cambridge: Cambridge University Press.

Wenger, E., McDermott, R. and Snyder, W.M. (2002) *Cultivating Communities of Practice*, Boston, MA: Harvard Business Press.

Woodman, R.W., Sawyer, J.E. and Griffin, R.W. (1993) 'Toward a theory of organizational creativity', *Academy of Management Review*, **18**: 293–32.

7 Science based entrepreneurship through the lens of sectoral and national innovation systems

French academic start-ups developing hydrogen and fuel cell technologies

Arman Avadikyan

Introduction

The necessity to reduce greenhouse gas emissions in energy production, to diversify energy resources and liberalization trends in the energy sector have motivated market oriented innovative approaches to develop sustainable technologies. In this context academic start-ups (ASUs) may be an important driver for creativity to bring science based breakthrough environmental innovations to the market and make use of opportunities that otherwise would go unnoticed. In fact, ASUs, although far more risky than other means for research valorization (e.g., collaborative research schemes, licensing of patented university inventions), attract a great deal of policy attention in considering academic entrepreneurship as a means to connect more closely scientific creativity and market opportunities. Traditionally, several arguments justify public support for ASUs. They concern first information / knowledge asymmetries and R&D market imperfections (Hall, 2002; Schneider and Veugelers, 2010; Takalo and Tanayama, 2010). A second argument in favor of public support is that, while developing breakthrough technologies, ASUs can generate social benefits that are in excess of the private returns they can capture (Nelson, 1959). From a macro policy perspective ASUs can critically contribute to overall creativity of the economic system by acting as engines in the industry exploration-exploitation discovery cycle (Gilsing and Nooteboom, 2006). In this respect ASUs can be seen as a key resource in the collective creativity process both through failure and success (McGrath, 1999). Third, when the environment is taken into account, there is the necessity, because of the double externality problem (Jaffe *et al.*, 2005), to coordinate innovation and environmental policies to reduce negative external costs. In this context, ASUs have the potential through breakthrough innovations to contribute to technological regime shifts and improve the social efficiency of environmental policies.

This chapter combines the sectoral and national systems approaches (Pavitt, 1984; Nelson and Rosenberg, 1993; Lundvall, 1998; Malerba, 2005) to analyze their roles in the innovation process of ASUs working on hydrogen and fuel cell technologies (HFCs). We focus our analysis on the French HFC innovation system and ASUs created and operating in France. Our aim is to explore the interactions between national and sectoral patterns of innovation as the basis for the empirical analysis of French HFC ASUs. Although ASUs are seen as critical to develop new knowledge and new markets, their opportunities and challenges are not neutral to the sectoral and national contexts within which they operate. We argue that both the dynamics of innovation patterns in the energy sector and factors related to the French innovation system have shaped the development processes of HFC ASUs.

The second section focuses on the sectoral environment of HFCs and on technological/market uncertainties faced by ASUs. The third section describes the evolution of the French science and technology (S&T) policy environment related both to HFCs and ASUs. The next section is devoted to the analysis of five HFC ASU cases. The final section concludes.

Sectoral opportunities and technology/market uncertainties for HFCs

The analytical framework of sectoral innovation systems explains why sector specific characteristics can lead to differences in terms of technology innovation opportunities and orient creativity. Referring to Schumpeter several authors distinguished between *Schumpeter Mark I* and *Schumpeter Mark II* technological regimes (Winter, 1984; Klevorick *et al.*, 1995; Breschi *et al.*, 2000). The former is characterized by low concentration of innovative activities, small size innovators, high entry and exit of innovators and disruptive innovations, whereas in the latter concentration of innovative activities is higher, innovators are of larger size, entry is lower, stability in the ranking of innovators is greater and innovations are mainly of an incremental type.

The liberalization of energy markets and energy security/environmental concerns over the past 30 years have initiated new *stimuli* in favor of both new business models based on decentralized electricity generation within more competition oriented markets and sustainable technologies for energy production and use. These trends contributed to some degree to transforming actors' perceptions on the variation and selection processes related to potentially radical innovations both in the electricity and transportation sectors (Sine and David, 2003; Markard and Truffer, 2006). One consequence was the broadening of scientific/technological paths and the diversification of both public and private actors involved in the innovation process.

Within this context, HFCs because of their promise to mitigate CO_2 emissions and their ability to be operated as decentralized energy systems in combination with renewable energy sources attracted growing interest as a new technological alternative. The market potential of HFCs is related to the broad range of possible uses in transportation (cars, auxiliary power unit), stationary (combined heat and power, backup power) and portable (battery charging, power unit) applications.

FCs rely on the electro-chemical reaction of combining hydrogen with oxygen in a catalytic process to produce direct electric current and heat. There are several types of FCs which can be distinguished by their operating temperature and the electrolyte used. Proton Exchange Membrane (PEM) FCs can be considered as particularly suited for entrepreneurial initiatives since they do not require massive investments in their early development stages. The FC technology represents however only the conversion part of the HFC value chain. Complementary parts are hydrogen production, distribution and storage. Promising advances on the PEMFC have also opened new opportunities along the hydrogen value chain. As a result new innovative firms, traditionally not part of the hydrogen production industry, have engaged in hydrogen technologies (Avadikyan *et al.*, 2003; Suurs *et al.*, 2010). Entrepreneurial firms focus their efforts on small scale decentralized hydrogen production technologies based on bio fuel reforming and water electrolysis. Research activities are also focused on high pressure or solid state hydrogen storage.

Progress in HFCs is dependent on research in a varied set of fundamental and applied scientific disciplines such as electrochemistry, composite materials, physics, nano and biotechnologies and energy engineering. The critical role played by S&T in the field of HFCs is evidenced by Hassan (2005) through the expansion trends of S&T outputs related to HFCs between 1990 and 2000 both in terms of publications and patents. An examination of the intensity of science (publications) and technology (patents) linkages related to HFCs during the same period, shows also the strong cognitive relations between these outputs in several main sub-fields, which may explain "the fact that researchers and inventors were inclined to solve more and more similar problems on both sides" (p. 233). Therefore, the science intensity of HFCs, the close linkage in this domain between S&T, optimistic expectations on their economic potential along with market oriented innovation policies have offered scientists, since the beginning of the 1990s, opportunities to value and transform research results on HFCs developed in universities into innovative applications through several technology transfer paths including the creation of entrepreneurial ventures.

Being however in their emerging phase along the innovation process, HFCs continue to be dominated by significant technological and market uncertainties. It is unclear which HFC technical alternatives will prevail (Hellman and van den Hoed, 2007). Risks during this emergence phase are the dominance of some trajectories to the detriment of others. Particularly in the creative phase of technologies fierce competition may lead to rapid obsolescence of S&T advances.

The challenges relate also to future and uncertain markets. The systemic nature of the HFC value chain which requires the integration of numerous specialized technological subsystems, the need for complementary assets and infrastructural investments constitute severe adoption barriers for HFCs. Furthermore HFCs must substitute for mainstream products and processes which define the standard in the valuation of emerging technologies. Although HFCs have a variety of potential applications there are uncertainties about whether, when and which markets will adopt them. The prospects for an application are likely to change as the technology and the markets for HFCs continue to develop. For instance, at the beginning of the 2000s first FC experiments in niche transport markets were highly diversified, but the one application which recently had the highest growth rate was the auxiliary power unit (APU) (Adamson and Jerram, 2009). Although market experimentation remains critical to support the technology market matching process and niche markets can be a stepping stone for the diffusion of breakthrough innovations (Agnolucci and McDowall, 2007; Hendry *et al.*, 2007) technology and prototype developments are still based on expectations and the technology/market selection environment remains in flux.

Another critical challenge for HFCs start-ups during this phase is the uncertainty related to the improvement potential of competing technologies (Nair and Ahlstrom, 2003). HFCs, both in terms of fuels and conversion technologies, are in fact competing with many alternatives that might also significantly contribute to emission reduction and diversification of energy sources. For instance, if significant progress in HFCs and the influential role of some HFC companies had created at the beginning of the 1990s shared optimistic expectations and had raised the interest of researchers, industrial actors, investors as well as policy makers during one decade, this wave of interest subsided progressively during the 2000s following a collective learning process that took better into account the development challenges and weaknesses of HFCs (e.g. their efficiency). The initial rise of interest in HFCs coincided with that of other clean energy sectors such as wind and solar. Contrary however to HFCs these sectors continued to attract significant investment. Also, if initially HFCs were presented as an alternative solution to electric batteries avoiding some of their technical barriers, investments in batteries again showed a significant rebound in the automotive sector considering their progress in portable applications.

Positioning of ASUs within the French HFC innovation system

The national system of innovation played a critical role in supporting the structuring of the French HFC industry. Allocation of public resources, technological policies, private-public collaboration patterns, institutional features of public research establishments and ASU support mechanisms all contributed to the emergence and development of the French HFC industry.

Several contributions pointed to the distinctive features of the French innovation system, through both its enduring traits and recent transformations (Mustar and Larédo, 2002; Vavakova, 2006). With regard to the energy innovation system three main elements can be put forward to explain the structuring of the French technology/industrial policy model. The first is the extensive role played by large technology programs supported by public intervention for instance in nuclear energy. The second trait concerns the creation of mission oriented public research institutes such as the *Commissariat à l'Energie Atomique* (*CEA*) in nuclear energy. The third trait relates to the industrial policy which privileged "energy champions" such as *AREVA*, *EDF* and *GDF*.

The priority given to nuclear energy in order to reduce the energy dependence of France since the 1970s explains why many energy related decisions including those on HFCs are today linked in France to the scientific, technological and industrial competences that were developed in this field by national public and private actors. Today nuclear power still remains a cornerstone of the French environmental and energy security policies and is considered as an opportunity to reinforce its role in reducing greenhouse gas emissions. Nevertheless, the environmental policies that were initiated following the Kyoto protocol and the alignment with European energy objectives led France progressively to diversify its R&D towards alternative energy technologies. HFCs have been an important component of this diversification trend (Meritet, 2007; IEA, 2010).

Although HFCs had been intensely investigated in France during the period 1960–1975, R&D efforts were afterwards significantly reduced. Compared for instance to Germany which acted as a leader in Europe in HFC investments, France had accumulated during the 1990s a significant gap. It was the influential role of the French automotive industry which at that time was experiencing difficulties in developing battery electric vehicles that motivated the decision to consider HFCs as a priority alternative energy technology and to support the emergence of a French HFC industry.

This orientation motivated the integration of HFC programs within the more general national innovation policy trend which at that time began to initiate several successive reforms (Laws on Innovation promulgated in 1999 and 2005) that privileged new market oriented public support mechanisms based on networking between public-private actors in priority technological areas, the creation of ASUs, competitive schemes in research funding and regional clustering.

Through this new policy orientation, HFC programs gave a new impulse to R&D projects associating teams from public research establishments and industry. Given the optimistic prospects and the political and industrial priorities on HFCs, several spin-off firms, including those by large companies, were created. Both *AREVA* and *Air Liquid* launched in 2001 their own respective spin-offs. Conceived initially as exploratory programs they supported a wide area of technologies and long term breakthrough trajectories.

With respect to HFC ASUs, most of them were launched following the introduction in 1999 of ASU support mechanisms. These mechanisms included several measures namely (1) a new civil servant status for public researchers to allow them, after a selective review process, to create ASUs, become stakeholders or offer scientific consultancy, (2) a national contest supporting the emergence and creation of high tech start-ups, (3) the establishment of incubators and finally (4) sectoral public seed capital funds which take minority interests in selected start-ups with high growth potential irrespective of their origins (Mustar and Wright, 2010). This science based entrepreneurship policy is based on a systemic approach where public intervention covers each stage to address "failures" from the lab to the development phase. The different measures are complementary and each acts as a sequential selection mechanism during the support process.

From 1999 to 2008, 5 HFC ASUs were created with the support of the national contest. Among the seed funds, one is dedicated to energy and environmental technologies with institutional and industrial investors also contributing to the fund. From 2003 to 2010, this fund supported 19 start-ups, with only one of them engaged in HFCs. Interviews with ASUs show that selection by the national contest and support by incubators contributed to reducing ASUs obstacles in the search for investors particularly for those lacking personal contacts. All could access either pre-seed capital and/or incubation funds in their creation and first development phases. Nevertheless HFC ASUs stressed also the fact that this general technology entrepreneurship policy tended to abstract from the individual and sectoral specificities of HFC companies.

Although other innovative SME support mechanisms (project based subsidies and grants) financed HFC projects by ASUs, the most important support came from HFC national programs based on competitive schemes and structured around private-public R&D cooperation projects.

During the initial explorative stage of these programs the competition field could be considered as open. All French forerunner companies and a large number of public research establishments participated in these programs. The development of HFCs was however progressively dominated by a few private and public actors. Although no less then 100 public laboratories participated in the programs only a few were stable participants through time. Espacially those which had maintened a research activity throughout the 1980s could take advantage of these programs and exploit their expertise. As a mission oriented R&D laboratory, the *CEA* acted as the central actor both in terms of R&D efforts and as an orchestrator across private-public projects. Clustering initiatives evidence also the high concentration of activities in the *Grenoble* region, representing 70 percent of HFC patents filed in France (DRIRE, 2008). This result is not surprising if one considers that the most active HFC actors (e.g. *CEA*, other research establishments, *Air Liquide*) are located in this region. Also, progressively corporate spin-offs acquired a dominant position within the programs because of their larger capacity, in terms both of diversity of their R&D activities and their networking with a larger set of private and public research establishments.

Although ASUs could benefit from these programs, they were confronted with several challenges. Applying within the framework of competitive programs they were not sure to get support for their activities. Particularly, the dominant position of corporate spin-offs made access to these programs progressively more difficult for ASUs. Some ASUs participated also in projects which were not directly related to the development of their core technologies. This reflects the dilemmas faced by ASUs in deciding whether to participate in time consuming side-projects for finance's sake, network or experience or to reject them in order to remain focused on their core technologies. Also because of technological uncertainties some ASUs could not always finalize their projects in due time and prolonged them without financial compensation. Another critical factor was the narrowing down of technological options pursued by the programs. In fact, after an initial phase of technological exploration there was a shift towards a more focused approach. Although support on PEMFC was relatively stable, it appears that the focus on hydrogen technologies shifted progressively towards R&D on high temperature electrolysis and on solid state hydrogen storage reflecting more long term opportunities connected with nuclear energy.

More recently, after a decade during which HFCs were high on the S&T policy agenda, the technical difficulties and environmental weaknesses associated with their development motivated a shift of public funds towards other alternative energy technologies. Although support through these programs was in the past critical for ASUs, the number of accepted projects progressively decreased and made public support even more uncertain and difficult to get for ASUs.

French ASUs involved in HFC technologies

In this section we examine five ASU cases involved in HFCs and provide insights into their development processes. Our analysis is based on interviews conducted with founders, managers and researchers of three ASUs and on Internet information sources. For anonymity's sake, companies are renamed A, B, C, D and E. Our study insists on three aspects that emerged as critical to the activities of ASUs. They relate to HFC technological and market uncertainties, ASUs interactions with public research establishments and with industrial actors.

Technological and market uncertainties

ASUs may have an advantage in being first to exploit scientific results but their creation often represents only a first step to demonstrate the feasibility of a technological concept. The novelty of the technology means generally a high level of uncertainty regarding the possibility of replicating laboratory attributes in prototypes (Maine and Garnsey, 2006). Furthermore, technologies developed must be incorporated into wider systems and need complementary innovations. Technological risks can therefore be particularly high if ASUs focus on a single technology without being able to switch to alternative options.

Although the generic nature of ASUs' scientific background may confer them the flexibility to transfer their knowledge to different applications, investments in prototypes are generally made prior to the existence of an observable demand and it can therefore be challenging to decide when to invest in which application (Hellman and van den Hoed, 2007). Experimentation about new market-technology combinations to reduce market uncertainties can turn out to be expensive and resources can become scarce when different applications require unrelated and costly investments. Dividing resources across many applications and long lead market times can thus impede the rapid development of ASUs.

Several of the ASUs studied were confronted with significant challenges and uncertainties during their development phase. They were obliged to reframe their business models, their strategies and to abandon or to realign some of their activities either because initial opportunities became irrelevant or because they discovered new ones during their development process.

The case of company B reflects the difficulties when developing proof of technological and market concepts. The ASU was motivated by the promising results obtained during a three year Ph.D. project on a high temperature multi-fuel hydrogen generation process. Shortly after its creation in 2001, a business plan was established reflecting how the company perceived its technology/market development paths with the belief of providing market ready products in ten years' lead time. However, first development projects revealed that the new process was inappropriate for the application it was meant for, that is on-board automotive multi-fuel reformers. The company continued nonetheless to focus on the same technology by targeting several other niche markets for automotive (APU) and decentralized stationary (FC CHP – combined heat and power) applications with alternative designs. These developments brought into light other deficiencies making the original opportunity technically unprofitable. Although the company diversified its potential markets, its development strategy focused on a single technical process which turned out in the end to be unsustainable. As one of the academic co-founders said, progress along ASU's technology required long term R&D that the company could not afford because of insufficient financial resources. These failures forced the ASU to abandon its technology and to reorient its activities towards HFC R&D services, technical assistance and consultancy. Due to lack of promising perspectives the ASU finally ceased its activities.

Company A successfully developed its technology but is still confronted with market uncertainties. Created in 1997, it was initially conceived as an R&D sub-contractor for HFCs. This strategy allowed the ASU to extend its technological exploration activities and to benefit from synergies between different projects. Three technological projects were more particularly investigated. Nevertheless these technologies were at different maturity stages. The decision to industrialize the most mature project, the one which was perceived as the most easy to develop (PEM electrolyzer) since several subsystem components were already standardized and easily available in the market, obliged the firm – because of financial constraints – to put other projects on stand-by. This orientation meant however that the ASU was confronted with a highly competitive electrolyzer niche market at the international level and with difficulties in finding customers to achieve a sustainable business path for its product line.

Companies D and E, beyond focusing on HFC markets, developed alternative applications for their technologies. Company D conceived its innovative ceramic membranes as a platform technology. During its research activities the ASU explored several potential applications. Several of them were abandoned during the process. Finally, besides centering the main application of its technology on PEM electrolyzers for hydrogen production the company retained the possibility of using its membranes for water disinfection. Since its creation in 2005, efforts have been focusing on the development of pilot equipments. More recently a pre-industrialization plan was initiated. The company therefore has not yet embarked upon commercially exploiting its technological assets.

One of the best positioned ASUs, company E can be differentiated from the others by the length of research activities that were conducted within its parent research organization (PRO) already before 1990 on its magnesium based solid-state hydrogen storage technology. When the company was founded in 2008, the innovation had benefited from the accumulation of a significant knowledge base. The proof of concept was therefore strong and the ASU needed just one year to deliver its prototype after its creation. Furthermore thanks to an industrial experienced management team the company elaborated a clear strategic business plan. The company combined capabilities both in the production of powdered magnesium hydrides and hydrogen storage and targeted existing markets where there was already an increasing demand for hydrogen storage and magnesium as a material (in industrial uses and chemical applications) as well as long term markets (related to renewable energies). The company adopted furthermore an international market positioning and supplied first industrial clients in Europe and Asia in a very short timeframe.

Finally, company C, contrary to all other ASUs developing stationary applications, centers its activities on portable applications based on its innovative FC components. The capability to market its products depends on complementary innovations such as for instance miniaturized hydrogen storage devices. Through several collaborative R&D projects the company has been actively involved in the development of these complementary innovations. In order to sustain its long term portable FC product development, the company has acted since its creation in 2003 as a R&D sub-contractor, as a supplier of its innovative FC components in niche applications and as a provider of FC test materials and educational kits. Although the low capital intensity of portable FCs allowed the ASU to be involved in several demonstration projects, it remains still uncertain whether or not FCs will be able to compete with advanced battery technologies in portable applications.

In the case of ASUs that were created earliest the long lead times to develop and market HFC technologies rendered them highly vulnerable. Technological and market uncertainties affected heavily their activities and for some of them threatened their survival obliging them to review their ambitions or even cease their activities. It appears that some of the ASUs spun out too early without even a preliminary proof of concept or without clearly framing technological and market opportunities and uncertainties which characterize the HFC sector. Most ASUs spent the largest part of their capital on technology development and testing various assumptions related to various applications. Several ASUs searched for alternative applications in order to hedge against market risks related to HFCs. But none has until now generated sustainable returns from sales.

Interactions with public research establishments

Academic R&D represents an important source of creativity for high tech start-ups to improve their core technologies (Cohen *et al.*, 2002; Santoro and Chakrabarti, 2002). Nevertheless, because of their limited resources, ASUs can have difficulties diversifying their academic connections (Mohnen and Hoareau, 2003; Laursen and Salter, 2004). Through four ASU cases Johansson *et al.* (2005) observed that their university relations are characterized by a small number of strong ties to PROs. Such limited strong ties, although decisive for the transfer of complex knowledge, create a high degree of dependence and vulnerability for ASUs and make them difficult to substitute. Focusing on the comparison between ASUs and non-ASUs, Colombo *et al.* (2010) showed that the association with high quality research universities contributes positively to the growth of ASUs, whereas their influence on the growth of non-ASUs is found to be far less significant. These results are justified by several of ASUs' attributes that facilitate the detection, absorption and exploitation of academic knowledge. Furthermore, because of their science orientation and the social contacts of their founders, ASUs can be advantaged in receiving support from their PROs and in motivating academics to get involved in their ventures as members or consultants.

Access to public research establishments is considered by all ASUs studied as being critical in order to improve their innovativeness. Initially, the motivation relates to the need to access equipment and facilities of their former establishments. Most of the ASUs in fact first operated within the premises of their PROs. The cases evidence also that although PROs contributed to lowering ASUs' financial needs in their emergence phase, they rarely supported them financially, except in case D where the PRO had an explicit ASU creation policy with dedicated resources. Academic research continues also to play a catalytic role after the creation of ASUs since they build their business on the findings that come out of their university collaborations. The cases reveal therefore a clear division between research conducted by universities and development efforts by ASUs. Historical relations, geographic proximity, knowledge specificity and resource constraints explain the strong ties of ASUs to a limited number of research establishments and mainly to their PROs and their networks.

Another common trait is that for all ASUs collaborative links with universities are made possible through public funding based on national and European R&D programs. These programs, beyond providing financial support, create an environment where it becomes easier to find mutual interests between ASUs and research establishments, to articulate priorities and to develop joint projects. They allow also the financing of doctoral students and common investments in research equipments.

Beside these common elements, one observes also differences between ASUs. For instance, an important characteristic of ASU D relates to the central position of its PRO in the French HFC innovation system. The resources and scientific reputation of its PRO helped the ASU participate in ambitious scientific projects and diversify its academic partners. Companies D and E located in the *Grenoble* region benefited from the dense scientific network on HFCs. In the case of company E, the reputation of the academic inventor, his boundary spanning role by keeping positions both within the university and the ASU as a scientific advisor and his capability to leverage university and laboratory resources were key in nurturing the innovativeness of the ASU.

Compared to other ASUs, company A was not involved before its creation in HFC projects and its university relationships had been based on the founder's personal connections. The venture was motivated by the founder's belief that his background as a former well-known scientist, his research work related to HFCs and his detailed knowledge on research projects conducted in France could position the ASU as a key intermediary actor transforming academic research into innovative HFC technologies. Shortly after its creation the ASU established strong links with three research establishments, which had been at the basis of the three technologies developed by the ASU. These projects critically influenced the strategic orientation of the company and served also as a basis for recruitment of young researchers.

ASUs have been nevertheless also confronted with challenges in keeping connections with their PROs or in involving researchers more strongly in their venture. In the case of company B its founders were the most actively engaged in HFC research within their PRO. The ASU could benefit initially from the research facilities of its PRO which also participated in several R&D projects through the national HFC R&D program. However, even if one of the founders of the ASU retained his academic post and informal relations in the beginning facilitated cooperative projects, the human resource base of the PRO dried up progressively after the spin-out. It became therefore more difficult for the start-up to rely on its PRO.

Although company B was involved in projects with other research establishments, these were peripheral to its activities and were mainly motivated by financial and networking purposes and did not contribute to the core product of the company.

Company A reveals another aspect related to the fragile balance between the interests of ASUs and those of university researchers. In fact, after focusing its business on the most promising technology, the ASU's commercial orientation and conflicts between the new manager and initial employees created tensions within the company. Two of them resigned from the ASU and divested the company from critical competences. This created the need to involve more actively in the ASU the main academic researcher who had contributed to the electrolyzer project. But the scientist had also his own priorities in terms of teaching and research in other fields which competed increasingly for his limited time. Therefore he was not eager to be more intensely involved in the ASU. Company A tried to diversify its collaborations with other research establishments, but it became more difficult to run in parallel the development of its core product and to launch new research projects due to its limited human and financial resources.

All ASUs evidence the critical role played in their innovation activities by universities and particularly by their PROs. This dependence on their former establishments is reflected in the need to access equipment and facilities but more fundamentally to develop additional knowledge in terms of research for their innovativeness. Thus even after the creation of ASUs university research remains the main source for their business developments. Nevertheless, their capacity to maintain a strong connection to the scientific world rests on public support. Therefore the downsizing of HFC programs might in the future seriously affect the capability of most HFC ASUs to establish R&D arrangements with universities.

Industrial partnerships

Since the beginning of the 1990s the HFC industry has been characterized by an impressive number of R&D partnerships and collaborations. HFC ventures interact both with original equipment manufacturers (OEMs) and energy utilities to convince them to incorporate these technologies into their products and business models (Jacquier-Roux and Bourgeois, 2002; Brown *et al.*, 2007).

For innovative start-ups, a key dimension in the emerging phase of technologies is partnering with a diversified set of industrial actors to reduce uncertainty around their activities and get support in the development and diffusion of their technologies (Birley, 1985; Gomes-Casseres, 1997). Partnerships with large industrial actors can allow ASUs to access financial resources, complementary assets and innovations (Teece, 1986), managerial advice (Dushnitsky and Lenox, 2005) and eventually facilitate market entry (Acs *et al.*, 2001).

For incumbent firms, whose R&D activities are dominated by existing products, collaboration with ASUs provides access to science based innovations. By partnering with ASUs they can avoid investments in duplicative R&D projects (Gans and Stern, 2003), benefit from their curiosity driven creativity improving thereby their absorptive capacity and innovative performance (Ahuja and Katila, 2001; Rothaermal, 2002). Even taking options on failing ventures can constitute for incumbent firms a learning experience to the extent that it offers new technological insights or points at market unattractiveness.

However, ASUs can also be confronted during their search for partners with several challenges related for instance to adverse selection problems because of technological uncertainties (Gans and Stern, 2003). The ability of ASUs to signal their quality through scientific achievements and patenting (Shane and Stuart, 2002), the social capital of scientific founders, of their PROs or the network relationships of TTOs (Di Gregorio and Shane, 2003) can play a key role in attracting leading partners. Colombo and Piva (2008) suggest that even under a weak appropriability regime, ASUs' propensity to resort to partnerships remains high because of financial needs and the necessity for managerial expertise, complementary assets and distribution channels.

Partnerships with industrial actors were critical in all our ASU cases. Collaborations had been initiated for some ASUs, within their PROs, even before their creation and represented a stepping stone for the spin-out decision. In the case of company B, research on its technology had been possible within the PRO when a French car manufacturer decided to finance a Ph.D. project. Although the automotive company had priority to exploit the patents filed, its interest on the technology dampened since it perceived the technology too early stage to develop further. This left the researchers with the opportunity to exploit the patents themselves. In the case of company E, its creation had been motivated by the enduring cooperation between the academic inventor and a high tech SME specialized in magnesium production processes interested in the alternative applications of the metal. The collaboration was supported both through bilateral and European projects on solid hydrogen storage. The ASU was created on the basis of several patents licensed from the PRO to the new ASU and the transfer of magnesium production competences already possessed by the SME.

The fact that ASUs develop breakthrough technologies based on advanced scientific competences makes their technology and market potential more difficult to evaluate, making therefore the necessity to find support for their innovations much more acute. A common trait in all companies was therefore short term collaborative projects with several incumbent companies, through either bilateral R&D contracting or through national or European private-public projects. All ASUs except company D also searched for external partners as investors and shareholders to support their development and pre-industrialization efforts.

The decision by company D to remain independent is explained by the fact that investments in portable applications were not capital intensive. Instead of relying on external investors, the company searched for R&D joint ventures with other high tech start-ups developing complementary technologies or new to the market applications to commercialize its FCs. This strategy led however the spin-out to be struck into very specific potential niche applications.

In the case of company A, its development activities were first financed through initial capital by the founders and their employees. The decision to commercialize its most advanced technology required nonetheless restructuring the ASU's internal management in order to adopt a more industry oriented approach. This led the scientific founder to retire voluntarily from the ASU and to hire an experienced manager as the new CEO. His role was to attract industrial partners for financial support and to facilitate access to complementary competences. The ASU tried also to initiate a large scale demonstration project with other industrial actors. The difficulties of bringing together a diversity of stakeholders with different interests and the necessary financial resources hindered however the launch of the project. Despite being acquired by a French family owned investment company in the renewable energy and environmental sectors and agreements with several foreign gas equipment companies to distribute its technology, the ASU could not until now establish a viable customer base.

Concerning company B, beyond personal capital by the founders and their family, its development activities were supported by taking advantage of financial opportunities existing at the regional level. The ASU was supported by two regional financial intermediaries responsible for backing up local economic development activities. After failing to develop its technology, thanks to the founders' entrepreneurial initiatives at the local level and the business reorientation on HFC engineering assistance, the ASU obtained a second round of funding by a regional company involved in the exploitation of renewable energy resources. Despite launching several small HFC related projects, this minority stake proved hardly to be sufficient to allow the ASU to recover from its failure.

Company C secured first financial support for its development by business angels through the network of its incubator. The ASU's development efforts during three years attracted the venture capital fund of one of Europe's leading energy utility companies aiming to diversify its portfolio of start-ups developing alternative energies. This corporate investment brought changes within the management board of the ASU through the involvement of a new experienced representative of the corporate venture and an independent director to prepare and support the company for a profitable exit. Being associated with a leading energy company involved in both gas and wind power installations provided the ASU with the opportunity to access critical assets and resources making it possible to develop its technology through a systemic approach.

Finally, company E was the only ASU to obtain financial backing by three different sources shortly after its creation (public seed capital, financial intermediaries and corporate seed capital from a large energy OEM company). The investments it attracted were by far the highest and also the most diversified compared to other ASUs. This success can be explained by three factors. First, the technology of the company had already been developed to a state of near market readiness. Another decisive factor was the personal relationships and reputation of the scientist and the industrial experience of the co-founders. A third factor relates to the ASU's access to the extensive network of industrial and financial actors structuring the French HFC cluster. The perceived high technology potential and the credibility of the founders and of the management team possessing significant industry specific, technical and managerial experience through their previous positions provided company E with the ability to attract in a very short time frame significant financial resources and to speed up its development. A noticeably larger second round funding after the development of the first prototype was secured from two other institutional investors which gave the company the possibility of initiating its pre-industrialization and production plant.

French HFC ASUs resorted both to collaborations for joint R&D projects and partnerships for acquiring capital funds and access to distribution channels. Collaborative R&D projects indicate that HFCs are still in their demonstrative phases. Engaging in collaborative projects with established actors allowed ASUs to further develop, communicate and evaluate their innovations. From the perspective of established actors ASUs served as exploration engines. During this process some ASUs positioned themselves as competitors of established firms which developed in parallel, through their corporate spin-offs, similar technologies. Almost all ASUs relied also on OEMs and energy utility companies to access capital and try to bring their products to the market. Considering the limited market potential for HFCs in France, several ASUs searched for foreign partners and/or customers. ASUs differed also in terms of the market power of actors they have been involved with. Some are leading players in the energy market; others have only a regional reach or have a small market share. The strategy of established firms consisted in all cases in making optional positioning investments to diversify their renewable energy technology portfolio. The long process of market acceptance for HFCs and expectation dynamics on their uncertain potential make however the situation of all ASUs highly vulnerable since they strongly depend on market oriented investment incentives of OEMs and energy companies which, in the case of disruptive technologies, may be highly unstable.

Conclusion

Innovative start-ups working on HFCs face critical challenges in commercializing their technologies. They experience numerous technical and market uncertainties in shifting from R&D to production and sales activities. Commercialization projections of HFCs went consistently unrealized revealing the excessive optimism that motivated entrepreneurial initiatives from the mid-1980s and it is still questionable when and whether they will gain widespread market acceptance.

This article analyzed science based entrepreneurial activities in HFCs in France through five ASUs created between 1999 and 2008. Two systemic perspectives were mobilized to gain insights into the development trajectories of these firms.

The sectoral innovation system perspective focused on the specificity of HFCs. The creation of HFC ASUs was motivated by the opening up of windows of opportunities in the energy sector for science intensive breakthrough technologies related to environmental concerns. Most French ASUs were confronted during their development with critical challenges forcing some of them to reorient their activities or even abandon their ambitions. Due to technological and market uncertainties initial choices were revealed in some cases to be non viable. Some ASUs, due to their limited financial resources, had difficulties in reorienting their activities towards new directions. The transition to commercialization remains for all of them a risky process. Although external investments played a critical role in their development process, positioning investments by incumbent energy companies represent optional investments which may be withdrawn if initial opportunities do not materialize. None of the companies has in fact until now achieved a sustainable returns phase.

To complement the sectoral perspective, our analysis focused on the positioning of ASUs within the French innovation system. The French energy context contributed heavily to the structuring of the national innovation strategy on HFCs. The involvement of France in HFCs corresponded also to a period where several market oriented mechanisms were implemented to support science based entrepreneurship initiatives. These mechanisms and HFC R&D public-private programs have been important for all ASUs. The dependence of ASUs on public funds reveals however the innovation policy risk involved for most ASUs to shifts in national S&T trends. In fact, priority changes towards other alternative technologies and pessimistic prospects on HFCs led progressively to downscale public support for HFCs, which affected negatively the capability of most ASUs to maintain their R&D efforts.

From a general policy perspective, the HFC case bears evidence that the coordination of sectoral and national innovation perspectives can be a complex task in the case of potentially radical technologies. National innovation policy aims to provide the framework conditions to encourage variety both by the development of different technological options and by providing support to explorative and creative activities. The motives for supporting ASUs are here mainly explained in terms of "R&D market failures" since the technologies they develop are supposed to be an agent of change.

But an innovation policy may lead to ineffective results without taking into account the sectoral context. HFC ASUs are in fact facing a situation where the existing energy regime has not yet embraced HFCs and where market demand has yet to emerge. ASUs' innovations may therefore take a long time to materialize and penetrate the market, if at all. In that sense, national support mechanisms seem to have systematically underestimated the sectoral obstacles and technological and market barriers that HFC ASUs might face after their creation. One may therefore question whether most HFC ASUs have not been created too early and whether market oriented innovation policies such as ASU support mechanisms are the most appropriate ones in the case of HFCs which still require long term research efforts, heavy investments and sector level coordination among established and new actors.

References

Acs, Z.J., Morck, R.K. and Yeung, B., 2001. Entrepreneurship, globalization, and public policy. *Journal of International Management*, **7**(3), pp. 235–51.

Adamson, K.-A. and Jerram, L.C., 2009. 2009 Niche Transport Survey. *Fuel Cell Today*, August.

Agnolucci, P. and McDowall, W., 2007. Technological change in niches: Auxiliary power units and the hydrogen economy. *Technological Forecasting & Social Change*, **74**(8), pp. 1394–1410.

Ahuja, G. and Katila, R., 2001. Technological acquisitions and the innovation performance of acquiring firms: A longitudinal study. *Strategic Management Journal*, **22**(3), pp. 197–220.

Avadikyan, A., Cohendet, P. and Héraud, J.-A., 2003. *The Economic Dynamics of Fuel Cell Technologies*. New York: Springer.

Birley, S., 1985. The role of networks in the entrepreneurial process. *Journal of Business Venturing*, **24**(1), pp. 107–17.

Breschi, S., Malerba, F. and Orsenigo, L., 2000. Technological regimes and Schumpeterian patterns of innovation. *Economic Journal*, **110**(463), pp. 388–410.

Brown, J., Hendry, C. and Harborne, P., 2007. Developing radical technology for sustainable energy markets: The role of new small firms. *International Small Business Journal*, **25**(6), pp. 603–29.

Cohen, W.M., Nelson, R.R. and Walsh, J., 2002. Links and impacts: The influence of public research on industrial R&D. *Management Science*, **48**(1), pp. 1–23.

Colombo, M.G. and Piva, E., 2008. Strengths and weaknesses of academic startups: A conceptual model. *IEEE Transactions on Engineering Management*, **55**(1), pp. 37–49.

Colombo, M.G., D'Adda, D. and Piva, E., 2010. The contribution of university research to the growth of academic start-ups: An empirical analysis. *Journal of Technology Transfer*, **35**(1), pp. 113–40.

Di Gregorio, D. and Shane, S., 2003. Why do some universities generate more start-ups than others? *Research Policy*, **32**(2), pp. 209–27.

DRIRE, 2008. Analyse de l'environnement technologique des pôles de compétitivité par les brevets: Cas du pôle Tenerrdis, *DRIRE Rhônes-Alpes*.

Dushnitsky, G. and Lenox, M.J., 2005. When do incumbents learn from entrepreneurial ventures? Corporate venture capital and investing firm innovation rates. *Research Policy*, **34**(5), pp. 615–39.

Gans, J.S. and Stern, S., 2003. The product market and the market for "ideas": Commercialization strategies for technology entrepreneurs. *Research Policy*, **32**(2), pp. 333–50.

Gilsing, V. and Nooteboom, B., 2006. Exploration and exploitation in innovation systems: The case of pharmaceutical biotechnology. *Research Policy*, **35**(1), pp. 1–23.

Gomes-Casseres, B., 1997. Alliance strategies of small firms. *Small Business Economics*, **9**(1), pp. 33–44.

Hall, B., 2002. The financing of research and development. *Oxford Review of Economic Policy*, **18**(1), pp. 35–51.

Hassan, E. (2005) The evolution of the knowledge structure of fuel cells. *Scientometrics*, **62**(2), pp. 223–38.

Hellman, H.L. and van den Hoed, R., 2007. Characterising fuel cell technology: Challenges of the commercialisation process. *International Journal of Hydrogen Energy*, **32**(3), pp. 305–15.

Hendry, C., Harborne, P. and Brown, J., 2007. Niche entry as a route to mainstream innovation: Learning from the phosphoric acid fuel cell in stationary power. *Technology Analysis and Strategic Management*, **19**(4), pp. 403–25.

IEA, 2010. *Energy Policies of IEA Countries: France 2009 Review*. Paris: International Energy Agency.

Jacquier-Roux, V. and Bourgeois, B., 2002. New networks of technological creation in energy industries: Reassessment of the roles of equipment suppliers and operators. *Technology Analysis & Strategic Management*, **14**(4) pp. 399–417.

Jaffe, A.B., Newell, R.G. and Stavins, R.N., 2005. A tale of two market failures: Technology and environmental policy. *Ecological Economics*, **54**(2/3), pp. 164–74.

Johansson, M., Jacob, M. and Hellström, T., 2005. The strength of strong ties: University spin-offs and the significance of historical relations. *Journal of Technology Transfer*, **30**(3), pp. 271–86.

Klevorick, A., Levin, R., Nelson, R.R. and Winter, S., 1995. On the sources and significance of inter-industry differences in technological opportunities. *Research Policy*, **24** (2), pp. 185–205.

Laursen, K. and Salter, A., 2004. Searching high and low: What types of firms use universities as a source of innovation? *Research Policy*, **33**(8), pp. 1201–15.

Lundvall, B., 1998. Why study national systems and national styles of innovation? *Technology Analysis and Strategic Management*, **10**(4), pp. 407–21.

Maine, E. and Garnsey, E., 2006. Commercializing generic technology: The case of advanced materials ventures. *Research Policy*, **35**(3), pp. 375–93.

Malerba, F., 2005. Sectoral systems of innovation: A framework for linking innovation to the knowledge base, structure and dynamics of sectors. *Economics of Innovation and New Technology*, **14**(1/2), pp. 63–82.

Markard, J. and Truffer, B., 2006. Innovation processes in large technical systems: Market liberalization as a driver for radical change? *Research Policy*, **35**(5), pp. 609–25.

McGrath, R.G., 1999. Falling forward: Real options reasoning and entrepreneurial failure. *Academy of Management Review*, **24**(1), pp. 13–30.

Meritet, S., 2007. French perspectives in the emerging European Union energy policy. *Energy Policy*, **35**(10), pp. 4767–71.

Mohnen, P. and Hoareau, C., 2003. What type of enterprise forges close links with universities and government labs? Evidence from CIS 2. *Managerial and Decision Economics*, **24**(2/3), pp. 133–46.

Mustar, P. and Larédo, P., 2002. Innovation and research policy in France (1980–2000) or the disappearance of the Colbertist state. *Research Policy*, **31**(1), pp. 55–72.

Mustar, P. and Wright, M., 2010. Convergence or path dependency in policies to foster the creation of university spin-off firms? A comparison of France and the United Kingdom. *Journal of Technology Transfer*, **35**(1), pp. 42–65.

Nair, A. and Ahlstrom, D. 2003. Delayed creative destruction and the coexistence of technologies. *Journal of Engineering and Technology Management*, **20**(4), pp. 345–65.

Nelson, R.R., 1959. The simple economics of basic scientific research. *Journal of Political Economy*, **67**(3), pp. 297–306.

Nelson, R. and Rosenberg, N., 1993. Technical innovation and national systems. In: R.R. Nelson, ed. *National Innovation Systems: A comparative analysis*. New York, Oxford: Oxford University, pp. 3–38.

Pavitt, K., 1984. Sectoral patterns of technical change: Towards a taxonomy and a theory. *Research Policy*, **13**(6), pp. 343–73.

Rothaermel, F.T., 2002. Technological discontinuities and interfirm cooperation: What determines a start-up's attractiveness as alliance partner? *IEEE Transactions on Engineering Management*, **49**(4), pp. 388–97.

Santoro, M. and Chakrabarti, A.K., 2002. Firm size and technology centrality in industry-university interactions. *Research Policy*, **31**(7), pp. 1163–80.

Schneider, C. and Veugelers, R., 2010. On young highly innovative companies: why they matter and how (not) to policy support them. *Industrial and Corporate Change*, **19**(4), pp. 969–1007.

Shane, S. and Stuart, T., 2002. Organizational endowments and the performance of university startups. *Management Science*, **48**(1), pp. 154–70.

Sine, W.D. and David, R.J., 2003. Environmental jolts, institutional change, and the creation of entrepreneurial opportunity in the US electric power industry. *Research Policy*, **32**(2), pp. 185–207.

Suurs, R., Hekkert, M.P. and Smits, R., 2010. Understanding the build-up of a technological innovation system around hydrogen and fuel cell technologies. *ISU Working Paper* N° 09.10, Utrecht University.

Takalo, T. and Tanayama, T., 2010. Adverse selection and financing of innovation: Is there a need for R&D subsidies? *Journal of Technology Transfer*, **35**(1), pp. 16–41.

Teece, D.J., 1986. Profiting from technological innovation: Implications for integration, collaboration, licensing, and public policy. *Research Policy*, **15**(6), pp. 285–305.

Vavakova, B., 2006. Reconceptualizing innovation policy: The case of France. *Technovation*, **26**(4), pp. 444–62.

Winter, S.G., 1984. Schumpeterian competition in alternative technological regimes. *Journal of Economic Behaviour and Organisation*, **5** (3/4), pp. 287–320.

Part II

Economics of creativity, firms and organizations

8 Managing creativity in the firm

The fuzzy front end of innovation and dynamic capabilities

Patrick Cohendet, Jean-François Harvey and Laurent Simon

Introduction

The literature on innovation, particularly with regard to new product development, explores how firms manage their innovation process, the organization of their material, their financial and human resources, the tools they use, as well as the culture and leadership in place. To a large extent, this literature, which is essentially focused on the conditions of development of new processes and new products, pays only marginal attention to the emergence and generation of the creative ideas from which the innovation processes are undertaken. Creativity is generally confined to the black box of R&D which is supposed to produce new ideas with an innovative potential, or in the blind spot of what is happening before the R&D process takes shape. The focus is on the output of the exploration phases of firms, rather than on the processes within, before and around the R&D box. The processes of formation of ideas, the history of their emergence, the conditions of their growth and the complex routes of their transformation into potential new products are largely untold (e.g., Hansen and Birkinshaw, 2007). What really and essentially matters seems to be the selection process between competing ideas that will lead to new innovative processes and products.

However, there are recent repeated signs that this situation may dramatically change. In a context of open innovation, firms tend to pay increasing attention to the emergence, formation and consolidation of creative ideas whether these ideas are internal or external to the firm. To better assess the importance of this change of focus, we rely on the representation of the innovative process given by Koen *et al.* (2001). The authors deconstruct the innovation process into three distinct phases: the fuzzy front end (FFE), the new product and process development (NPPD) and commercialization. Each of these innovation process phases embodies a set of specific issues that calls for particular methods, procedures and management approaches. The last two phases (NDDP and commercialization) are well known, expertly studied and associated with a set of practices that is relatively well mastered and at a very advanced state of knowledge. Conversely, the first phase (FFE) is poorly documented and analyzed. For instance, since the transformation of new product development's design and engineering, which occurred during the 1990s, and the intensification of the focus on management rationalization, most firms now master the techniques and procedures related to the second phase (NDDP). This phase is commonly accompanied by the introduction of a traditional project management perspective. Similarly, many firms now routinely do sophisticated work in the third phase (commercialization). In short, the saturation of improvement efforts is evident in the intermediate and final phases of the innovation process and the first phase has received less attention in organizational science to date (Mainemelis, 2010).

From Schumpeter (1939) via the Oslo Manual (OECD, 1992) and to date, the central authors in the domain of innovation in firms still provide very few lessons on the FFE phase. For instance, as part of his stage-gate model, Cooper (2001) presents what he calls "discovery" as stage zero of the new product development process before the NPPD phase is entered. The discovery stage is described as the moment when the firm chooses which projects to pursue. More precisely, in the traditional vision that Cooper portrays, the first step of the process is generally the brainstorming phase. This phase comprises gathering the maximum number of ideas to solve a predetermined problem, or to reach a given objective. Subsequently, through a "funnel" process – shaped by a sequential "go–no go" procedure – the number of competing ideas is progressively trimmed. No-go ideas are generally discarded, and only a small number of go ideas pass the various gates before being transformed into an innovative output. In this sense, the ideas then generated are not only easily understood, but also offer an easily discernible value for decision makers. Within their small group, the decision makers prioritize these ideas according to their appropriateness logic. The selected ideas are sent to be transformed into distinct projects.

The traditional management vision considers new ideas as preformatted "black boxes" – originating from outside or inside the firm – containing pieces of knowledge that codified materials or procedures describe well. To a large extent, this perspective regards new ideas as given "objects" external to the firm's sphere but whose meanings are instantaneously accessible. The origin of these ideas and the history of the ways in which they have been built and formatted do not matter. Only their potential economic value, which guides each step of the stage-gate process of the managers' selection procedures, matters to the organization. In other words, all ideas have the same form. Like fruit, they are ready to be picked (e.g., Frese *et al.*, 1999).

In this contribution, we question this rather static and deterministic vision of the trajectory of ideas on which firm innovation is based. Based on repeated observations of ideation processes in innovative firms, such as Ubisoft Montreal, IREQ – Hydro-Quebec and Cirque du Soleil, we consider the ideation process as essentially an evolutionary process. Ideas do not "fall from the sky" equipped with a codebook or a grammar of interpretation and then undergo a funnelling process during which most of them are progressively discarded. Our work with innovative firms shows us that the real flavors of the fruits – the ideas generated – are not known *ex ante*. They do not obey the principle of the gradual reduction of their complexity, nor is their form a finished product. Instead, they constantly evolve and are constantly processed by many different stakeholders inside and outside the firm.

Frames of interpretation

In this evolutionary process, an idea emerges from a creating act that "uncovers, selects, re-shuffles, combines, synthesizes already existing facts, ideas, faculties, skills" (Koestler, 1964: 120). The history of past projects, of the routines, of the reservoir of knowledge accumulated by the firm does matter. The internal diversity, openness and absorptive capacity also contribute to the process by feeding the organization with new, complementary ideas and/or new elements of knowledge (Cohen and Levinthal, 1990; Zahra and George, 2002). This history contributes to shaping the emergence and identification of new ideas in the firm. Indeed,

> the evolution of new ideas often entails a dynamic transition: when first proposed, new ideas are often rejected because they are perceived as weird, inappropriate, unworkable, or too risky, but these same ideas may later result in an outcome that the social context accepts as useful and breakthrough.
>
> (Mainemelis, 2010: 558)

In this sense, Witt (2009) emphasizes that through a more or less accidental displacement of attention to something not previously noted, new elements are introduced into a given context with which they were not previously associated. This first phase of the ideation process is a "generative operation" that produces new (re-)combinations of elements.

Then, once a new idea – at this stage often called "an opportunity"[1] – has emerged in the organization, a maturation phase may develop. This phase cannot be reduced to the mere evaluation of the idea's economic potential under the principles of a final go-no go decision process. It is a phase of repeated and multiple feedback loops throughout the firm. The small group of agents at the origin of the idea must exercise considerable efforts to alert other organization actors or communities to convince them of their new idea's relevancy, usefulness and potential. From these interactions, an initial idea is frequently modified, combined with other ideas, contextualized, tested, discarded or reinforced. New ideas change from a very specialized knowledge localized in one community into procedural knowledge encompassing numerous firm agents' perceptions. We assist with the deepening, the enriching and the technical validation of the idea before its rhetorical broadening. The grammar of interpretation – the codebook or the dictionary (in the firm language) – that equips the idea at the end of the maturation period is thus not an attribute that is given *ex ante*, but results from the continuous iterations between different communities in the firm. Witt (2009) calls this phase, which serves the new (re-)combination that is integrated into a newly emerging or a more general, already existing concept and context, "interpretative operation."

Once equipped with a codebook and a grammar of usage, ideas are ready for the organization's selection process, which is often a go-no go structure. However, contrary to the rather traditional vision, the idea does not generally die if it does not "pass the gate." It is noteworthy that the small groups of agents who worked to nurture, reinforce and install the idea in the context of the firm, contribute to the formation of a "creative slack" for further expansion of the firm's potential creativity (Cohendet and Simon, 2007). The creative slack notion refers intentionally to the organizational slack notion proposed by Penrose (1959). The latter suggested that firms always have some stock of unused, or underused, resources (e.g., knowledge, relationships, reputation, managerial talent, physical assets, etc.) that inevitably accumulate in the course of developing, producing and marketing any given products or services. In Penrose's view, these unexploited or underexploited productive resources are the primary factors determining the extent and the direction of growth, which is firms' dominant motivation, and is only limited by their administrative capacity.

This creative slack is partly distributed in the firm's formalized, codified knowledge base and partly in the cognitive functioning of and interactions between its diverse communities. The ideas in the slack are also constantly nurtured and enriched by the knowledge acquired in the firm's actual innovative projects (learning by searching) and exploitation processes (learning by doing). This knowledge acquisition is due to the active cognitive functioning of the diverse communities' members, who have memorized – thanks to the knowledge that their fellow members contribute – parts of the learning during innovative projects. Although it is well known that firms have extreme difficulties with memorizing what has been learnt during a project, the interest of communities in this issue is that they rather easily memorize the routines their members practice (Simon, 1981; Nelson and Winter, 1982; Cohen *et al.*, 1996). The firm's creative potential relies on its ability to recombine within bits of the creative slack that these heterogeneous, formal and informal structures hold (Drazin *et al.*, 1999).

Consistent with Staw's (1990) theory of creativity, the evolutionary trajectory of ideas in a given firm depends inherently on its ability to facilitate the emergence of ideas, to nurture and develop them, to choose the right moment to use them in innovative projects, or to actively "store" them as creative slack. In our view, these abilities correspond to the "dynamic capabilities" defined by Teece in his seminal 2007 article. His argumentation maintains that "dynamic capabilities can be disaggregated into the capacity: (1) to sense and shape opportunities and threats, (2) to seize them, and

(3) maintain competitiveness through enhancing, combining, protecting, and, when necessary, reconfiguring the business enterprise's intangible and tangible assets" (Teece, 2007: 1319). Beside the clear specification of dynamic capabilities as processes, such a disaggregation emphasizes the purposes of dynamic capabilities rather than what they are and what they allow for.

By making organizational capabilities dynamic, easy to remove and to mobilize inside and outside the organizational boundaries, these organizational processes help firms to achieve, in a first step, an internal evolutionary fitness. This internal fitness is a necessary condition to ensure the overall evolutionary fitness in tune with the external environment. Put differently, communities inside or outside the firm generate and play with ideas to make them grow to the point where they are understandable to and exploitable by the firm.

Throughout this chapter, we try to better understand the first phase of the innovation process or FFE. Inspired by Teece (2007), our work on innovative firms has led us to present the FFE from three steps that are critical for generating ideas within firms:

1. Sensing, which represents the appraisal of a genuine (re-)combination of knowledge from inside and outside the firm that allows the discovery of a new opportunity.
2. Seizing, which represents the translation and seduction processes that bring this new opportunity to the attention of many actors from different communities, who are thus likely to play an active role in its evolution and enrichment as a new idea.
3. Reconfiguring, which represents the harmonization of the new idea with the firm's grammar in order to turn it into an appropriate concept.

The process's illustration in Figure 8.1 primarily serves analytical purposes. In reality, it is less linear and more recursive; that is, agents and their diverse communities usually move back and forth between the various steps.

Sensing the opportunity

The first step, sensing the opportunity, refers to the firm's creative capability that brings together actors belonging to a community to actively develop knowledge (Brown and Duguid, 1991; Wenger, 1998; Sawhney and Prandelli, 2000). In other words, the firm's creative capability depends on the absorptive capacity (Cohen and Levinthal, 1990) of its respective communities. As we pointed out, actors are members of a community which gives them the ability to integrate knowledge from their environment, pass it around in order to ponder, investigate and get beyond the gross and the obvious and thus assess its value in relation to their practice (Amin and Cohendet, 2004).

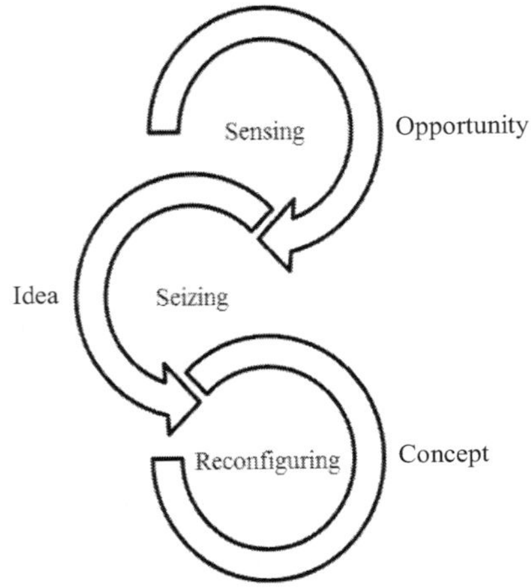

Figure 8.1 The three steps of the FFE phase

In so doing communities provide the *creative slack* necessary for firms to sustain their innovative power (Cohendet and Simon, 2007).

> [Since] creativity does not occur at a single point in time but is rather the outcome of a series of interconnected events and undertakings, [it has to be] examined as a form of connectivity, the ability to make connections between heterogeneous materials.
>
> (Styhre and Sundgren, 2005: 47)

In this sense, firms must not rely solely on their own employees' current knowledge in order to maintain strong creative capabilities. They must also bank on their employees belonging to internal/external communities, where it is possible to absorb new knowledge and to deepen existing knowledge (Chesbrough, 2003; Cohendet and Simon, 2007). As illustrated by numerous authors (e.g., Drazin *et al.*, 1999; Flynn *et al.*, 2003; Hargadon and Bechky, 2006), the generation of new ideas within firms depends heavily on the communities associated with the firm, rather than only on the firm's formal R&D activities.

Hence, an actor makes a discovery through the interactions, either inside or outside the firm, within his community. He manages to identify a technological, procedural or business opportunity that leads him to wonder "what if?" An opportunity is thus defined as "a business or technology gap [that] exists between the current situation and an envisioned future in order to capture competitive advantage, respond to a threat, solve a problem, or ameliorate a difficulty" (Koen *et al.*, 2002: 7). Although the idea remains unclear in his mind, the actor feels that "there is a difference that could make a difference."

Triggered by accident or voluntarily, the actor then brings the opportunity to the attention of his community (of knowledge or of practice), which generally represents a place where there is a climate of psychological safety that encourages the free expression of new ideas (Edmondson, 1999; Baer and Frese, 2003). This community creates a relatively homogeneous network of actors around the opportunity and tries to make sense of it by giving it a meaning and updating it in relation to its common practice. The community gives the actor a chance to share his insight without investing too much time and energy in a detailed explanation. The members of one community share epistemic standards, which blur the differences between distinctively different views. Such an opportunity is therefore usually context-sensitive. Along with his peers, the actor can thus dig into the opportunity by concentrating on the cognitive tasks and creative assets that fit his community.

At this point, the evolution of the opportunity shows striking similarities with the concept of "meme," where a belief, a behavior or a style extends from one individual to another within the same culture (Dawkins, 1976). The "meme" acts as a unit of cultural, symbolic or practical value which can be transmitted from one mind to another through the use of writings, dialogues, rituals or any other inherited and reproducible phenomenon. The opportunity then takes the appearance of genes in the sense that it gradually modifies the methods and enters the mindsets of the community. Put differently, the opportunity, like a "meme," behaves like a virus – that moves through a population, taking hold in each person it infects (inspired by Gladwell, 2000).

The community allows the adoption of a broader vision of the opportunity through an iterative process in which its members deepen and/or transform the knowledge about it. This is where, ultimately, the analytical systems available to firms are found with which to learn about, identify, screen, shape and calibrate new, creative arrangements. These arrangements are part of the history of knowledge specific to the community, the creative slack which submits the opportunity to a relevance test in relation to the community's practice and the opportunity's evolution.

Seizing the idea

The second step, seizing the idea, alludes to the heuristic processing and cognitive setting of the idea within the organization. This step allows us to emphasize the need to specifically analyze the role and practices of active creative units, communities in the early innovation phases. As demonstrated by Cohendet *et al.* (2006), even before an idea can be patented and become economically viable, the "ideators" are usually asked to participate in a collective invention phase. This time, several of the communities superimposed on more traditional organizational structures play a crucial role. After avoiding other cognitive positions and various types of reasoning in order to develop the knowledge around the opportunity, it is time to introduce these features to the development agenda. Thus, this phase is characterized by an exchange of knowledge between a multitude of actors from different backgrounds, who thus operate at a considerable cognitive distance (Nooteboom, 2000), and whose skills and talents complement one another. Diversity obviously plays a crucial role at this stage. Hence, the success of an innovation depends on the linkages built around the opportunity in order to make the network of heterogeneous actors aware of it as a new business idea (Fleming *et al.*, 2007). Here, the idea "consists of a high-level view of the solution envisioned for the problem identified by the opportunity" (Koen *et al.*, 2002: 7).

To some extent illustrated by other authors on this subject (e.g., Hellström and Hellström, 2002; Van Dijk and Van de Ende, 2002; Bakker *et al.*, 2006; Grandadam *et al.*, 2011), we need to stress the significance of the actions that are pursued by the community that initially carries the opportunity for the firm in order to promote it and raise it to the state of a new idea that can be adopted by other communities, which are stakeholders in relation to the idea itself and within the firm in general. However, it takes time for a group of distinct communities to create a "world" of their own where they feel comfortable with a new idea. Nevertheless, members of these communities are not actually attached to their routines in a passive and insular manner. They are quite willing to change very quickly if they are able to find an engaging purpose for what they are offered. While their habits may change, they have a very reasonable and instinctive appreciation of the risks posed by a new idea (Crozier and Friedberg, 1977).

Sometimes called creative entrepreneurs (Napier and Nilsson, 2006) or corporate entrepreneurs (Augsdorfer, 2005), the community members from whom the opportunity comes try to keep it within the firm by finding allies in other communities who have a stake in it. In order to do so, they develop and use specific processes that allow the deployment of undertakings that enable other communities to better understand what led to the discovery of the opportunity and what is likely to come from it. Behind this pragmatic approach lies a rhetoric that is deployed idiosyncratically in a number of communities in order to bring the creative idea within the firm's knowledge trajectories. Usually, during a first phase, the connections between communities occur "below the surface", as "bootleg activities" (Augsdorfer, 2005). In a second phase, depending on the "ripeness" of the idea, the main actors bring the idea to the knowledge of management, initiating an evaluating phase.

From an interpretative perspective, emphasis is placed on the communication aspect associated with the presented opportunity. A common vision – shared by a variety of communities within the firm – can only be developed through a negotiation process, because the parties do not approach the opportunity from the same angles. Therefore, instead of instinctively selling the opportunity, it must be built into an idea that reaches consensus among the various communities. In so doing, from the previous activities of sensemaking (i.e., the development of a meaningful framework in order to understand the idea), activities of sensegiving (i.e., various communities' influence on the construction of the idea) are orchestrated to protect each community's identity in respect of the idea being built (Gioia and Chittipeddi, 1991; Boland and Tenkasi, 1995). A gradual process – evolutionary, not revolutionary – is needed to create a win-win situation and, accordingly, communicate and integrate the communities' viewpoints. Consequently, the various communities can offer their own interpretation of the idea, rather than passively moving to the innovation process's intermediate phase. The organization of the different perspectives provides the real potential for the idea development (Midler, 2004).

This phenomenon can also be explained from the perspective of the actor-network theory (ANT) which describes how some ideas become successful innovations and others do not (Amblard *et al.*, 2005). A new idea is not labeled as innovative for its intrinsic qualities or because it is superior to competing ideas. Its emergence requires the presence of spokespeople who will try to recruit or convert actors and sponsors, throughout a process that Callon and Latour call "translation" (Callon and Latour, 1990; Callon, 1999; Akrich *et al.*, 2006), which is a process that allows "the understanding between universes that were initially strangers to each other ... universes that are progressively reconfigured as they are connected" (Callon, 1999: 42–3). Every specialized community has its own criteria and expectations vis-à-vis the idea. When a community carries an idea, it must also operate a "translation" of the idea that is twofold. First, it needs to be translated to fit into the economic value chain of the firm or to advance towards a new one. Second, a translation must be operated to fit the idea into the production and administration routines of the firm or to renew them to adjust to the idea itself. This double translation allows the transition from the idea to a concept that is workable as a project.

The communication established between communities is partly transcribed and therefore materializes through permanent efforts to codify knowledge within the firm. The construction of a cognitive setting allowing the various communities to communicate can be understood by using translation as an example. The cognitive efforts that a community makes to interest another one (or several others) in its knowledge production activities (which, in turn, underlie the search for a common commitment, a voluntary adhesion and a common language and beliefs) correspond to enrolment mechanisms. These mechanisms were suggested by Callon (1999), who identifies the construction of an intermediary that supports the translation and the alignment of interests.

It is throughout this step that non-human actors (mediators) or boundary objects come into play and become indispensable for the consideration of the multiple issues related to the idea and support the alignment of the different perspectives. Given the cognitive distance existing between the various communities, the boundary objects always need to be used in a very simplistic manner for presenting the idea but also offer great opportunities to flesh out the idea. Indeed, this process represents a two-step evolutionary scenario. First, the efforts of translation and seduction require an unsophisticated reformulation of the idea. Second, the idea is not received passively by the various communities since each community has its own history of ideas. Thus, we assist with a mounting of the idea with preexisting ones. Hence, the activities of translation and seduction transform the idea through a process of oversimplification-complexification in order to support the appropriation of the idea by a collective of communities within the firm. For example, at Cirque du Soleil, an idea that comes from a group of artists must be fragmented in order to focus on its economic viability before being presented to the firm's community of managers. The idea is then supplemented by a set of codes from the managers' practice, which gives the idea an economic potential. Put differently, by enhancing the idea with their outlook, the community of managers help develop the idea into a concept that, from their perspective, is appropriable. For instance, when introduced to the community of technicians, the (simplified) focus is on the idea's technical aspects, and so forth.

The validation and collective dissemination of new knowledge can thus be interpreted as a process of the diffusion and progressive contagion of the communities. Within this framework, collective decision-making procedures can be regarded as processes of the construction of sense, the formation and the revision of collective beliefs, through constant negotiations. The behavior of choice is thus affected by the various social, cognitive and power structures in which the actors are embedded (March and Olsen, 1976). As presented by Cohen *et al.* (1996), this usually implies (i) learning a language within which to code successful routines, (ii) creating cognitive artefacts that can be diffused, and (iii) translating the high-level description contained in the cognitive artefact into actual practice as well as symbolic discourses, thus generating a new routine adapted to the new context. In short, "[ideas] are built up, torn down, combined, reshaped, modified, and upgraded. An idea may go through many iterations and changes as it is examined, studied, discussed, and developed in conjunction with other elements" (Koen *et al.*, 2002: 26–7). When adopting a more utilitarian approach, these iterations take place around the questions "why / why not?" from the various firm communities' perspectives.

In order to seize the idea in some way, the organization must make resources available to its communities (work time, materials, budgets and the use of equipment or other property). Doing so, the communities enhance the idea initially presented by transforming it into a concept that has the required legitimacy to reconfigure the firm, or part of the firm, for its development. Management's role appears critical at this point to make room for the exploration of ideas that do not offer crystal-clear answers to the "why / why not" examination and therefore they need to provide some degree of freedom to communities to elaborate on their ideas (Amabile, 1988; Shalley,

1991; Oldham and Cummungs, 1996; Mainemelis and Ronson, 2006). Managers must avoid strict evaluations leading to a final decision in order to offer communities a chance to develop their idea and show its merits can go a long way (Cummings, 1965; Staw, 1990; March, 1991; Augsdorfer, 1996, 2008). Indeed, as asserted by Mainemelis, "premature selection may extinguish ideas before their merits can be realized. To heighten creativity, organizations should delay idea selection and allow employees to elaborate on new ideas prior to deciding which ideas will be implemented" (2010: 565).

Reconfiguring for the concept

The third and final step, reconfiguring for the concept, points out that a new idea always disturbs the organizational configuration in place to a greater or lesser extent. Thus, for an opportunity to become an idea and a concept to be introduced as a project in the NPPD phase of the innovation process, the firm needs to reconfigure or change some of its parts. Indeed, any idea that is ready to be launched as a project generates some sort of creative destruction in the Schumpeter sense. The firm needs to anticipate the alignment and the dynamic realignment of its tangible and intangible assets to grasp the opportunity. The existing order or way of doing things needs to change to introduce the concept, which includes having a written and a visual description of it, as well as defining it broadly according to the "estimates of [its] market potential, customer needs, investment requirements, competitor assessments, technology unknowns, and overall project risk" (Koen *et al*., 2001: 51).

The actions following the various firm communities' decisions to subscribe to the new idea require a learning that is collective and transpires through the integration of procedural knowledge associated with the idea. Hence, the restructuring of formal structures – for example, modifications in the value chain, changes in the authority figures, transformation in the coordination mechanisms, variations in the balance between exploration and exploitation, etc. – and/or informal ones occurs: for example, changes in terms of the intensity and/or structure of interactions between the various stakeholders in the social network, fluctuations in the flows of knowledge in terms of content and intensity and variations in the degree of cooperation between various actors, allocation of financial and administrative support. This necessarily comprises unlearning.

The firm needs to remove some of its organizational routines or norms as well as the accumulated knowledge that could hinder the new idea. This will pave the way for the strengthening of meanings throughout the firm and for developing the competences necessary for the functioning specific to the introduced concept: the translation of the concept into a project formally acknowledged and supported by the hierarchy. In other words, the firm must abandon some of its past competences, since the results of previous learning are not only obsolete, but can also become a great source of inertia given a new organization with new challenges. This does not necessarily happen smoothly as, for instance, some of the targeted organizational routines may play an important role in the firm's current functioning. Having acquired legitimacy throughout the firm, such routines tend to perpetuate themselves and are very difficult to unlearn as they may establish a type of defense mechanism against the proposed change (Argyris, 1985). Thus, the competences of the past can become core rigidities that inhibit innovation (Leonard-Barton, 1992) – even more so if actors face difficult and uncomfortable issues, and are challenged in their established practices and beliefs. In this case, actors tend to draw their solutions from the repertoire of pre-existing competences, even if these are inappropriate for the actual concept introduction.

Firms must make room for new ideas and therefore focus on unlearning (Hedberg, 1981). In this sense, it is essential to do away with mental schemes and other collective patterns that can also become sources of inertia. Yet, this dimension is often overlooked in the innovation process. Given the required cognitive restructuring among and between communities for innovation to occur, unlearning is a very important step to success since it is the quality of the collective learning (and unlearning) that allows the evolution of an idea into a shared concept that can give rise to regulated action for its development. First, agents and their respective community need to unlearn some practices from work arrangements that prevail in the firm and, second, they need to create new representations about the way of doing things together. This requires a strong dose of trial and error, learning by doing, and improvization since most of the new representations can only come from the actual doing of things under the new paradigm. They cannot be taught a priori and framed independently of the knowledge that comes from the actual experience of working under the new system.

The learning (and unlearning) that needs to take place is predominantly reflective, taking a step back and understanding the new system that is developing, and experiencing new work situations and being opened to spontaneous learning opportunities (Argyris and Schön, 1978). Rather than simply disseminate knowledge in a formal manner, the success of an idea happens only when the concept itself can change the routines and collective understanding of the firm's activities. Individuals need to be able to think their experience (Dewey, 1933; Argyris and Schön, 1978; Schön, 1983), to construct and internalize a new interpretation of the meaning of their work, which guides them into action.

Furthermore, any sort of reconfiguration is imperfect. It is always problematic to identify what competences the actors are to mobilize in future circumstances whose contours remain blurred until the activities are actually launched. Limited by the inadequate knowledge associated with the activities that are to be accomplished, some communities within the firm may try to predict the impacts of individual actions within the upcoming system and for that reason tend to impose their preferred structural arrangements to others. In other words, the degree of uncertainty in emergent phases is so high that agents cannot anticipate the behavior of others. In fact, the agents' behavior remains largely opportunist and calculative. It is also deliberately procedural: the community will start a procedural process of convergence and formation of collective beliefs through a series of negotiation processes and continuous feedback until it stabilizes. The period of convergence (elimination of uncertainty) is generally a fertile period in the process of formation of collective beliefs. However, collective beliefs are more likely to converge rapidly when a "metacode" or "meta-design" between communities already exists (Fischer, 2000).

Yet, it is required to establish a compromise that is acceptable for the various communities and that gives them some sort of direction beforehand. Indeed, the combination of novelty and insecurity creates a reluctance to change. Any situation of change conveys its attendant uncertainties and concerns: fear for the agents losing some of their influence or power, doubt about their ability to master new instruments of labor, fear of cognitive obsolescence, job competition, etc. Therefore, when change is growing in many different directions, it becomes an even more important source of uncertainty and anguish. It is thus difficult for individuals to establish causal relationships between their actions and the modifications that need to be made (Lant and Mezias, 1992).

Seeking to move towards the concept development and its potential exploitation, the agents involved at this stage ask themselves "go / no go?" before moving to the NPPD phase. Mostly, this is when the hierarchical authorities in place consider the organization's strategic priorities in order to authorize the transformation of the concept into an official project.

Conclusion

According to a recent survey, 70 percent of executives from various sectors regard innovation as one of the three main growth drivers for their firm (Barsh *et al.*, 2007). However, 65 percent of them are not confident about their decisions to stimulate innovation (Barsh *et al.*, 2007). This large discrepancy between the approaches and the perceived relevance of the methods regarding supporting and accelerating innovation is a major challenge for firms' economic development and competitiveness. A better understanding of the steps that form the dynamic capabilities of the firm's communities may be a good starting point to tackle this challenge. In this chapter, we have suggested that unfolding the black box of the "fuzzy front end" phase and better capturing the potential of creativity could lead to significant improvements in the ways firms manage their innovative processes.

Questions remain regarding the ability of the firm management to create contexts within which and around which opportunities can grow into ideas and turn into concepts developed as official projects. The three steps' management principles differ greatly from the more traditional ones that support the NPPD and commercialization phases. Firms that adopt a closed and rigid idea generation vision are left with a system in which the upstream steps are heavily biased by the expectations of the downstream activities. Yet, too many scholars and practitioners still make no distinction between the approaches for the FFE phase and those for the NPPD and commercialization phases. These scholars therefore assume that knowledge is always concentrated in a specific place (R&D) rather than scattered across and outside the firm (interacting knowing communities). Yet, it is not so much a matter of managing what is "known" but grasping the "unknown" with the latter often expanding faster than the former (Hatchuel and Le Masson, 2007).

In opposition to what a majority of studies on new product development and innovation management advance, idea generation is no longer the preserve of a select group of actors, part of the R&D department or strategic business units. In fact, new ideas come first and foremost from employees (41 percent) and from business partners and customers (over 35 percent) while the R&D department accounts for just over 15 percent (IBM, 2006). Firms must therefore remain opened to the maximum of knowledge and depart from the closed and rigid model presented by most authors. The innovative firm has procedural features that are established gradually and are constantly reconfigured but also less formal ways of communicating an appreciation for new ideas (Gist, 1989; Schein, 1992; Shalley and Perry-Smith, 2001; Zhou, 2003).

Creative ideas take time to construct (Wallas, 1926; Campbell, 1960; Russ, 1993; Csikszentmihalyi, 1997; Mainemelis, 2002). They are progressively, but also iteratively and collectively, revealed, enhanced, nurtured, interpreted, enacted and codified. Adopting an idea is like listening to music; we are attracted by the melody, and as we get to know its meaning, we learn the lyrics. Therefore, in today's creative economy, ideas are not simply picked up from a basket of ideas and readily transformed into commanding projects. Instead, the first phases of the innovation process are chaotic and unpredictable (Backman *et al.*, 2007, Crawford and Di Benedetto, 2008). This is where we observe the dynamic capabilities of a firm's communities – a space which offers the most potential for improving firms' innovation process (Zhang and Doll, 2001; Verworn *et al.*, 2008). In other words, this is where we find the greatest challenges of creating value since today's business world requires firms to explore and continuously take advantage of all the diverse knowledge areas they can access and to not only exploit a set of accumulated knowledge.

In terms of management, the above development opens large avenues of research: there are still important divergent views about the practical implications of managing creativity in firms: On the one side, many authors argue that the three steps presented in this chapter and summarized in Table 8.1 should not fall under the strict control of management and should develop with the greatest freedom possible to explore new ideas and options around a

Table 8.1 The fuzzy front end of innovation and dynamic capabilities

	Sensing the opportunity	*Seizing the idea*	*Reconfiguring the concept*
Inquiry	What if?	Why / Why not?	Go / No go?
Knowledge	Internal and external to the firm	Mostly internal to the firm	Internal to the firm
Process	Sensemaking	Sensegiving	Taking action
Stakeholders	An homogeneous group of actors	An heterogeneous network of communities	Work groups and project teams
Leaders	Experts	Creative entrepreneurs	Managers
State of codification	Sparks and sketches	Manifesto	Codebook

set of specific issues (e.g., Woodman *et al.*, 1993; Amabile, 1998; Augsdorfer, 2008). Among them, Augsdorfer stresses that "the management lesson is to leave creative processes in the underground world and formal processes in the official world [and that] any attempt to mix the two worlds simply does not work" (2008: 47). On the other side, other authors (Hargadon and Sutton, 1997; DeFillippi *et al.*, 2007; Hansen and Birkinshaw, 2007) suggest that managers should be more aware of those processes and could potentially be in charge of managing the creative, upstream phases of the process of innovation, but that these phases should not be managed under the same controlling rules as the rest of the organization. For instance, in the creative phases, people are more sensitive to intrinsic incentives than extrinsic incentives, failures could be normal (provided that the teams can rebound from failures), the interpretation of the stage-gate processes may be different, more focused on exploration and learning – divergence – than on selection and actualization of ideas – convergence, etc. In this contribution, we clearly suggested developing further research to assess the cognitive, social and organizational dynamics of the FFE. A first perspective calls for a significant redefinition of the dynamic capabilities of the firm – based on intra- and inter-communities learning. In a second perspective, such a view also invites reconsidering the roles of the managers dedicated to innovation and organizational flexibility. On the one hand, they should be more focused on the instauration of connective, open to the outside and exploratory contexts. On the other hand, they should pay more attention to emerging novelty and weak signals from the internal communities, providing them with a wisely balanced mix of weak prescriptions – to partially shape and orient collective interests and curiosities aligned with the firm's strategic issues – and timely support for the unfolding, growth, validation and enactment of new ideas.

Notes

1 We are aware that the use of the term "opportunity" may be confusing, especially to those who are familiar with the work of Williamson (1975, 1985) but we wanted to follow on the work of Koen *et al.* (2001). To be clear, what is meant here is the spark that precedes the emergence of an idea.

References

Amabile, T.M. (1988) "A model of creativity and innovation in organizations," *Research in Organizational Behavior* **10**: 123–67.

Amabile, T.M. (1998) "How to kill creativity," *Harvard Business Review*, September: 77–87.

Amblard, H., Bernoux, P., Herreros, G. and Livian, Y.-F. (2005) *Les nouvelles approches sociologiques des organisations*. Éditions du Seuil. Paris.

Akrich, M., Callon, M. and Latour, B. (2006) *Sociologie de la traduction: Textes fondateurs*. Presses de l'École des Mines. Paris.

Amin, A. and Cohendet, P. (2004) *Architectures of Knowledge: Firms, Capabilities and Communities*. Oxford University Press. Oxford.

Argyris, C. (1985) *Strategy, Change, and Defensive Routines*. Pitman. Boston.

Argyris, C. and Schön, D. (1978) *Organizational Learning: A Theory of Action Perspective*. Addison-Wesley. Reading.

Augsdorfer, P. (1996) *Forbidden Fruit: An Analysis of Bootlegging, Uncertainty, and Learning in Corporate R&D*. Avebury. Aldershot.

Augsdorfer, P. (2005) "Bootlegging and path dependency," *Research Policy*, **34**(1): 1–11.

Augsdorfer, P. (2008) "Managing the unmanageable," *Research Technology Management* **51**(4): 41–7.

Backman, M., Börjesson, S. and Setterberg, S. (2007) "Working with concepts in the fuzzy front end," *R&D Management* **37**(1): 17–28.

Baer, M. and Frese, M. (2003) "Innovation is not enough: Climates for initiative and psychological safety, process innovations, and firm performance," *Journal of Organizational Behavior* **24**: 45–68.

Bakker, H., Boersma, K. and Oreel, S. (2006) "Creativity (ideas) management in industrial R&D organizations: A crea-political process model and an empirical illustration of Corus RD&T," *Creativity and Innovation Management* **15**(3): 296–309.

Barsh, J., Capozzi, M. and Mendonca, L. (2007) "How companies approach innovation," *McKinsey Quarterly*, October.

Boland, R.J. and Tenkasi, R.V. (1995) "Perspective making and perspective taking in communities of knowing," *Organization Science* **6**(4): 350–72.

Brown, J.S. and Duguid, P. (1991) "Organizational learning and communities of practice: A unified view of working, learning and innovation," *Organization Science* **2**(1): 40–56.

Callon, M. (1999) "Le réseau comme forme émergente et comme modalité de coordination: le cas des interactions stratégiques entre firmes industrielles et laboratoires académiques," in M. Callon, P. Cohendet, F. Eymard-Duvernay *et al.* (eds.) *Réseau et Coordination*. Economica. Paris : 13–64.

Callon, M. and Latour, B. (1990) *La Science telle qu'elle se fait*. Éditions La Découverte. Paris.

Campbell, D.T. (1960) "Blind variation and selective retention in creative thought as in other knowledge processes," *Psychological Review* **67**: 380–400.

Chesbrough, H. (2003) *Open Innovation: The New Imperative for Creating and Profiting from Technology*. Harvard Business Press. Boston.

Cohen, M.D., Burkhart, R., Dosi, G. *et al.* (1996) "Routines and other recurring action patterns of organizations," *Industrial and Corporate Change* **5**(3): 653–98.

Cohen, W.M. and Levinthal, D.A. (1990) "Absorptive capacity: A new perspective on learning an innovation," *Administrative Science Quarterly* **35**(1): 128–52.

Cohendet, P. and Simon, L. (2007) "Playing across the playground: Paradoxes of knowledge creation in the videogame firm," *Journal of Organizational Behaviour* **28**: 587–605.

Cohendet, P., Créplet, F. and Dupouët, O. (2006) *La gestion des connaissances: Firmes et communautés de savoir*. Economica. Paris.

Cooper, R.G. (2001) *Winning at New Products*. Basic Books. New York, 3rd edition.

Crawford, C.M. and Di Benedetto, C.A. (2008) *New Products Management*. McGraw-Hill. New York.

Crozier, M. and Friedberg, E. (1977) *L'acteur et le système*. Editions du Seuil. Paris.

Csikszentmihalyi, M. (1997) *Creativity: Flow and the Psychology of Discovery and Invention*. Harper. New York.

Cummings, L. (1965) "Organizational climates for creativity," *Academy of Management Journal* **8**: 220–7.

Dawkins, R. (1976) *The Selfish Gene*. Oxford University Press. New York.

DeFillippi, R., Grabher, G. and Jones, C. (2007) "Introduction to paradoxes of creativity: Managerial and organizational challenges in the cultural economy," *Journal of Organizational Behavior* **28**: 511–21.

Dewey, J. (1933) *How We Think*, D.C. Heath and Company. Boston.

Drazin, R., Glynn, M.A. and Kazanjian, R.K. (1999) "Multilevel theorizing about creativity in organizations: A sensemaking perspective," *Academy of Management Review* **24**: 286–307.

Edmondson, A. (1999) "Psychological safety and learning behaviour in work teams," *Administrative Science Quarterly* **44**: 350–83.

Fischer, G. (2000) "Social creativity, symmetry of ignorance and meta-design," *KnowledgeBased Systems Journal* **13**(7–8): 527–37.

Fleming, L., Mingo, S. and Chen, D. (2007) "Collaborative brokage, generative creativity, and creative success," *Administrative Science Quarterly* **52**(3): 443–75.

Flynn, M., Dooley, L., O'Sullivan, D. and Cormican, K. (2003) "Idea management for organizational innovation," *International Journal of Innovation Management* **7**(4): 417–42.

Frese, M., Teng, E. and Wijnen, C.J.D. (1999) "Helping to improve suggestion systems: Predictors of making suggestions in companies," *Journal of Organizational Behavior* **20**: 1139–55.

Gioia, D.A. and Chittipeddi, K. (1991) "Sensemaking and sensegiving in strategic change initiation," *Strategic Management Journal* **12**: 433–48.

Gist, M.E. (1989) "The influence of training method on self-efficacy and idea generation among managers," *Academy of Management Review* **42**: 787–805.

Gladwell, M. (2000) *The Tipping Point: How Little Things Can Make a Big Difference*. Little Brown & Company. Boston

Grandadam, D., Cohendet, P. and Simon, L. (2011) *The knowledge-creating firm in a creative economy*, Working Paper, MosaiC HEC Montréal.

Hansen, M.T. and Birkinshaw, J. (2007) "The innovation value chain," *Harvard Business Review* **85**(6): 121–30.

Hargadon, A.B. and Bechky, B.A. (2006) "When collections of creatives become creative collectives: A field study of problem solving at work," *Organization Science* **17**(4): 484–500.

Hargadon, A.B. and Sutton, R.I. (1997) "Technology brokering and innovation in a product development firm," *Administrative Science Quarterly* **42**: 716–49.

Hatchuel, A. and Le Masson, P. (2007) "Shaping the unknown and the emergent," *European Academy of Management* **21**.

Hedberg, B. (1981) "How organizations learn and unlearn," in P.C. Nystrom and W.H. Starbuck (eds.) *Handbook of Organizational Design*. Oxford University Press. London: 8–27.

Hellström, C. and Hellström, T. (2002) "Highways, alleys and by-lanes: Charting the pathways for ideas and innovation in organizations," *Creativity and Innovation Management* **11**(2): 107–14.

IBM (2006) *Expanding the Innovation Horizon: The Global CEO Study 2006*. IBM Business Consulting Services.

Koen, P., Ajamian, G., Burkart, R. et al. (2001) "Providing clarity and a common language to the fuzzy front end," *Research Technology Management* **44**(2): 46–55.

Koen, P., Ajamian, G., Boyce, A. *et al.* (2002) "Fuzzy front end: Effective methods, tools and techniques," in P. Belliveau, A. Griffin and S. Somermeyer (eds.) *The PDMA Toolbook for New Product Development*. John Wiley & Sons, Inc. New York: 5–35.

Koestler, A. (1964) *The Act of Creation*. Hutchinson & Co. London.

Lant, T.K. and Mezias, S.J. (1992) "An organizational learning model of convergence and reorientation," *Organization Science* **3**: 47–71.

Leonard-Barton, D. (1992) "Core capabilities and core rigidities," *Strategic Management Journal* **13**: 111–26.

Mainemelis, C. (2002) "Time and timelessness: Creativity in (and out of) the temporal dimension," *Creativity Research Journal* **14**: 227–38.

Mainemelis, C. (2010) "Stealing fire: Creative deviance in the evolution of new ideas," *Academy of Management Review* **35**(4): 558–78.

Mainemelis, Ç. and Ronson, S. (2006) "Ideas are born in fields of play: Towards a theory of play and creativity in organizational settings," *Research in Organizational Behavior* **27**: 81–131.

March, J.G. (1991) "Exploration and exploitation in organizational learning," *Organization Science* **2**: 71–87.

March, J.G. and Olsen, J.P. (1976) *Ambiguity and Choice in Organizations*. Universitetsforlaget. Bergen.

Midler, C. (2004) *L'auto qui n'existait pas*. Dunod. Paris.

Napier, N.K. and Nilsson, M. (2006) "The development of creative capabilities in and out of creative organizations: Three case studies," *Creativity and Innovation Management* **15**(3): 268–78.

Nelson, R.R. and Winter, S. (1982) *An Evolutionary Theory of Economic Change*. Harvard University Press. Cambridge.

Nooteboom, B. (2000) *Learning and Innovation in Organizations and Economics*. Oxford University Press. Oxford.

OECD (1992) (1997 revised edition) *OECD Proposed Guidelines for Collecting and Interpreting Technological Innovation Data – Oslo Manual*. OECD. Paris.

Oldham, G.R. and Cummings, A. (1996) "Employee creativity: Personal and contextual factors at work," *Academy of Management Journal* **39**: 607–34.

Penrose, E. (1959) *The Theory of the Growth of the Firm*. Oxford University Press. New York.

Russ, S.W. (1993) *Affect and Creativity: The Role of Affect and Play in the Creative Process*. Lawrence Erlbaum Associates. Hillsdale.

Sawhney, M. and Prandelli, E. (2000) "Communities of creation: Managing distributed innovation in turbulent markets," *California Management Review* **42**(4): 24–54.

Schein, E.H. (1992) *Organizational Culture and Leadership*. Jossey-Bass. San Francisco.

Schön, D.A. (1983) *The Reflective Practitioner: How Professionals Think in Action*. Basic Books. New York.

Schumpeter, J.A. (1939) *Business Cycles: A Theoretical, Historical, and Statistical Analysis of the Capitalist Process*. McGraw-Hill. New York and London.

Shalley, C.E. (1991) "Effects of productivity goals, creativity goals, and personal discretion on individual creativity," *Journal of Applied Psychology* **76**: 179–85.

Shalley, C.E. and Perry-Smith, J.E. (2001) "Effects of social-psychological factors on creative performance: The role of informational and controlling expected evaluation and modelling experience," *Organizational Behavior and Human Decision Processes* **84**: 1–22.

Simon, H.A. (1981) *The Sciences of the Artificial*. MIT Press. Cambridge.

Staw, B.M. (1990) "An evolutionary approach to creativity and innovation," in M. West and J.L. Farr (eds.) *Innovation and Creativity at Work: Psychological and Organizational Strategies*. Wiley. Chicester.

Styhre, A. and Sundgren, M. (2005) *Managing Creativity in Organisations: Critique and Practices*. Palgrave Macmillan. New York.

Teece, D.J. (2007) "Explicating dynamic capabilities: The nature and microfoundations of (sustainable) enterprise performance," *Strategic Management Journal* **28**(13): 1319–50.

Van Dijk, C. and Van Den Ende, J. (2002) "Suggestion systems: Transferring employee creativity into practicable ideas," *R&D Management* **32**(5): 387–95.

Verworn, B., Herstatt, C. and Nagahira, A. (2008) "The fuzzy front end of Japanese new product development projects," *R&D Management* **38**(1): 1–19.

Wallas, G. (1926) *The Art of Thought*. Harcout Brace. New York.

Wenger, E. (1998) *Communities of Practice: Learning, Meaning, and Identity*. Cambridge University Press. New York.

Williamson, O. (1975) *Markets and Hierarchies: Analysis and Antitrust Implications*. Free Press. New York.

Williamson, O. (1985) *The Economic Institutions of Capitalism*. Free Press. New York.

Witt, U. (2009) "Propositions about novelty," *Journal of Economic Behavior and Organization* **70**: 311–20.

Woodman, R., Sawyer, J. and Griffin, R. (1993) "Toward a theory of organizational creativity," *Academy of Management Review* **18**(2): 293–321.

Zahra, S.A. and George, G. (2002) "Absorptive capacity: A review, reconcepualisation, and extension," *Academy of Management Review* **27**(2): 185–203.

Zhang, Q. and Doll, W.J. (2001) "The fuzzy front end and success of new product development," *European Journal of Innovation Management* **4**(2): 95–112.

Zhou, J. (2003) "When the presence of creative coworkers is related to creativity: Role of supervisor close monitoring, development feedback, and creative personality," *Journal of Applied Psychology* **88**: 413–22.

9 Appropriation strategies and endogenous technological regime

Towards a dynamic theory of the role of patents

Jérôme Freyermuth, Patrick Cohendet and Julien Pénin

Introduction

Taking the opposite of the standard vision of intellectual property developed within the frame of the Arrow model (1962), most empirical studies show that the primary role of patents is not necessarily to exclude imitators and to secure monopoly positions. Patents are flexible instruments that can serve a multitude of different strategies. Among others, in line with the work of Chesbrough (2003), patents often have a role of coordination of innovative activities; they help to structure open innovation (Cohendet and Pénin, 2011). Moreover, empirical studies also highlight that the use of patents is not uniform across firms and across sectors (Levin *et al.*, 1987; Cohen *et al.*, 2000).

Also, Bah and Le Bas (2011) mobilized the literature on sectoral systems of innovation to develop the concept of sectoral systems of intellectual property (SSIP), a concept which illustrates that the optimal use of intellectual property is contingent to a set of elements which themselves depend largely on the sector in which the company operates. In parallel to this work, Pénin (2011) studied the determinants of firms' optimal patenting strategy and argued that the technological regime (Nelson and Winter, 1982) plays a central role.

In these analyses, the technological regime is generally regarded as fixed and does not depend on firms' strategies. However, economic literature has long emphasized that the dynamics of innovation affects the needs and strategies of the actors, and vice versa (Dosi, 1982; Zuscovitch, 1986; Teubal *et al.*, 1991; Callon, 1999).

In this context, this chapter aims to deepen the analysis of the relationship between firms' appropriation strategies and innovation dynamics. In particular, we focus on the characteristics of the technological regime as a central determinant of firms' patenting strategies but – and this is the originality of our work – we consider those characteristics as evolving endogenously over time. Technological regimes are built by the strategies of the

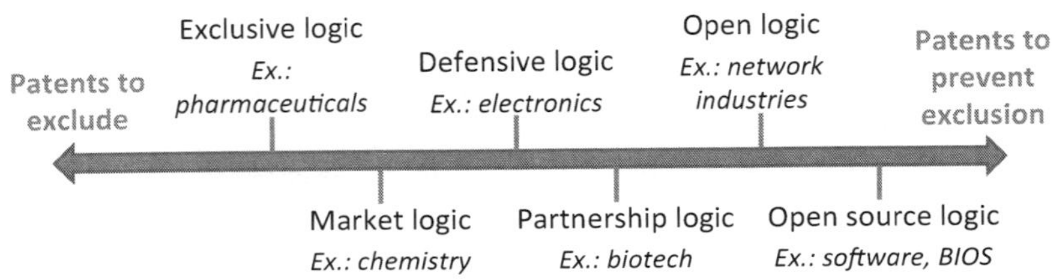

Figure 9.1 The management of patents: a wide spectrum of uses

actors of the innovation process, and therefore the optimal patenting strategies change over time, depending on the evolution of the technological regime. Several case studies in various sectors support this evolving vision of firms' appropriation strategies (in life sciences, software, creative industries, etc.).

The next section summarizes the multiple roles of patents put forward in the literature. The third section offers an explanation for this multiplicity of roles by tying firms' optimal use of their patent portfolio with the technological regime of the sector in which they operate. The next section introduces dynamics in the reasoning by presenting the work of Zuscovitch (1986) and Callon (1999), which led to considering the evolution of technological regime as endogenous. The final section attempts to link this endogenous evolution of technological regime and firms' optimal patenting strategy.

The many roles of patents put forward in the literature

We show in this section that, although patents are traditionally seen only as instruments of exclusion, there is a wide spectrum of use for firms. The latter can use patents in a pure logic of exclusion or, at the other end of the spectrum, in pure open source logic. We will show in the next section that patenting strategies are indeed widely contextual and depend on the technology used in the sector in which the company operates (Figure 9.1 illustrates these two points, namely the multiple roles of the patent and the importance of sectoral context).

Exclusive logic

Following Arrow (1962), the traditional view considers innovation as an individual and solitary activity which generates significant positive externalities (innovation is similar to the production of information). Hence, the lone inventor must primarily be able to protect his invention. This vision leads to focusing essentially on the incentives to innovate (rather than on the issue of coordination of the innovative activities) and on the role of patents as an instrument of exclusion mostly used to prevent imitation and to ensure a monopoly position to the holder (generally a manufacturing firm inventor-producer-distributor). This exclusive use of patents is found mainly in the pharmaceutical industry.

In contrast with this paradigm, the contemporary vision of the innovation process considers the production of knowledge and innovation as essentially a collective activity that requires collaboration and interactions between many different actors (Nelson and Winter, 1982; Chesbrough, 2003). Similarly, knowledge is seen as largely contextual and difficult and costly to imitate. Within this renewed framework, the innovation process is disintegrated, modularized (we leave the paradigm of the producer-inventor-distributor fully integrated firm) and the key issue for innovation actors is less to protect themselves from imitation than to interact and exchange with partners. The role of patents is therefore rather to contribute to include the various actors in the process, to facilitate their interactions under incomplete information, rather than to exclude (Cohendet and Pénin, 2011). This coordination role of patents, according to the context, gives rise to very different logics of use.

Market logic

Increasingly, patents ensure the market coordination of the innovation process. The combined properties of protection and knowledge disclosure are indeed essential for the market exchanges of technology (Rivette and Kline, 2000; Arora *et al.*, 2001). On the one hand the disclosure of the invention allows sellers to advertise their technology. On the other hand the protection dimension secures the disclosure by preventing free riding behaviors (which would consist of using the published invention without compensating the inventor). In other words, the combined properties of protection and knowledge disclosure solve the Arrow paradox (1962). Consequently, patents contribute to the development of markets for technology in which technology suppliers and technology buyers meet. In this context, patents sustain the rise of companies specializing in R&D and knowledge production (the so-called "fabless" firms) (Grindley and Teece, 1997; Arora and Merges, 2004).

Defensive logic

In a logic similar to the market one, patents are not always sold on markets for technology but are traded against other patents in cross-licensing agreements (they are somehow bartered). Patents are indeed very often used as bargaining chips in a defensive logic (they contribute to secure freedom to operate). Firms amass thousands of patents for the sole purpose of being able to protect themselves when they are accused of infringement. In this case they can cross-license their own patents, thus obtaining the right to continue to manufacture and sell their own products. This defensive use of patents is typical of sectors in which the technology is complex, such as electronics, where companies often face problems of freedom to operate (Grindley and Teece, 1997).

Partnership logic

Patents can also be used in an explicitly collaborative logic. Within open innovation frameworks, when firms have a vital need of partners, patents can be a way to signal competences and negotiate formal collaboration agreements. In this case, patents play a role very early in the innovation process and this role is not confined to merely allocating existing resources (through the development of markets for technology, for example). They help to structure more or less formal collaborative arrangements (research network, research consortium, "joint venture," informal exchanges, etc.), thus contributing to a collective dynamics of knowledge production. In this partnership logic patents are likely to act upstream of the collaboration (they signal the skills of potential partners and protect their intangible assets) during the negotiations (they formalize the intangible contributions of each partner), and downstream of the collaboration (they facilitate the sharing of the results via co-patenting).

Open (and open-source) logic

An extreme case of collaborative use of patents is when a firm uses its patent in order to free the patented technology, i.e., to facilitate its use by other firms or to make it available to other firms. Many examples of open use of this type occur, for instance, in network industries, where the purpose of most firms is to benefit from network effects, to impose a standard and to ensure compatibility of its products with other firms. In these cases, it is essential for the company, not to exclude imitators, but rather to facilitate the adoption of its technology. To do this, firms are encouraged to grant licenses on a nonexclusive basis and at a very reasonable price. Another example of open use of this type is "open source" projects which occur, for instance, in the software sector. In this case, participants use patents in a "copyleft" logic, i.e., in order to make sure that the technology and its subsequent derivatives are made available to all (Pénin and Wack, 2008).

Towards sectoral systems of intellectual property (SSIP)

Because a patent is a flexible instrument that can be used in many different ways, choosing the optimal patenting strategy for the firm is a central issue. When should firms rely on exclusive strategies versus more collaborative ones? This choice is largely influenced by the technological regime of the sector (Nelson and Winter, 1982). Obviously, the nature of the firm (its size, experience, etc.) and the nature of the competitive regime might also influence the patenting strategy. For instance, small firms are less able to

Table 9.1 Technological regime and firms' optimal patenting strategies (Pénin, 2011)

*Characteristics of technological regime**	*Impact on patenting strategy*
Tacit knowledge base	**Makes imitation and communication harder** *Favors the non-market coordination role of patents*
Codified knowledge base	**Facilitates imitation and technology exchanges** *Favors an exclusive or a market coordination role of patents*
Complex (versus simple) technology	**Reduces the freedom to operate** *Favors a defensive use of patents*
Modular (versus integrated) technology	**Increases the division of labor of inventive work** *Favors the coordination role of patents*
Existence of network effects	**Increases the gains linked to the diffusion of the technology** *Favors an open and non-exclusive use of patents*

Note: The characteristics in this table are not ordered by importance. The effect of each characteristic must be interpreted *ceteris paribus*.

rely on exclusive strategies because they need to collaborate with holders of complementary assets. Nonetheless, those conditions matter less than the nature of the technology.

The technological regime of a given industrial sector can be defined as the specific combination of factors inherent to the technology used by the actors in this sector (Nelson and Winter, 1982; Winter, 1984). The dimensions of the technological regime are the existence of technological opportunities, the cumulativeness of technical advances, the properties of the knowledge base, the degree of modularity, the existence of network effects (also affected by demand) and the degree of complexity of technology (Breschi *et al.*, 2000).

Most studies on technological regimes studied the impact on the pace of innovation in a given sector. For instance, they used technological regimes in order to explain why different sectors exhibit different innovation rates (example: Schumpeter mark I versus mark II). However, we claim here that the technological regime also strongly affects actors' strategy. In particular, the use of patents depends on certain dimensions of the technological regime, as detailed in Table 9.1.

Codified versus tacit knowledge base

The more the knowledge base is codified the easier it is to transfer knowledge, either voluntarily (through collaborative and/or market arrangements) or unintentionally (through externalities) (Teece, 1986). So, all things being equal, when the knowledge base is highly codified firms use their patent in an exclusive logic to prevent imitation or in market coordination logic to grant licenses on markets for technology. Conversely, the more the knowledge base is tacit, the more difficult it is to communicate and exchange knowledge, thus leading firms to use their patents in a logic of coordination or even of collaboration.

The complex versus simple nature of technology

This dimension affects the freedom to operate of patent holders. When the technology is simple the patent holder has usually few problems of freedom to operate. For instance, in cosmetics and pharmaceuticals there is usually an almost perfect bijection between a molecule and a drug. Also the owner of the patented molecule often has a monopoly on the drug market. This is not the case in complex technologies, where patent holders may not have the right to use the patented technology. For example, in electronics, very often patents give to their owner a right to exclude but not a right to use the technology because there are so many overlapping patents that using the technology would imply infringing patents held by third parties (Grindley and Teece, 1997). Also, ultimately, in the case of a simple technology firms tend to use patents in an offensive way, to exclude other firms and to enjoy a monopoly position, whereas in the case of complex technology they tend to use it in a defensive logic, to avoid being excluded (Kingston, 2001).

The existence of network effects

Network industries, that is to say those where network effects are important, are characterized by a very specific use of patents. Indeed, as the value of the goods produced increases with the number of users, compatibility and standard issues are central. However, to facilitate the establishment of a standard and encourage the compatibility of different technologies it is often necessary to open these technologies and disseminate them widely. Strategies based on secrecy and exclusion are rarely relevant to impose a standard (in all cases they carry a high risk). So, all things being equal, in the case of strong network effects, firms are encouraged to use their patents in an open, collaborative and inclusive logic, in order to foster the emergence of a standard compatible with their technology.

The integrated versus modular nature of the technology

In connection with the previous point, the modular nature of technology also affects firms' patenting strategy. Simply put, when a technology is highly modular, actors have a vested interest in producing it in a decentralized way in order to benefit from specialization gains. In this case, when companies are largely interdependent, because each produces only a small subpart of the product, issues of interoperability and compatibility are again central.

Also, firms have strong incentives not to use their patents in an exclusive way, but rather in logic of coordination, in order to smooth interactions and collaboration.

This analysis, still very exploratory, allows us to highlight the main determinants of the optimal patenting strategy for a firm and to explain why the use of patents varies across industrial sectors (pharmaceuticals, electronics, network industries, chemicals, cosmetics, software, aeronautical, etc.). In the next section we introduce dynamics in this reasoning by presenting the works of Callon and Zuscovitch which both focused on the dynamics of knowledge and technology and which we consider as being largely complementary.

Introducing dynamics in the reasoning

The Callon model

Callon (1999) made an important contribution to the analysis of the dynamics of innovation by showing that the properties of knowledge are not fixed and exogenous, but change over time depending on the strategies adopted by the different actors. By considering that the properties of knowledge are functions of the state of the network in which this knowledge circulates (emerging versus stabilized networks), Callon thus shows that the possibility of replication and appropriation of new knowledge is highly dependent on the context of its creation. For instance, he points out that very often in order to duplicate a scientific statement one needs to reproduce the entire experiment. The fundamental error, says Callon, is to believe that codified and easy to duplicate knowledge contains within itself the attributes of its reproduction.

Callon's view leads to a conception of knowledge in terms of rivalry and appropriability quite different from that of the standard Arrow model. First, the non-rival property of knowledge, which implies a marginal cost close to zero, is endogenous; it is the outcome of a long process of construction by all the actors of the innovation process. For Callon, when knowledge is elaborated it evolves first in emerging networks and it is characterized by a high cost of reproduction (the reproduction of the whole environment in which it is created). It is almost useless outside the community which has produced it. Knowledge becomes non-rival only when the investments needed to create a consolidated network are performed, i.e., the creation of a common language, of a standard and of all the infrastructures in which it can flow with an almost zero cost of reproduction.

Second, the degree of appropriation of knowledge is also endogenous. In emerging networks, knowledge is easily appropriable; its reproduction is comparable to the reproduction of a physical good because, again, it is not just information that is reproduced but the whole environment of the discovery. In addition, the network that allows knowledge to flow is not yet developed, meaning that there is little possibility of spillovers. It is only when the network is consolidated that knowledge becomes close to information and is less easily appropriable.

Finally, in emerging networks, knowledge is very specific. The generality of knowledge does not follow from its intrinsic usefulness but depends on its level of diffusion. And this diffusion, for Callon: "depends entirely on the existence of relays, translators and mediators who, while following their own interests, invest a lot of energy to integrate, adapt and create compatibility" (Callon, 1999, p. 37). The generality of knowledge is therefore built with the evolution of the network in which it flows.

In Callon's analysis, the properties of knowledge and technology are not given but depend on the decisions of the actors of the economy who influence the evolution of the network in which knowledge flows. Knowledge will take the character of a public good only if investments are made so as to establish a consolidated network. The passage from an emerging network to a consolidated one does not follow a natural trajectory. It is not automatic. The strategic choice of agents in terms of disclosure, retention and protection of knowledge is a crucial element of this dynamic. Configuration and reconfigurations of the network are endogenous.

The contribution of Callon (1999) on the dynamic properties of knowledge provides a framework to understand the varying roles of patents according to the evolution of technological development. However, one major limitation of this model is its lack of characterization of intermediate configurations. It considers only the two extremes cases of the network which is either emerging or consolidated. Nothing is said on the cases in between these two extremes while it is likely that these intermediary configurations play a critical role in the dynamics and evolution of the network. This is why it may be interesting to complete this analysis with the work of Zuscovitch on the dynamics of technological change.

Zuscovitch's view of technical change

Zuscovitch's analysis put forward the influence of irreversibility and learning as the main forces of technological evolution. According to Zuscovitch (1986), "the dynamics of technology development is a cumulative process which operates on a wide variety of innovations." In line with the evolutionary approach, Zuscovitch thus envisages the dynamics of technical development as a cumulative and irreversible process in which the regular emergence of diversity is accompanied by a selection process. He places learning at the heart of this dynamics by considering that agents must always learn the new properties of technologies. This permanent learning of new characteristics ensures the gradual emergence of new products and processes and new markets, also often accompanied by a standardization process and increasing returns.

According to this logic, Willinger and Zuscovitch (1993) operate a sequential cutting of the technology development process depending on the type of irreversibility at work. Zuscovitch distinguishes the irreversibility according to both a static and a dynamic point of view. In the first one, the irreversibility comes from a techno-economic determinism defined by the existing paradigm, while in the second it corresponds to a cumulative process of learning. From this distinction, Zuscovitch defines technological evolution in three stages:

The pre-paradigmatic stage

In this first stage, the choices of agents are constrained by the paradigm in place. Indeed, during the early phases of emergence, the new technology is confronted with the performance of the old one. The unknown is compared to the known in a rational exercise of cost-benefit analysis. The constraints of the old technological regime thus force producers of new knowledge to always demonstrate the superiority (or the potential for superiority) of the new technological trajectory. Anderson and Tushman (1990) call this initial phase the "era of ferment." Different technological trajectories are typically experienced, but only one, according to the principle of selection, will emerge as a dominant design, thus opening the door to a new paradigm. The irreversibility in this stage comes therefore from the selection process between competing technologies.

The stage of self-organization

This stage is driven by a cumulative irreversibility induced by the learning process and the phenomena of lock-in. Past choices determine to a great extent future choices through a process of accumulation of technical knowledge. Thus, the choice of a technology will determine not only its own evolution but also the evolution of competing technologies that will be limited in their scope of application. Technology is subject to increasing returns of adoptions but also to a broadening of its scope of application. In other words, technology does not spread evenly but evolves along its successive replications. The key point here is the process of co-evolution between the body of knowledge and the diffusion of the technologies.

The paradigmatic stage

The paradigmatic stage is characterized by an allocative irreversibility due to the standardization of technical solutions. The new knowledge base is crystallized in a set of concepts, language and "technology module" enabling its adaptation to a variety of situations. Standardization is found not only at the level of the technical solution, but is also at work in the incentives and information system. Indeed, the search for increasing returns becomes the norm and the finding of technical solutions can no longer exceed the natural boundaries of the technology which is used. Solutions at the periphery, which are much less profitable, are thus progressively abandoned to the benefit of the standard solution, thus reinforcing the paradigm.

The evolution of the technological regime and firms' optimal patenting strategy

Combining Callon and Zuscovitch analysis to understand the dynamics of the technological regime

So far, we have explained the use of patents as a function of the technological regime in a static view. However, and this is the whole message of both Callon and Zuscovitch, the innovation process constantly changes the technological properties of the system. This constant evolution, among others, leads to a reconsideration of the conditions of appropriation of knowledge. Indeed, if we accept the fact that appropriation strategies are largely dependent on the technology, we need to be able to characterize the dynamics and evolution of technology in order to understand firms' optimal patenting strategies. Figure 9.2 shows how it is possible to combine the model of Callon and the work of Zuscovitch in order to understand the dynamics of the technological regime. In particular, the work of these two authors allows us to put forward three phases in the evolution of technological regime.

Pre-paradigmatic phase and emerging networks

This stage is characterized by the emergence of a new technical principle and the new body of knowledge associated with it. New technological options are constantly explored outside the existing paradigm, but their value and purpose are still largely uncertain. Because of the irreversibility due to the technological selection process, each organization producing a new principle desires that the others also adopt the same technological trajectory. But knowledge in this emerging configuration is strongly appropriable (it is difficult to reproduce outside their context of production). Hence, major investments are needed in order to create the environment necessary for this operation of dissemination. Given the high uncertainty that characterizes this phase, the risk is much higher for firms that others do not pay any attention to their technology than to be opportunistically imitated.

In order to expand the network in which knowledge can flow firms must hence develop strategies of sharing and translation (Callon, 1999). The objective is to reduce the degree of ownership of that knowledge or, in other words, to start a process of knowledge disappropriation. Indeed, new knowledge, to be useful, must be able to escape from the circle in which it was created in order to provoke an emulsion outside the network. If nobody cares about it, it is doomed to disappear, regardless of its intrinsic value. Thus, through sharing strategies deployed by producers of new knowledge, relationships will be forged between worlds that previously did not communicate and new

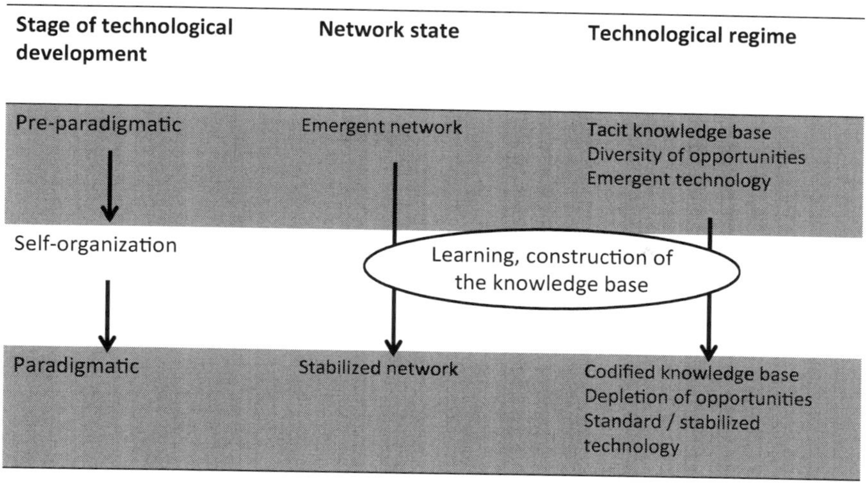

Figure 9.2 Combining Callon and Zuscovitch analysis to understand the dynamics of the technological regime

opportunities will take shape for new players in the network. During these multiple efforts of translation, networks evolve, interconnect and reconfigure themselves. Production and diffusion of knowledge converge into one similar process.

At this stage, the technological regime consists mostly of a tacit knowledge base coupled with high technological opportunities. Specifically, the tacitness of the knowledge base does not necessarily come from an inability to explain explicitly the "recipe" of the technology but from the characteristics of the environment (instrument, infrastructure, etc. ...) in which the knowledge was created. For example, a scientific publication is an output codified, but understanding this publication often requires a majority of tacit knowledge encapsulated in the elements that have served the scientific study.

Auto-organization phase and intermediary configurations

While in the pre-paradigmatic stage strategies of des-appropriation dominated, in this phase the system has adopted a technological trajectory, thus involving agents in a process of exploration and learning characterized by cumulative irreversibility. It is during this phase that the need for coordination is essential. Indeed, in the consolidating process of the network, the knowledge base is gradually shifting from a highly tacit, contextual and specific state to a highly codified and generic state. However, the need for coordination appears (and is possible) only when the properties of the knowledge base are in an intermediate state. Indeed, it is necessary that

..e actors have a minimum capacity of understanding and replication of knowledge in order to exchange and to be able to coordinate the process. As soon as the network is consolidated the need for coordination disappears, given the public nature of knowledge.

The fundamental impulse in this intermediary stage is the learning process. New opportunities constantly arise as the knowledge base expends and is replicated. Thus the exploration of new technology increases its scope of application, thereby pushing the boundaries of the network away. At this intersection of emerging and consolidated networks, the strategic choices of actors largely determine the switch to one or the other state of the network. But as we shall see in the next section, the technological regime will also lead the players' strategies.

Paradigmatic phase and consolidated networks

This phase is characterized by stabilized networks where knowledge tends to take the properties of a public good. The space of strategies is known in advance and knowledge can be replicated at very low cost anywhere in the network. Opportunities are exhausted and every innovation is easily imitated because of a strongly codified and generic knowledge base. The technology converges towards a standard and its evolution is largely governed by the agenda of problems to solve. The innovation process is hence characterized by micro-innovations imitable at low cost.

At the intersect of the different phases

The former analysis clearly shows that these three phases encapsulate the evolution of the technological regime in terms of knowledge base (codified / tacit) and opportunity conditions. However, as shown in Figure 9.3, this process of innovation also affects the degree of modularity and complexity of the technology, which is largely built as the network evolves. Indeed, modularity is not given. It is an outcome, via the standardization process, of the paradigmatic phase. Yet, the trend towards modularity already starts during the intermediary, self-organization phase, when knowledge is sufficiently replicated in the network in order to allow the coordination of joint learning of novelty and thus the generation of new opportunities. Complexity is also the outcome of the technological evolution that takes place at the intersection of the self-organization and paradigmatic phases. The progression of technological development is indeed an incremental step where each new innovation is bringing gradually more complexity. New connections between existing techniques are found, thus bringing increased complexity. And, modularity often emerges in order to simplify this development process of a complex technology.

Hence, the properties of modularity and complexity are not given a priori but change with agents' decisions, i.e., they are endogenous to the process.

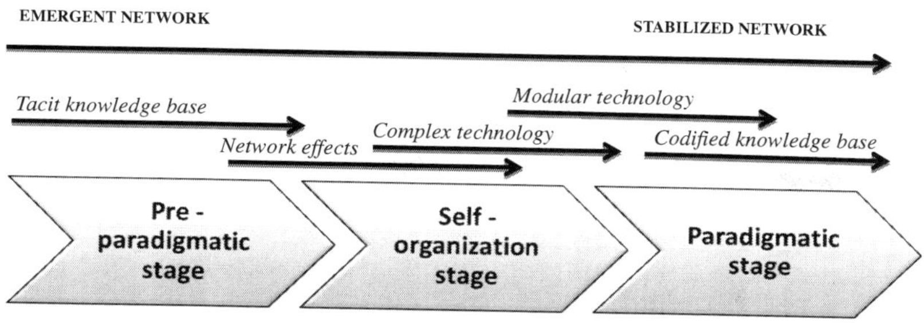

Figure 9.3 The evolution of technological regime

But this is not the case of network effects, which are largely an intrinsic characteristic of the technology. Their impacts on agents' decisions occur early in the process and continue until a critical mass of users is reached (thus often leading to lock-in).

Dynamics of technological regime and optimal patenting strategies

Previous sections showed how the optimal use of patents is shaped by the characteristics of the technological regime and how technological regimes evolve in time according to the choices made by the actors of the innovation process. Thus, it remains to connect the two and to analyze optimal patenting strategies in a dynamic setting.

The general dynamics of optimal patenting strategies is displayed in Figure 9.4. This dynamic is influenced by two opposing forces: The process of dis-appropriation which is necessary in emerging phases; and the process of appropriation necessary in consolidated phases. However, this tension becomes visible only if one takes a sufficiently large detachment in order to observe technological development since the emergence of a technology until the exhaustion of its opportunities.

In the emerging phase, the actors want to be understood by the other stakeholders and thus they develop for this purpose sharing strategies (Callon, 1999) that may result, for instance, in open-source logics of use of patents. Firms need to translate their knowledge and to share it in order to expand the network and create a dynamics of innovation. Many scholars have recently reported such a process of free sharing of knowledge during the early stages of an innovation process. Nuvolari (2004) for example described cooperation between engineers during the development of the steam engine in the mines of Cornwall (England). Meyer (2003) also documented this phenomenon in the case of the first microcomputers (developed within the *Homebrew Computer Club*, which saw the birth of the first Apple computer) and the first planes.

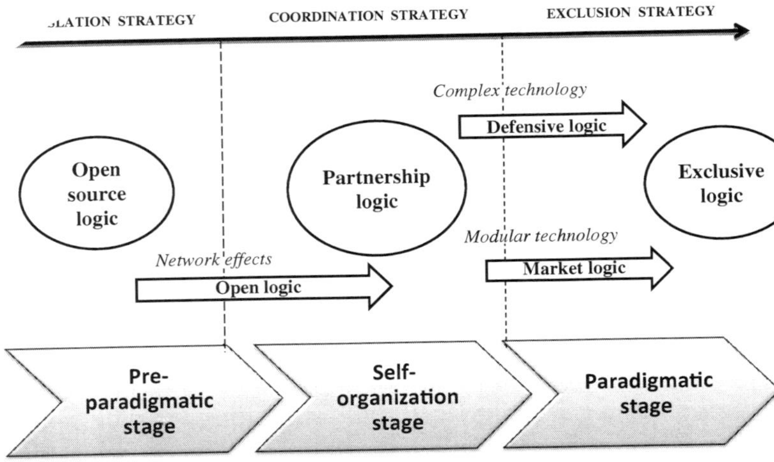

Figure 9.4 Patenting strategies in a dynamic setting

The next stage is characterized by coordination strategies. We hence find here all the patenting strategies that enter the open innovation paradigm (Chesbrough, 2003). These strategies are driven by the need to build a common body of knowledge necessary to standardize the technology. Network effects can lead to a more or less open use of patents at the beginning of the stage in order to take advantage of better technology diffusion. Also, at the end of this intermediary phase both market and defensive use may begin to emerge. The development of microcomputers provides a nice example of this evolution of appropriation strategies. As noted above, an open-source type of strategy has marked their first moments. Yet, as learning and codification progressed, patents have been increasingly used in a market and defensive logic. This is clearly illustrated by the practices of cross-licensing that are widely used today in this sector.

Finally, in the latest phases of technical development, an exclusive logic will tend to dominate. The knowledge base becomes highly codified and is easily controlled by all the actors. It is hence very easy to reproduce the micro-changes that are introduced by competitors. In this case, firms have therefore an interest in protecting their innovation which otherwise could be easily be imitated by a low-cost competitor.

To summarize, appropriation strategies differ greatly depending on the dynamics of technological development. This implies that differences in technological regimes across sectors (and therefore differences in firms' strategies) do not come from intrinsic differences in the technology but from a difference in the level reached in the technical development.

Conclusion: Towards a process of co-evolution between firms' appropriation strategies and the characteristics of the technological regime

We have proposed an interpretative model that combines the works of Callon (1999) and Zuscovitch (1986) to show how technological regimes are progressively built by the actors of the innovation process and how, in turn, they affect actors' strategies. Our analysis therefore conveys a dynamic process of co-evolution of firms' strategies (particularly their patenting strategies) and the characteristics of technological regimes. However, this contribution also opens broad avenues of research: for instance, if the results clearly show that patents are critical instruments for coordinating innovative activities between firms, how do patents contribute in practical terms to "including" all the heterogeneous stakeholders of the innovation process, particularly in a context of open innovation? Also, how do firms re-evaluate their intellectual property strategy periodically when contexts change (for instance, a coordination strategy that was successful in emerging phases may have to be changed to a strategy of exclusion as the situation becomes more stable)?

References

Anderson P., Tushman M. (1990), "Technological discontinuities and dominant designs: A cyclical model of technological change," *Administrative Science Quarterly* **35**, 604–33.

Arora A., Merges R. (2004), "Specialized supply firms, property rights and firm boundaries," *Industrial and Corporate Change* **13**, 451–75.

Arora A., Fosfuri A., Gambardella A. (2001), *Markets for Technology: The Economics of Innovation and Corporate Strategy*, Cambridge, Mass, MIT Press.

Arrow K. J. (1962), "Economic welfare and the allocation of resources for invention," in R.R. Nelson (ed.), *The Rate and Direction of Inventive Activity: Economic and Social Factors*, Princeton, Princeton University Press: 609–25.

Bah M.A., Le Bas C. (2011), "Un nouveau cadre d'analyse des fonctions du brevet: les systèmes sectoriels de propriété intellectuelle," in P. Corbel and C. Le Bas (eds.), *Les nouvelles fonctions du brevet*, Paris, Economica.

Breschi S., Malerba F., Orsenigo L. (2000), "Technological regimes and Schumpeterian patterns of innovation," *Economic Journal* **110**, 388–410.

Callon M. (1999), "Le réseau comme forme emergente et comme modalité de coordination: le cas des interactions stratégiques entre firmes industrielles et laboratoires académiques," in M. Callon, P. Cohendet, N. Curien *et al.*, *Réseau et Coordination*, Paris, Economica.

Chesbrough H. (2003), *Open Innovation: The New Imperative for Creating and Profiting from Technology*, Boston, Harvard Business School Press.

Cohen W. M., Nelson R. R., Walsh J. (2000), "Protecting their intellectual assets: Appropriability conditions and why US manufacturing firms patent (or not)," NBER working paper 7552.

Cohendet P., Pénin J. (2011), "Patents to exclude vs. include: Rethinking the management of intellectual property rights in a knowledge-based economy," *Technology Innovation Management Review*, December, 12–17.

Dosi G. (1982), "Technological paradigm and technological trajectories: A suggested interpretation of the determinants and directions of technical change," *Research Policy* **11**, 147–62.

Grindley P., Teece D. (1997), "Managing intellectual capital: Licensing and cross-licensing in semiconductors and electronics," *California Management Review* **39**, 8–41.

Kingston W. (2001), "Innovation needs patent reform," *Research Policy* **30**, 403–23.

Levin R. C., Klevorick K., Nelson R. R., Winter S. (1987), "Appropriating the returns from industrial research and development," *Brooking Papers on Economic Activity* **3**, 783–820.

Meyer P.B. (2003), "Episodes of collective invention," *BLS working papers* 368.

Nelson R. R., Winter S. G. (1982), *An Evolutionary Theory of Economic Change*, Cambridge, Harvard University Press.

Nuvolari A. (2004), "Collective invention during the British Industrial Revolution: the case of the Cornish pumping engine," *Cambridge Journal of Economics* **28**, 347–63.

Pénin J. (2011), "Le brevet d'invention comme instrument de coordination de l'innovation ouverte," in P. Corbel and C. Le Bas (eds.), *Les nouvelles fonctions du brevet*, Paris, Economica.

Pénin J., Wack J-P. (2008), "Research tool patents and free-libre biotechnology: A unified perspective," *Research Policy* **37** (10), 1909–21.

Rivette K. G., Kline D. (2000), *Rembrandts in the Attic: Unlocking the Hidden Value of Patents*, Boston, Harvard Business School Press.

Teubal M., Yinnon T., Zuscovitch E. (1991), "Networks and market creation," *Research Policy* **20**, 381–92.

Willinger M., Zuscovitch E. (1993), "Efficience, irréversibilité et constitution des technologies," *Revue d'Economie Industrielle* **65**, 7–22.

Winter, S. G. (1984). "Schumpeterian competition in alternative technological regimes," *Journal of Economic Behavior and Organization* **5**, 287–320.

Teece D. (1986), "Profiting from technological innovation: implications for integration, collaboration, licensing and public policy," *Research Policy* **15**, 285–305.

Zuscovitch E. (1986), "The economic dynamics of technologies development," *Research Policy* **15**, 175–86.

10 Estimating the impact of a potential process innovation and the optimal strategy of licensing

Weifang Lou

Introduction

It is a well-established notion that innovation plays a vital role in the modern world in boosting economic growth and improving consumers' well-being. The task of accurately evaluating a potential innovation is strategically important. Nevertheless, the economics literature consists mainly in *ex post* evaluations of product innovations based on the observed economic changes in the marketplace (e.g. Hausman 1997; Hausman and Leonard 2001; Nevo 2003; Petrin 2002). Studies on emerging process innovations are rare. The main difficulty of this study is that a new technology embodied in process innovation does not lead to directly marketable products, which can be evaluated through investigating the pattern of substitution among alternative products in the market. The second difficulty is due to the lack of data – no observed prices and sales in a market of varieties of similar technologies. At the early stage of an innovation, only experimental information on the technology is available. In response to these challenges, this chapter proposes a practical approach where a process innovation is evaluated *ex ante* by its application to the final product marketed to consumers.

The economics literature on innovation evaluation is still pretty small. Several economists have developed estimation frameworks measuring the welfare impact from technological progress or the introduction of new goods. For instance, Bresnahan (1986) estimates the demand for computing at the industry level and measures the gains from the improvement of computing in financial services. Trajtenberg (1989) uses a nested multinomial logit demand model when evaluating the benefits from innovation in the Computed Tomography (CT) scanners market. Hausman (1997) applies a three-level demand model to evaluate new products in the ready-to-eat cereal industry. The models used in these studies are tractable in terms of construction and computational ease. However, they ignore some critical issues such as the effect of new goods on other related products or the implication of consumer heterogeneity for estimation. Another stream of literature related to this chapter is technology licensing.[1] Existing studies in this literature build on a theoretical framework, and empirical analysis to investigate the benefits from licensing a particular innovator is rare.[2] The most relevant theoretical paper is Doganoglu (2004), who models the transfer of a drastic cost-reduction or quality-improving technology for a differentiated product market within a Cournot oligopoly.

A few studies have looked into the pattern of innovation for the digital camera market. Song and Chintagunta (2003) model the adoption process of new goods by heterogeneous and forward-looking consumers. Zhao (2007) uses the pattern of new goods adoption by consumers and their interaction with camera manufacturers to explain price fall. Carranza (2010) analyses the decision on product introduction through a full equilibrium model of demand and supply. Xiao (2008) employs a standard logit model when measuring the welfare benefits from introducing the 'Easy-to-Use' storage system by Sony and the paper suffers from the independence from irrelevant alternatives (IIA) property. Lou *et al.* (2012) investigate in particular demand dynamics in the US digital camera market and compare estimation results from different model specifications.

This chapter estimates the potential implication of a recently developed process innovation named the Aluminium Metal Injection Moulding (ALMIM) technique. The technology was developed recently by Liu, Sercombe and Schaffer in the School of Engineering at the University of Queensland (Liu *et al.* 2008). We simulate the post-adoption prices and sales of the products with new characteristics and assess the potential benefits to consumers and producers. Like Petrin (2002), we apply the random coefficients structural model proposed by Berry, Levinsohn and Pakes (Berry *et al.* 1995, henceforth 'BLP'). Unlike Petrin and other earlier papers, we do not introduce a new product but alter the characteristics of the existing products as made possible by ALMIM. Moreover, existing studies use observations after an innovation is adopted and thereby are restricted to adopted innovations. By simulation, we consider a wide range of options on the way of introducing the new technology. We further examine the variation in the potential gains extractable by the ALMIM inventor. This is the first attempt to empirically tackle the problem of licensing. The proposed approach also avoids the criticism of becoming too engineering or firm focused as in the conventional technical cost model, which is typically utilized in the engineering science literature to assess new technologies (e.g. Kirchain and Field 2001).

Our results show that the potential gain from ALMIM varies dramatically depending on the superiority of the new technology and who adopts ALMIM. Higher benefits can be achieved if quality improvement is high and cost appreciation is small. When the single-adopter strategy is taken, the value of ALMIM is maximized by introducing ALMIM to the largest camera maker. The total gain increases substantially if increasing the number of adopters but the benefit to each single firm declines. Unlike existing welfare studies, we find that consumers benefit less than producers for all scenarios investigated. The result may be driven by the small cross-elasticities estimates, suggesting that each digital camera maker has substantial market power in pricing their products and can earn higher profits from a superior technology.

The rest of the chapter begins with an introduction to the way ALMIM can be applied to the digital camera industry as well as the corresponding designs of simulation scenarios. The third section presents the model specification and the welfare change measures. The fourth section reports the estimated demand and supply system and discusses the simulation results and the fifth section considers the potential gain to the inventor. The last section concludes.

The application of ALMIM to the digital camera industry

ALMIM is a new manufacturing technology that applies aluminium materials to a metal manufacturing technique called the metal injection moulding process (MIM).[3] MIM is superior for gaining high precision and quality standards when fabricating small and complex products while aluminium has some attractive mechanical properties such as light weight, stiffness and wear-resistance. ALMIM combines the advantages and is most appropriate for producing small and complex components of integrated final products (Liu *et al.* 2008). This makes ALMIM appealing to many fast growing industries in recent years, such as new consumer electronics. ALMIM can be utilized to improve the quality of existing cameras in two aspects (Liu *et al.* 2008): to reduce sizes at potentially lower cost and to add an extra metal feature. The corrosion resistance, strength and light weight of aluminium would improve the quality of digital cameras desired by photographers.

Transferring the new technology for use in mass production of digital cameras can take two ways, licensing to an intermediate ALMIM manufacturer or to digital camera maker(s). In both cases, the adoption of ALMIM leads to higher quality digital cameras at potentially lower costs, hence potential higher profits shared by the inventor and the adopter(s). As a consequence, the evaluation problem can be tackled by using the observed and simulated data on digital camera purchases to calculate the profit changes to digital camera makers. The predicted profit changes in the digital camera market are upper bounds on the potential gains extractable by the ALMIM inventor. The benefits further depend on the extent of quality improvement, the amount of cost appreciation and which firms and how many firms use ALMIM.

Considering the cost of ALMIM production, the capital cost incurred for investing in ALMIM production varies significantly from below US$100,000 to well above US$1,000,000. Necessarily, ALMIM equipment includes a moulding machine, a debinding furnace and a sintering furnace.

The prices of some sample equipment are given in the appendix (Table 10.A3). Aluminium also is more expensive than commonly used plastic, due to higher material, tooling or finishing costs. These measures suggest that an ALMIM body camera may cost more than a comparable plastic one. To allow for the cost appreciation, we increase the marginal costs of ALMIM digital cameras by a set of values ranging from zero to a maximum of US$3.0 in the counterfactual scenarios.[4]

The potential cost and benefit can also differ substantially depending on who is the adopter(s). On the one side, ALMIM provides its adopter with the opportunity to obtain a certain degree of market power and potential benefits may vary depending on the firm's capitalization, market share, brand loyalty, technological capacity and human resources. On the other side, adoption by one firm also affects the market power of their rivals, influencing their competitors' sales, profitability and new technology investment decisions. As a result, the profitability of all firms would be affected with the arrival of the innovation, no matter who in fact adopts it. This results in divergences among firms in their willingness-to-pay for acquiring the new technology. In other words, the benefits extractable by the ALMIM inventor can also differ as the composition of the adopters changes. Consequently, we consider two licensing options taken by the inventor: the single-adopter strategy and the multiple-adopter one. In the multiple-adopter case, we assume all firms have the right of adoption and will choose to adopt ALMIM whenever they are better off.

Considering different factors outlined above, we end up with a set of hypothetical scenarios. These counterfactual scenarios differ from one another by a three-level specification:

1) Improve the attributes of products in two separate ways, i.e. smaller size and metal body replacement. In either case, quality change is of one dimension with all other features fixed. Changes for the counterfactual scenarios are assumed as follows:

    a) Reduce the sizes of adopting firms' products by 10 per cent or 20 per cent, respectively.[5]
    b) A dummy feature for 'metal' attribute. The hypothetical marginal value for the new attribute (β_{metal}) is set at 0.02 and 0.04, equivalent to an increase in the value of cameras by 0.2MP and 0.1MP resolutions, respectively.[6]

2) Specify the firms that adopt ALMIM at a time by changing both the number of firms and the particular firm(s) that adopt(s) the new technique:

    a) Set the number of adopting firms to be either one or all of the six firms.[7]
    b) Alter the firms and run simulation separately for each case.
    c) In multiple-adopter cases, we first set all six firms to be the adopters. If an adopter is found to be worse off as a result of adoption, we change it back to be a non-adopter and rerun the simulation.

3) Modify the marginal cost of products involving the new technology by adding the cost increment, ranging from US$0 to US$3.0.

The above counterfactual scenarios resolve the problem of a lack of data from the market post adoption. Simulation results give guidance to adopters in determining their willingness-to-pay for ALMIM. While firms have much clearer insight on the actual production cost and quality improvement, the exact value of ALMIM can be updated if supplemented with this information.

Model specification

The estimation of demand and supply is based on the model proposed initially by BLP. The model is appealing for the current study as it allows the estimation of demand and supply using the characteristic space approach while also accommodating problems such as consumer heterogeneity, price endogeneity and dynamic issues.

The demand side

Let us suppose there are J^t distinct camera models marketed in period t and each product is indexed by $j = 1, 2, \ldots, J^t$. The characteristics that define each camera model j include its fixed quality measures $x_{j,k}$ ($k = 1, 2, \ldots, K$, including brand dummy), the time-varying price P_j^t and the age of the model in the market $x_{j,age}^t$. Let us suppose further that there are I^t consumers/households in market t and the utility of household $i \in I^t$ from consuming product j is

$$\begin{aligned} U_{i,j}^t &= \alpha \ln(y_i^t - p_j^t) + \sum_{k=1}^{K} (\beta_k + \sigma_k v_{i,k}) x_{j,k} + (\beta_{age} + \sigma_{age} v_{i,age}) x_{j,age}^t + \xi_j + \varepsilon_{i,j}^t \\ &= \delta_j + u_{i,j}^t + \varepsilon_{i,j}^t, \end{aligned} \tag{1}$$

where

$$\delta_j = \sum_{k=1}^{k} \beta_k x_{j,k} + \beta_{age} x_{j,age}^t + \xi_j;$$

$$u_{i,j}^t = \alpha \ln(y_i^t - p_j^t) + \sum_{k}^{K} \sigma_k x_{j,k} v_{i,k} + \sigma_{age} x_{j,age}^t v_{i,age}$$

In (1), ξ_j captures product-specific characteristics observable to consumers but unobservable or immeasurable to researchers, and $\varepsilon_{i,j}^t$ represents the idiosyncratic shock to utility. The vector $v_i^t = (y_i^t, v_{i,1}, v_{i,2}, \ldots v_{i,K}, v_{i,age})$ defines a consumer's characteristics, capturing his idiosyncratic tastes that deviate from the population means (β_k and β_{age}). The extent of taste variations around their means are measured by σ_k and σ_{age}, respectively. An outside non-purchase alternative is introduced and its utility is specified as

$$U_{i,0}^{t} = \alpha \ln(y_i^t) + \sigma_0 v_{i,0} + \xi_0 + \xi_{i,0}^{t}. \tag{2}$$

Conditional on the realization of v^t, the probability of consumer i choosing product j out of choice set J^t becomes

$$\Pr_{i,j}(v)^t = \frac{\exp\left(\alpha \ln(y_i^t - p_j^t) + \sum_{k=1}^{K} (\beta_k + \sigma_k v_{i,k}) x_{j,k} + (\beta_{age} + \sigma_{age} v_{i,age}) x_{j,age}^t + \xi_j\right)}{1 + \sum_{l=1}^{J^t} \exp\left(\alpha \ln(y_i^t - p_l^t) + \sum_{k=1}^{K} (\beta_k + \sigma_k v_{i,k}) x_{l,k} + (\beta_{age} + \sigma_{age} v_{i,age}) x_{l,age}^t + \xi_l\right)} \tag{3}$$

Given the joint probability density function $P(v^t)$, the aggregated probability of consumers choosing camera model j, or the market share of model j is obtained by integration:

$$s_j(p^t, x^t, \xi; \alpha, \beta, \sigma) = \int \Pr_j(v^t) P(v^t) d v^t, \tag{4}$$

where p^t is the price vector, x^t represents the collection of all observable characteristics and ages $(x_{j,1}, \ldots x_{j,K}, x^t_{j,age}, j = 1, \ldots J^t)$, β and σ are the collections of all beta and sigma parameters, ξ is the vector of all unobservable characteristics, which will form the error term required for the GMM estimator.

The supply side

The marginal cost of producing a camera j, mc_j^t, is modelled as a semi-log hedonic function of a set of broadly available quality measures, χ_j, a time trend, t and unobservable cost characteristics, ϖ_j,

$$\ln(mc^t) = \gamma \chi_j + \lambda t + \varpi_{j,}, \qquad j \in J_f^t, \tag{5}$$

In the cost equation, γ is a vector of coefficients measuring the marginal effect of each observable characteristic on the logarithm of the product's marginal cost and λ captures the cost reduction due to technological progress. In equilibrium, each multi-product firm f has a set of J_f^t number of first-order conditions (FOC) that take the form:

$$s_j(p^t, x^t, \xi) + \sum_{l \in J_f^t} (p_l^t - mc_l^t) \frac{\partial s_l(p^t, x^t, \xi)}{\partial p_j^t} = 0, \qquad j = 1, 2, \ldots J^t. \tag{6}$$

Combining the FOC equation with the above cost equation (5), we can write a supply side moment condition:

$$\varpi = \ln(p^t - \Delta(p^t, x^t, \xi)^{-1} s(p^t, x^t, \xi)) - \gamma\chi - \lambda t. \tag{7}$$

The term $\Delta(\cdot)^{-1}s(\cdot)$ is the markup estimated jointly with the demand side equations, where $\Delta(p^t, x^t, \xi)$ is a $J^t \times J^t$ matrix and each element in the matrix is given by

$$\Delta_{j,l}(p^t, x^t, \xi) = \begin{cases} \frac{-\partial s_l(p^t, x^t, \xi)}{\partial p_j^t}, & l, j \in J_f^t \\ 0, & \text{otherwise} \end{cases}. \tag{8}$$

The cost side moment condition given by (7) is used jointly with the demand side condition (4) to generate the objective function in the GMM estimation routine. By minimizing the GMM objective function, estimates for both demand and supply equations are derived simultaneously.[8]

Welfare change measure

A firm will adopt a process innovation if the new technology brings more profits than the existing technologies. Following the standard approach, we measure the change in producer surplus for each multi-product firm as the total sum of profit differences of all products manufactured by the same firm:

$$\Delta\pi_f = \sum_{j \in J_f} (p_j^1 - mc_j^1) M s_j^1 - \sum_{j \in J_f} (p_j^0 - mc_j^0) M s_j^0 \tag{9}$$

In the equation, p_j^0 and s_j^0 signal respectively the observed price and market share of each product *j*. The marginal cost before technology change is estimated using the existing market observations and is denoted by mc_j^0. The corresponding price, sales and cost variables after change are denoted by p_j^1, s_j^1 and mc_j^1, respectively. The market size M is assumed to remain fixed. The total change in producer surplus as a result of technological progress is then calculated as the sum of profit changes $\Delta\pi_f$ over all competing firms $f = 1,2,\ldots,F$ in the product market, including both adopting and non-adopting firms.

$$\Delta\Pi_{f=1}^F = \sum_{f=1}^F \Delta\pi_f, \quad \forall f = 1, 2, \ldots, F. \tag{10}$$

Since the product characteristics prices change as a result of adopting ALMIM, consumer welfare will also differ. In this study, the Hicksian Equivalent Variation (*EV*) is utilized to measure the change in consumer surplus. Let us suppose the equilibrium price vector prior to and post innovation are p^0 and p^1, respectively. The corresponding utility levels an individual i can achieve at the two time periods are $U_i^0(p^1, y_i)$ and $U_i^1(p^1, y_i)$, where y_i represents the consumer's disposable income. Then, the Hicksian EV_i measure can be calculated as[9]

$$EV_i^1 = e_i(p_i^0, U_i^1(p^1, y_i)) - e_i(p_i^1, U_i^1(p^1, y_i)). \tag{11}$$

McFadden (1981) and Small and Rosen (1981) show that for logit models, the EV^1 can be obtained by calculating:

$$EV_i^1 = \frac{U_i^1 - U_i^0}{\alpha_i}, \tag{12}$$

where α_i represents the marginal utility of income for consumer i. Given the individual level EV_i^1 estimates, the total consumer welfare change measure is obtained by an integral over the mass of the consumers in the market. With our utility specification, (12) can be written as:

$$EV_i^1 = \int \frac{\ln\left[\sum_{j \in J_t} \exp(U_{i,j}^1)\right] - \ln\left[\sum_{j \in J_t} \exp(U_{i,j}^0)\right]}{\alpha_i} dP_v^*(\nu), \tag{13}$$

where $P_v^*(\nu)$ is population distribution function and $\ln\left[\sum_{j \in J_t} \exp(U_{i,j}^t)\right]$ is the logit inclusive value which indexes the utility achievable by individual consumer i in period t. Given the prices and characteristics of products prior and post innovation adoption, the logit inclusive values are calculated separately and in turn to obtain the change in consumer surplus. The total change in consumer surplus is then calculated as the total sum of the measures of all consumers.

The results

Data and estimation of demand and supply

The data for digital camera demand and supply estimation are comprised of two parts: a data set purchased from the NPD market research company and a supplemented product characteristic data set from web sources. A total of 4253 observations (351 models) are included, covering more than 83 per cent of total US point-and-shoot (P&S) camera unit sales during the sample period (January 2003 to May 2006). Each model is characterized by its brand name, resolution, optical zoom range, digital zoom range, LCD, weight and three dimensional size measure. Figure 10.1 illustrates

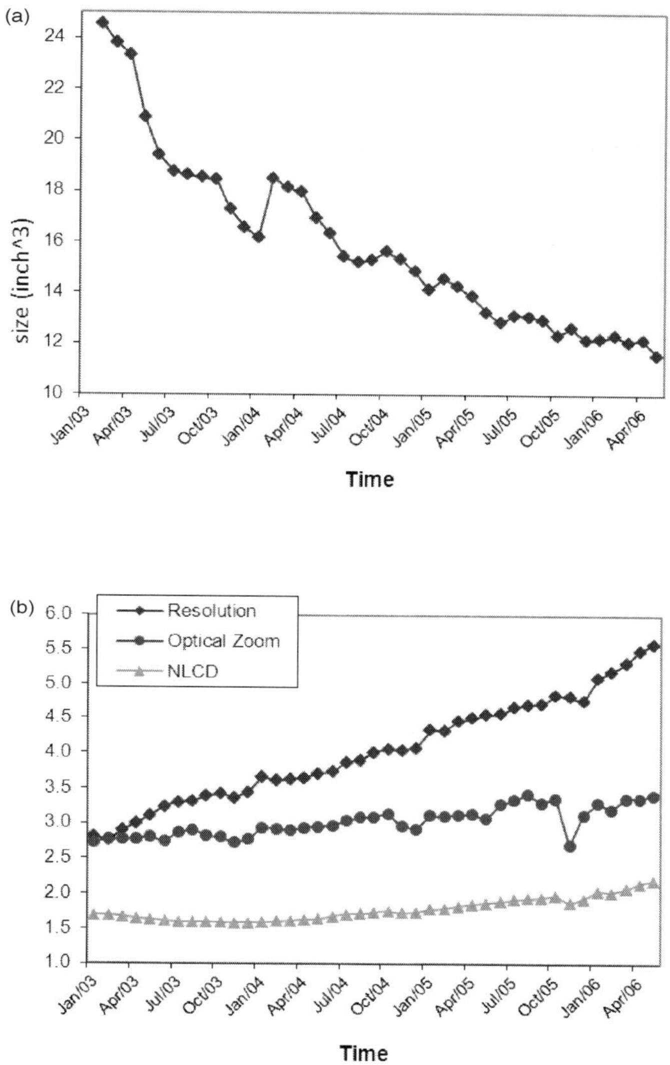

Figure 10.1 Evolution of camera characteristics (January 2003–May 2005)
a. Camera size
b. Resolution, optical zoom range and size of LCD

Table 10.1 Model-level total sales statistics

Firm	*No. observation*	*Mean*	*25%*	*75%*	*Max*
Canon	62	122,698	13,415	218,208	752,598
Fujifilm	30	65,012	3,015	71,731	352,404
Kodak	56	111,623	10,982	194,971	473,176
Nikon	31	106,304	15,142	126,214	575,509
Olympus	49	77,940	22,682	106,739	408,940
Sony	48	147,265	15,626	263,334	482,869
Market	276	108,666	13,757	169,203	752,598

the evolution of weighted average value of the main product characteristics. Table 10.1 documents the sales statistics. We define the last month of the sample period as the baseline situation prior to innovation adoption. There were 143 distinct models marketed at the time.

Table 10.2 and Table 10.3 report the estimated demand and cost parameters from equations 3 and 5. On the demand side, the price coefficient is 0.324 and the marginal utility of age is negative and strongly significant (−0.2253), suggesting that the longer the camera is in the market, the lesser consumers are willing to pay for it. The reported coefficients on the brand dummies range from −0.76 to 0.27, measuring consumers' subjective value on each brand of digital cameras relative to the sample mean. The estimates show that consumers prefer Sony and Canon to other brands while Fujifilm is the least favoured brand. The coefficient of Nikon is positive but is statistically insignificant. Consumers respond favourably to image resolution and optical zoom range, size of LCD screen and weight. The effect of size is significantly negative (−0.2577), suggesting that the majority of P&S consumers prefer a smaller camera, other things being equal. The random effects for most attributes are negligible. However, there exists a substantial variation in the marginal values of digital zoom range and the age variable, suggesting we can reject the hypothesis that the random coefficients model is equivalent to the standard logit model.

The marginal cost estimates in Table 10.3 show most variables have the expected sign and are statistically significant. The trend parameter is −0.0122, implying an overtime decreasing production cost. This is in accordance with the dramatic technological progress in the digital camera industry during that time. Positive coefficients are obtained for resolution, optical zoom and LCD size, indicating the higher these features the more costly for firms to produce. The parameter for weight is positive and significantly different from zero, demonstrating that producing cameras with more robust material and extra components is costly. This is intuitive given larger weight signals some favoured unobservable components or higher quality of the camera. It also justifies results on the demand, where we have positive weight coefficients. On

Table 10.2 Demand estimates

Variables	*Parameter estimate*	*Standard error*	*t-statistic*
Alpha: term on price			
ln(*y*-*p*)	0.3240*	0.1157	2.80
Beta: mean coefficient			
Constant	−7.4278*	0.5955	−12.47
Resolution	0.3825*	0.1187	3.22
LCD	0.2472	0.1427	1.73
Opt. zoom	0.2492*	0.0297	8.39
Size	−0.2577*	0.1250	−2.06
Weight	0.0232	0.0621	0.37
Dig. zoom	1.5229*	0.5736	2.65
Age	−0.2253*	0.0130	−17.33
Canon	0.2401*	0.0620	3.87
Fujifilm	−0.8256*	0.0879	−9.39
Kodak	−0.3068*	0.0705	−4.35
Nikon	0.0984	0.0861	1.14
Olympus	−0.2324*	0.0572	−4.06
Sony	0.5277*	0.0627	8.42
Sigma: standard deviation of beta			
Constant	0.8424	1.1050	0.76
Resolution	0.0372	0.3615	0.10
LCD	−0.2922	0.2031	−1.44
Opt. zoom	0.0052	0.0987	0.05
Size	−0.1809	0.1037	−1.74
Weight	0.0774	0.0529	1.46
Dig. zoom	1.5235*	0.5015	3.04
Age	0.0950*	0.0058	16.38

Notes: 1 * Coefficient significantly different from zero at the 1 per cent level.
2 Time dummies are included in the estimation but their parameter estimates are not reported for the sake of space.

the contrary, the cost is negatively related to camera size, suggesting smaller cameras associated with extra input (cost), other things being equal.

Simulation for the adoption of ALMIM

Simulation results for different scenarios are organized according to how ALMIM improves camera quality. The welfare effect of metal camera body is very similar to the effect of camera size reduction, therefore only

Table 10.3 Cost estimates

Variable	*Parameter*	*Standard error*	*t-statistic*
Canon	1.7735*	0.2125	8.35
Fujifilm	1.7951*	0.2103	8.54
Kodak	1.7158*	0.2132	8.05
Nikon	2.1474*	0.2006	10.70
Olympus	1.9107*	0.2120	9.01
Sony	1.6834*	0.2218	7.59
Ln(resolution)	0.4333*	0.0958	4.52
Ln(LCD)	0.4187*	0.0457	9.16
Opt. zoom	0.0191*	0.0084	2.27
Ln(size)	−0.6601*	0.0394	−16.75
Ln(weight)	1.4230*	0.0791	17.99
Dig. zoom	−0.0114	0.1618	−0.07
Trend	−0.0122*	0.0019	−6.42

Note:* Coefficient significantly different from zero at the 1 per cent level.

the size simulation is presented fully below while other details are given in the appendix. Table 10.4 reports the profit changes when a single P&S camera manufacturer adopts the ALMIM technology. There are six pairs of columns, where the first column of each pair is the adopter's profits and the second is the aggregate profits of the remaining five non-adopting manufacturers. Two scales of size reduction are considered (10 per cent and 20 per cent), corresponding to the upper and lower panels of the table, respectively. The simulation assumes an increase in marginal costs due to the adoption of ALMIM as listed in the first column in Table 10.4.

Reducing product sizes can possibly restrict other characteristics of digital cameras which further influences consumers' benefits. Therefore, the results reported here can be considered as the upper bounds for welfare estimates. As expected, ALMIM adopters' gains, no matter who the adopter is, are positively related to the scale of size reduction but negatively to cost appreciation. For any single adopter, the higher the rise of marginal cost, the smaller the benefit that a firm can achieve. For a 10 per cent size reduction, any of the six adopters wins positive gains when the unit cost rises by no more than US$1.5. Beyond this point, some firms start incurring losses. Larger brands tend to profit more from new technology than smaller brands, and Canon is most profitable and enjoys US$1.6 million profits even when cost appreciation is 2.5. On the contrary, Fujifilm has the smallest impact and the gain that Fujifilm could obtain is less than one-tenth of that of Canon. If the marginal cost increases by US$3, no firm can benefit from ALMIM.

Also, benefit estimates are larger if the quality improvement is higher. According to Table 10.4, reducing the size by 20 per cent will bring profits

Table 10.4 Change in producer surplus with one adopter (annualized profit changes in millions of dollars)

Unit cost rise	*Adopter*	*Five non-adopters*	*Adopter*	*Five non-adopters*	*Adopter*	*Five non-adopters*	*Adopter*	*Five non-adopters*	*Adopter*	*Five non-adopters*	*Adopter*	*Five non-adopters*
	CANON		*FUJIFILM*		*KODAK*		*NIKON*		*OLYMPUS*		*SONY*	
Size reduce by 10%												
0.0	**10.342**	−3.284	**0.934**	−0.536	**3.440**	−1.805	**3.391**	−1.716	**2.920**	−1.514	**5.016**	−1.941
0.5	**8.572**	−2.499	**0.635**	−0.354	**2.567**	−1.332	**2.552**	−1.236	**2.252**	−1.100	**3.739**	−1.257
1.0	**6.812**	−1.717	**0.339**	−0.172	**1.702**	−0.860	**1.719**	−0.758	**1.589**	−0.688	**2.470**	−0.575
1.5	**5.064**	−0.938	**0.046**	0.008	**0.845**	−0.391	**0.894**	−0.283	**0.931**	−0.278	**1.208**	0.104
2.0	**3.326**	−0.163	**−0.244**	0.187	**−0.006**	0.076	**0.076**	0.189	**0.279**	0.130	**−0.045**	0.780
2.5	**1.599**	0.610	**−0.532**	0.364	**−0.848**	0.540	**−0.736**	0.659	**−0.369**	0.535	**−1.290**	1.452
3.0	**−0.117**	1.379	**−0.816**	0.541	**−1.684**	1.003	**−1.542**	1.126	**−1.012**	0.938	**−2.528**	2.123
Size reduce by 20%												
0.0	**21.315**	−6.699	**1.900**	−1.089	**7.018**	−3.671	**6.909**	−3.487	**5.950**	−3.078	**10.168**	−3.919
0.5	**19.515**	−5.906	**1.596**	−0.904	**6.129**	−3.190	**6.055**	−3.000	**5.271**	−2.657	**8.876**	−3.229
1.0	**17.727**	−5.116	**1.296**	−0.720	**5.248**	−2.712	**5.209**	−2.516	**4.596**	−2.239	**7.592**	−2.542
1.5	**15.950**	−4.330	**0.998**	−0.538	**4.375**	−2.235	**4.369**	−2.034	**3.926**	−1.824	**6.316**	−1.858
2.0	**14.184**	−3.547	**0.702**	−0.356	**3.509**	−1.761	**3.537**	−1.555	**3.262**	−1.410	**5.048**	−1.177
2.5	**12.429**	−2.767	**0.410**	−0.176	**2.650**	−1.290	**2.711**	−1.079	**2.603**	−0.999	**3.787**	−0.499
3.0	**10.684**	−1.990	**0.121**	0.003	**1.799**	−0.820	**1.892**	−0.605	**1.949**	−0.590	**2.535**	0.177

Table 10.5 Size of digital cameras in cubic inch by brand (May 2006)

Firm	*Average sizes*	*Standard error*
Canon	13.927	1.799
Fujifilm	9.260	0.486
Kodak	12.028	0.547
Nikon	10.709	0.648
Olympus	10.903	0.695
Sony	8.378	0.734
Total	11.330	0.541

to all adopters even when the cost increment is at US$2.5. Interestingly, the size of benefits to Canon is more than double that of the second largest brand Sony in each counterfactual case. Examining observations of the two brands suggests two possible explanations. First, Canon has 62 distinct models in the data set that take advantage of the innovation, much more than that of Sony at 48. Second, the average size of Canon P&S cameras is 14 cubic inches; over 60 per cent larger than that of the contemporarily marketed Sony P&S cameras (see Table 10.5).

Table 10.6 lists the EV measures of change in consumer surplus corresponding to scenarios presented in Table 10.4. Unexpectedly, the predicted potential gains to consumers are much smaller than to producers, contrary to most existing studies (e.g. Petrin 2002, Xiao 2008). The reverse may be driven by the small cross-elasticities estimates among different digital camera models, particularly after accounting for the product ageing effect.[10] This further suggests that the digital camera firms have considerable market power to extract the surplus gains from consumers. Higher benefits can be achieved by consumers if a larger brand adopts ALMIM and cost appreciates less. For instance, the maximum estimated total benefit to consumers is US$1.126 million when Canon can reduce camera sizes by 10 per cent with no cost increment. For the same size reduction, consumers are better off in all single-adopter cases when cost appreciates by up to one dollar. Negative EV is obtained for Fujifilm as the single adopter at US$1.5 cost increment. But if cost increases by US$2.5 or more, consumers are always worse off. Alternatively, when ALMIM allows firms to produce cameras at 20 per cent smaller sizes, all EV estimates in Table 10.6 are positive.

In multiple-adopter cases, by assumption, a camera maker will choose to adopt only if it earns higher profit. The results are reported in Table 10.7, where non-adopting firms are reported by numbers with underlines. It is shown that for small cost increments, i.e., up to US$1.50, all firms can benefit from adoption and act as adopters. Total benefits are much higher compared to any comparable single-adopter cases. The value for each single firm is positive but differs substantially, ranging from US$0.165 million

Table 10.6 Change in consumer surplus with one adopter (in millions of dollars)

	The single adopter					
Unit cost rise	*Canon*	*Fujifilm*	*Kodak*	*Nikon*	*Olympus*	*Sony*
Size reduced by 10%						
0.0	1.126	0.127	0.486	0.410	0.360	0.554
0.5	0.898	0.084	0.358	0.301	0.270	0.386
1.0	0.671	0.041	0.231	0.193	0.180	0.219
1.5	0.445	−0.002	0.105	0.086	0.091	0.052
2.0	0.221	−0.044	−0.020	−0.020	0.002	−0.113
2.5	−0.002	−0.086	−0.145	−0.126	−0.086	−0.277
3.0	−0.224	−0.127	−0.268	−0.231	−0.173	−0.441
Size reduced by 20%						
0.0	2.305	0.258	0.989	0.833	0.733	1.119
0.5	2.074	0.214	0.859	0.722	0.640	0.949
1.0	1.843	0.170	0.729	0.612	0.549	0.780
1.5	1.614	0.127	0.600	0.503	0.458	0.612
2.0	1.386	0.084	0.472	0.395	0.368	0.445
2.5	1.159	0.041	0.345	0.287	0.278	0.278
3.0	0.934	−0.001	0.219	0.180	0.189	0.113

for Fujifilm to US$5.4 million for Canon. When an extra cost of US$2.0 is incurred, four firms will be better off acting as adopters. Surprisingly, the predicted profit change for Sony as a non-adopter is still negative (−0.199), but it could lose more by adopting ALMIM. Only the largest brand Canon can gain from ALMIM when cost appreciates by US$2.5 and no firm will adopt it if the marginal cost increases by US$3.0 or more.

Table 10.7 shows that the changes in consumer surplus also increase and almost triple the estimates in the comparable single-adopter cases. For a 10 per cent size reduction, EV can reach as high as over 3 million dollars, about three times the maximum EV in Table 10.4. As the marginal cost goes up, the consumer welfare gain drops continuously and quickly. In contrast to the comparable single-adopter cases, though, higher EV measures are predicted for all multiple-adopter cases. The results demonstrate that increasing the number of adopters leads to more intense competition which benefits consumers.

When comparing to the predictions in the single-adopter cases, the profit gains to each firm in the multiple-adopter cases become much smaller. To have a closer look at the causes of this, the upper and lower panels of Table 10.8 list the average price changes for different scenarios. It shows that all single adopters raise their product prices after adoption. The simulated

Table 10.7 Welfare change measure with multiple adopters

Quality improve	*Change in producer surplus*								*Change in Consumer Surplus*	
	CANON	*FUJIFILM*	*KODAK*	*NIKON*	*OLYMPUS*	*SONY*	*ADOPTERS TOTAL*	*NON-ADOPTERS TOTAL*	*Market Mean*	*Market Median*
Size reduced by 10%										
0.0	7.093	0.468	2.149	1.969	1.512	1.878	15.069	-	3.028	3.151
0.5	6.254	0.316	1.697	1.553	1.225	1.339	12.383	-	2.273	2.414
1.0	5.419	0.165	1.248	1.139	0.939	0.801	9.711	-	1.517	1.723
1.5	4.587	0.015	0.803	0.728	0.654	0.265	7.052	-	0.762	1.077
2.0	3.391	0.074	0.154	0.148	0.253	−0.199	3.946	−0.125	0.172	0.533
2.5	1.599	0.057	0.198	0.164	0.090	0.101	1.599	0.610	−0.002	0.057
3.0	-	-	-	-	-	-	-	-		
Size reduced by 20%										
0.0	14.603	0.936	4.332	3.957	3.036	3.725	30.589	-	6.099	6.347
0.5	13.755	0.782	3.875	3.537	2.745	3.183	27.878	-	5.344	5.588
1.0	12.910	0.630	3.420	3.119	2.457	2.644	25.18	-	4.589	4.862
1.5	12.068	0.479	2.969	2.704	2.169	2.106	22.495	-	3.835	4.160
2.0	11.230	0.329	2.520	2.291	1.883	1.570	19.823	-	3.080	3.480
2.5	10.395	0.180	2.075	1.881	1.599	1.035	17.164	-	2.326	2.834
3.0	9.563	0.032	1.633	1.473	1.315	0.501	14.517	-	1.572	2.176

Note: The numbers with underlines indicate the firms are non-adopters.

Table 10.8 Average price changes by adopter(s)

Unit cost rise	*Canon*	*Fujifilm*	*Kodak*	*Nikon*	*Olympus*	*Sony*
Single adopter						
Size reduced by 10%						
0.0	0.449	0.133	0.246	0.279	0.234	0.547
0.5	1.286	0.946	1.034	1.102	1.055	1.354
1.0	2.124	1.761	1.823	1.925	1.875	2.162
1.5	2.963	2.576	2.613	2.749	2.697	2.971
2.0	3.802	3.392	3.403	3.574	3.520	3.780
2.5	4.642	4.209	4.195	4.399	4.343	4.591
3.0	5.554	5.115	4.992	5.187	5.273	5.343
Size reduced by 20%						
0.0	0.939	0.268	0.504	0.570	0.477	1.260
0.5	1.776	1.082	1.292	1.392	1.298	2.069
1.0	2.614	1.897	2.081	2.215	2.118	2.879
1.5	3.452	2.712	2.871	3.039	2.940	3.689
2.0	4.292	3.528	3.661	3.863	3.763	4.500
2.5	5.132	4.345	4.453	4.688	4.586	5.312
3.0	6.041	5.244	5.246	5.427	5.521	5.685
Multiple adopter						
Size reduced by 10%						
0.0	−0.074	−0.216	−0.146	−0.203	−0.176	−0.118
0.5	0.762	0.559	0.658	0.572	0.608	0.679
1.0	1.599	1.334	1.463	1.346	1.392	1.477
1.5	2.435	2.108	2.268	2.121	2.176	2.274
2	3.486	−0.316	3.173	3.217	3.142	−0.691
2.5	4.642	−0.163	−0.117	−0.264	−0.200	−0.228
3	-	-	-	-	-	-
Size reduced by 20%						
0.0	−0.144	−0.442	−0.305	−0.421	−0.367	−0.156
0.5	0.692	0.332	0.499	0.353	0.416	0.640
1.0	1.528	1.106	1.304	1.127	1.199	1.436
1.5	2.364	1.880	2.108	1.901	1.983	2.232
2.0	3.200	2.654	2.912	2.675	2.766	3.029
2.5	4.036	3.428	3.716	3.449	3.550	3.825
3.0	4.929	4.464	4.567	4.561	4.359	4.152

Note: The numbers with underlines indicate the firms are non-adopters.

price increments are higher for larger brands. Camera prices also go up as the costs of ALMIM cameras rise. For the multiple-adopter cases, ALMIM camera prices initially drop if the size is reduced by 10 per cent but the marginal cost does not change. As the cost rises, prices also increase for all adopters but at a much smaller magnitude relative to single-adopter cases. The predictions indicate that increasing the number of adopters minimizes the market power of adopting firms and facilitates competition.

The potential gain to the ALMIM inventor

As a non-producer, the ALMIM inventor can be rewarded through licensing the new technology to the potential users and extracting part of the profit gains of the camera makers. Thus, the potential benefit to the inventor depends on the profitability of ALMIM and the likelihood of adoption by camera manufacturers. This is further subject to issues like the level of competition intensity, the new technology's superiority against the existing alternatives, as well as how the new technology is transferred (Kamien 1992; Kamien and Tauman 1986). In the case of the digital camera market, the anticipated profit changes to the top six digital camera brands have been simulated for a number of hypothetical scenarios. By examining the profitability of different digital camera makers under various occasions, the ALMIM inventor (or the ALMIM intermediate producer) can choose the number of adopting digital camera firms or the price of ALMIM in order to maximize his gain.

Using the simulation results reported in Tables 10.4 and 10.7 above, we can see the variations in the profits potentially extractable by the ALMIM inventor given different scenarios. In the single-adopter cases, it is shown that for any predetermined quality improvement and cost appreciation, extractable profits increase by selling to a larger brand and diminish as cost appreciates. For instance, if a 10 per cent size reduction can be achieved through ALMIM at no extra marginal production costs, the potential gains from ALMIM range from US$0.934 to US$10.342 million per annum depending on who is the adopter. The gains to all single-adopters decrease if higher replacement costs incur, with the median measure falling to US$1.71 million at a one-dollar cost increment and US$0.63 million at a US$2.5 cost increment. If the marginal cost appreciates by more than US$3.0, then the inventor can hardly benefit from the new technology as no firm is likely to gain through adoption. All reported profit estimates become positive when ALMIM allows for a 20 per cent size reduction and the gains are more than doubled, showing that a higher quality innovation is likely to generate higher benefits to the inventor.

If a multiple-adopter strategy is taken, the inventor can extract up to a total gain of over 15 million dollars when camera sizes can be reduced by 10 per cent without cost shifts. This amount falls rapidly if the marginal cost of ALMIM cameras increases, arriving at approximately US$7 million if the cost is increased by US$1.5 and down to only US$ 3.946 million by US$2.0.

If cost appreciation is too high, e.g. US$3.0 or above, the new technology turns out to be a failure as no firm will be able to afford it. However, when a larger magnitude of size reduction (20 per cent) takes places, the inventor can earn much more by introducing ALMIM to more camera brands than to any single brand.

In sum, the result shows that subject to the quality of the innovation and the extent of the market power of adopting firms, the potential gains to the ALMIM inventor differs subject to how the new technology is introduced. Our results are consistent with the findings in many theoretical papers such as Doganoglu (2004) and Poddar and Sinha (2004), who argue that the choice of optimal licensing depends on the nature and degree of innovation as well as the role of the players. We find that the profits to the inventor rise as the quality of ALMIM is higher and the number of adopters increases, which coincides with the conclusion of the theoretical paper by Doganoglu (2004) when product differentiation is allowed for. With different scenarios covered here, the *ex ante* study provides a guide to both the MIM manufacturing industry and the P&S digital camera industries. As manufacturers have more complete information regarding the exact breakdown costs, appropriate adjustment can be made directly following the estimation results.

Summary

This chapter proposes a tractable approach in making an *ex ante* evaluation for a technological innovation and investigating the optimal way of licensing the new technology by its owner. The approach is applied to evaluating a quality-improving manufacturing technique, the ALMIM process innovation, through its application in the digital camera market. While most present papers on innovation evaluation make *ex post* welfare analysis on the introduction of new goods after they have been marketed for a certain period of time, this study evaluates an emerging technological innovation before it is applied for mass production.

By simulating post-adoption prices and sales of the final marketed products where the new technology is applied to, the study circumvents the difficulties typically involved in and hence discouraging *ex ante* analysis on technological innovations in the empirical IO literature. Using the derived simulation results under various circumstances, we also examine the potential benefits that could be extracted by the owner of the innovation. As existing studies on technology licensing explore the issue using a theoretical framework, this is the first attempt that pursues the optimal way of transferring a technological innovation by its owner through an empirical investigation.

Notes

1 Built on the notable study by Arrow (1962), the literature of technology licensing has been extended in many different directions. Kamien (1992) makes an

excellent survey of earlier literature. For recent studies, see Hernandez-Murillo and Llobet (2006), Kabiraj (2005), Sen and Tauman (2007) and the literature cited there.

2 There are very few empirical studies in this literature, and the present studies are generally descriptive analyses on licensing contracts using historical or survey data; see e.g. Anand and Khanna (2000).

3 See Liu *et al.* (2008).

4 Though not reported here, we simulate for a wider range of cost increment (zero to US$5.00 per unit). By this, we consider the situations with unexpected high cost appreciation and allow for extra costs that may incur for maintaining other attributes of existing cameras in smaller sized camera bodies.

5 It is noted that reducing digital cameras' sizes can affect the products' other characteristics, such as shrinking the size of the LCD screen or limiting the optical zoom range.

6 See Appendix for more detailed explanation on the economic meaning of the metal feature.

7 Given a minimum scale requirement of ALMIM, the adoption of the innovative technique is more likely to benefit larger firms than smaller firms.

8 See Nevo (2000) and Lou (2009) for a detailed explanation of estimation and simulation procedure.

9 Since most digital camera products are at a price range between US$150 and US$300, the income effect is assumed away. Introducing the income effects will make the computation much more complicated. With this assumption, the two Hicksian welfare measures, i.e. the Equivalent Variation (EV) and Compensation Variation (CV) are equal.

10 The mean of sum of cross-elasticities for each brand is 1.62 without accounting for the age of the product. It is reduced by 60 per cent to 0.95 when the age variable is included in the demand model.

Appendix A: implications of a metal-body camera

Consider the case where ALMIM provides digital cameras with an extra 'metal' attribute. As the marginal value for 'Metal Camera' is unknown beforehand, two hypothetical measures (β_{metal}) are supplemented (0.02 and 0.04). The economic meaning can be interpreted through the ratio of coefficients in the demand model. For instance, the ratio $\beta_{resolution}/\beta_{metal}$ can be interpreted as consumers' willingness-to-pay (or tradeoffs) for metal attribute relative to resolution. The estimate for resolution is around 0.4. If β_{metal} were assumed to be 0.04, the ratio ($\beta_{resolution}/\beta_{metal}$) would be 10, indicating consumers are willing to give up 0.1 MP to have the metal body camera. Since the mean camera resolution is over 5.5 MP, the range 0.02–0.04 is considered a conservative estimate.

The changes in producer surplus with only one firm being the innovation adopter are reported in Table 10.A1. It is found that all firms could gain higher profits by adopting ALMIM if cost rises less than US$1.5. Fujifilm and Kodak start getting losses from the innovation if cost rises by US$2, but Canon and Sony would gain higher profits if the marginal cost was to rise by up to US$2.5. No firm will adopt if cost appreciates by more than US$3. Modifying the predictions of metal coefficient to a moderate level of

Table 10.A1 Change in producer surplus with one adopter of 'metal body' cameras (annualized profit changes in millions of dollars)

Unit cost rise	*Adopter*	*Five non-adopters*	*Adopter*	*Five non-adopters*	*Adopter*	*Five non-adopters*	*Adopter*	*Five non-adopters*	*Adopter*	*Five non-adopters*	*Adopter*	*Five non-adopters*
	CANON		*FUJIFILM*		*KODAK*		*NIKON*		*OLYMPUS*		*SONY*	
Marginal value for metal body cameral = 0.02												
0.0	**8.910**	−2.865	**1.064**	−0.638	**3.173**	−1.722	**3.359**	−1.771	**2.994**	−1.600	**6.328**	−2.638
0.5	**7.157**	−2.083	**0.766**	−0.456	**2.305**	−1.253	**2.521**	−1.297	**2.333**	−1.194	**5.051**	−1.960
1.0	**5.413**	−1.316	**0.473**	−0.278	**1.449**	−0.788	**1.701**	−0.824	**1.674**	−0.783	**3.791**	−1.283
1.5	**3.654**	−0.527	**0.174**	−0.082	**0.583**	−0.311	**0.865**	−0.346	**1.009**	−0.365	**2.513**	−0.592
2.0	**1.933**	0.237	**−0.112**	0.082	**−0.257**	0.148	**0.055**	0.124	**0.362**	0.035	**1.268**	0.076
2.5	**0.196**	1.024	**−0.404**	0.270	**−1.109**	0.618	**−0.769**	0.586	**−0.294**	0.453	**0.009**	0.762
3.0	**−1.497**	1.774	**−0.686**	0.437	**−1.930**	1.072	**−1.560**	1.060	**−0.926**	0.843	**−1.218**	1.420
Marginal value for metal body camera = 0.04												
0.0	**17.941**	−5.747	**2.149**	−1.288	6.401	−3.468	**6.777**	−3.568	**6.042**	−3.223	**12.752**	−5.301
0.5	**16.178**	−4.963	**1.848**	−1.101	**5.521**	−2.998	**5.930**	−3.091	**5.371**	−2.813	**11.461**	−4.613
1.0	**14.422**	−4.202	**1.552**	−0.927	**4.662**	−2.531	**5.105**	−2.620	**4.709**	−2.404	**10.194**	−3.946
1.5	**12.630**	−3.393	**1.244**	−0.738	**3.770**	−2.045	**4.240**	−2.124	**4.027**	−1.977	**8.886**	−3.238
2.0	**10.893**	−2.636	**0.956**	−0.560	**2.926**	−1.583	**3.430**	−1.657	**3.374**	−1.573	**7.632**	−2.573
2.5	**9.123**	−1.835	**0.653**	−0.378	**2.048**	−1.092	**2.584**	−1.167	**2.701**	−1.152	**6.341**	−1.871
3.0	**7.422**	−1.088	**0.372**	−0.199	**1.228**	−0.646	**1.789**	−0.710	**2.067**	−0.756	**5.115**	−1.218

Table 10.A2 Change in consumer surplus with one adopter of 'metal body' cameras (in millions of dollars)

Unit cost rise	*The single adopter*					
	Canon	*Fujifilm*	*Kodak*	*Nikon*	*Olympus*	*Sony*
Marginal value for metal = 0.02						
0.0	0.782	0.121	0.366	0.342	0.307	0.605
0.5	0.555	0.078	0.239	0.234	0.217	0.437
1.0	0.333	0.035	0.113	0.128	0.128	0.272
1.5	0.105	−0.007	−0.013	0.020	0.039	0.104
2.0	−0.115	−0.049	−0.137	−0.085	−0.048	−0.060
2.5	−0.340	−0.091	−0.261	−0.191	−0.137	−0.226
3.0	−0.557	−0.132	−0.383	−0.294	−0.222	−0.387
Marginal value for metal = 0.04						
0.0	1.571	0.244	0.737	0.690	0.618	1.216
0.5	1.341	0.201	0.608	0.581	0.527	1.047
1.0	1.120	0.158	0.482	0.473	0.438	0.881
1.5	0.887	0.114	0.354	0.363	0.346	0.710
2.0	0.668	0.072	0.229	0.258	0.259	0.546
2.5	0.438	0.029	0.102	0.149	0.168	0.376
3.0	0.223	−0.012	−0.020	0.046	0.083	0.216

0.04 increases the market power of the adopters and leads to dramatically higher gains to adopters. Higher valuation for the metal attribute by consumers also allows for higher marginal cost appreciation. EV measures for consumer surplus change are listed in Table 10.A2. A browse across the two panels shows clearly that the EV predictions depend largely on how much consumers value the new attribute. Higher benefits would be generated for consumers if more firms adopted ALMIM and the rises of marginal cost were low.

Welfare estimation for the scenario with multiple firms adopting the new technology is illustrated in Table 10.A3. The results exhibit a very similar picture to the comparable cases of size reduction reported in Table 10.7, in that the suppliers take the majority part of the gain from the adoption of ALMIM. Increasing the number of ALMIM adopters also generates higher benefits to consumers.

Table 10.A3 Welfare change due to 'metal body' cameras with multiple adopters

Quality improve	*Change in producer surplus*								*Change in consumer surplus*	
	CANON	*FUJIFILM*	*KODAK*	*NIKON*	*OLYMPUS*	*SONY*	*ADOPTERS' TOTAL*	*NON-ADOPTERS' TOTAL*	*Market mean*	*Market median*
Marginal value for metal=0.02										
0.0	5.182	0.593	1.809	1.886	1.571	3.463	14.503	-	2.507	2.482
0.5	4.348	0.441	1.359	1.470	1.285	2.923	11.826	-	1.752	1.847
1.0	3.517	0.292	0.915	1.061	1.000	2.384	9.169	-	1.005	1.269
1.5	2.686	0.140	0.467	0.646	0.715	1.846	6.498	-	0.242	0.691
2.0	1.833	0.087	0.275	0.167	0.398	1.267	3.665	0.362	**−0.318**	**0.023**
2.5	0.452	0.132	0.427	0.368	0.264	0.311	0.763	1.191	**−0.565**	**−0.408**
3.0	-	-	-	-	-	-	-	-	-	-
Marginal value for metal=0.04										
0.0	10.392	1.191	3.632	3.786	3.151	6.945	29.096	-	4.261	4.273
0.5	9.554	1.037	3.178	3.366	2.864	6.402	26.4	-	5.015	4.976
1.0	8.718	0.888	2.733	2.957	2.578	5.860	23.734	-	3.521	3.684
1.5	7.879	0.732	2.276	2.533	2.289	5.317	21.026	-	2.752	3.128
2.0	7.049	0.585	1.837	2.129	2.006	4.779	18.384	-	2.013	2.550
2.5	6.217	0.432	1.386	1.710	1.719	4.238	15.703	-	1.243	1.972
3.0	6.205	<u>−0.065</u>	<u>−0.207</u>	<u>−3.374</u>	1.618	4.214	12.037	−3.646	0.505	0.816

Note: The numbers with underlines indicate the firms are non-adopters.

Appendix B: major MIM equipment and prices

Device	*Price range*
Planetary or batch mixer with heater	$15,000 – $25,000
Continuous mixer with temperature control	$125,000 – $250,000
30 ton moulder	$25,000 – $60,000
80 ton moulder	$50,000 – $100,000
200 ton moulder	$90,000 – $150,000
Solvent debinding tank (controlled substance)	$400,000
Catalytic debinding furnace	$25,000 – up
Thermal debind furnace	$10,000 – $40,000
Laboratory-scale sintering furnace, batch	$25,000 – $50,000
Pilot-scale batch vacuum furnace	$200,000 – $250,000
Large-scale air furnace	$100,000 – $200,000
Large-scale batch vacuum furnace	$500,000 – $600,000
Continuous furnace (atmosphere)	$400,000 – $1,000,000

Source: Table 8.4 in German (2003, p. 109).

References

Anand, B. N. and Khanna, T. (2000) 'The structure of licensing contracts', *Journal of Industrial Economics*, **48**, 1: 103–35.

Arrow, K. (1962) 'Economic welfare and the allocation of resources for innovation', in R. R. Nelson (ed.) *The Rate and Direction of Inventive Activity*, Princeton, NJ, Princeton University Press: 609–26.

Berry, S., Levinsohn, J. and Pakes, A. (1995) 'Automobile prices in market equilibrium', *Econometrica*, **60**, 4: 889–917.

Bresnahan, T. F. (1986) 'Measuring the spillovers from technical advance: mainframe computers in financial services', *The American Economic Review* **76**, 4: 742–55.

Carranza, J. (2010) 'Product innovation and adoption in market equilibrium: the case of digital cameras', *International Journal of Industrial Organization*, **28**, 6: 604–18.

Doganoglu, T. (2004) 'Licensing of a drastic innovation with product differentiation and optimal number of licenses', unpublished manuscript, University of Munich.

German, R. M. (2003) *Powder Injection Molding: Design & Application*, Innovative Materials Solutions, State College, PA.

Hausman, J. (1997) 'Valuation of new goods under perfect and imperfect competition', in T. F. Bresnahan and J. G. Robert (eds) *The Economics of New Goods*, The University of Chicago Press, Chicago, IL, 207–48.

Hausman, J. and Leonard, G. (2001) 'The competitive effects of a new product introduction: a case study', *Journal of Industrial Economics*, **50**, 3: 237–63.

Hernandez-Murillo, R. and Llobet, G. (2006) 'Patent licensing revisited: heterogeneous firms and product differentiation', *International Journal of Industrial Organization*, **24**, 1: 149–75.

Kabiraj, T. (2005) 'Technology transfer in a stackelberg structure: licensing contracts and welfare', *The Manchester School*, **73**, 1: 1–28.

Kamien, M. (1992) 'Patent licensing', in R. J. Aumann and S. Hart (eds) *Handbook of Game Theory*, Elsevier Science, North-Holland, Amsterdam, 1, 331–54.

Kamien, M. and Tauman, Y. (1986) 'Fees versus royalties and the private value of a patent', *Quarterly Journal of Economics*, **101**, 3: 471–91.

Kirchain, R. E. and Field, F. (2001) 'Process based cost modeling: understanding the economics of technical decisions', *Encyclopedia of Materials Science and Engineering* **2**: 1718–27.

Liu, Z. Y., Sercombe, T. B. and Schaffer, G. B. (2008) 'Metal injection moulding of aluminium alloy 6061 with tin', *Powder Metallurgy* **51**, 1: 78–83.

Lou, W. (2009) 'Demand and supply of new durable goods and the implication of a potential technological innovation', PhD dissertation, La Trobe University.

Lou, W., Prentice, D. and Yin, X. (2012) 'What difference does dynamics make? The case of digital cameras', *International Journal of Industrial Organization* **30**, 1: 30–40.

McFadden, D. (1981) 'Econometric models of probabilistic choice', in Ch. Manski and D. McFadden (eds) *Structural Analysis of Discrete Data with Econometric Applications*, MIT Press, Cambridge, MA.

Nevo, A. (2000) 'A practitioner's guide to estimation of random coefficients logit models of demand', *Journal of Economics & Management Strategy* **9**, 4: 513–48.

Nevo, A. (2003) 'New products, quality changes and welfare measures computed from estimated demand systems', *Review of Economics and Statistics*, **85**, 2: 266–75.

Petrin, A. (2002) 'Quantifying the benefits of new products: the case of the minivan', *Journal of Political Economy*, **110**, 4: 705–29.

Poddar, S. and Sinha, U. B. (2004) 'On patent licensing in spatial competition', *The Economic Record*, **80**, 249: 208–18.

Sen, D. and Tauman, Y. (2007) 'General licensing schemes for a cost-reducing innovation', *Games and Economic Behaviour*, **59**, 1: 163–86.

Small, K. and Rosen, H. (1981) 'Applied welfare economics of discrete choice models', *Econometrica* **49**, 1: 105–30.

Song, I. and Chintagunta, P. (2003) 'A micromodel of new product adoption with heterogeneous and forward-looking consumers: application to the digital camera category', *Quantitative Marketing and Economics* **1**, 4: 371–407.

Trajtenberg, M. (1989) 'The welfare analysis of product innovations: with an application to computed tomography scanners', *Journal of Political Economy*, **97**, 2: 444–79.

Xiao, J. (2008) 'Technological advances in digital cameras: welfare analysis on easy-to-use characteristics', *Marketing Letters* **19**, 2: 171–81.

Zhao, Y. (2007) 'Why are prices falling fast? An empirical study of the US digital camera market', unpublished manuscript, Yale University.

11 Strategic alliances and knowledge flows between firms

Müge Özman and Patrick Llerena

One of the established results in the literature on strategic alliances stresses that firms learn from each other and, as such, alliances are important means through which knowledge flows in industries. The alleged positive relation between alliances and learning has resulted in a vast amount of research carried out to determine the conditions and contexts which facilitate learning processes. In this chapter, we focus on three factors which can be important: the duration, the scope of the alliance and the initial technological overlap between partners. Our results reveal that initial technological overlap has a positive relation with learning, and there is an inverted-U relation between learning and alliance duration.

Introduction

The impact of strategic alliances on organizational learning has been a fruitful field of study during the last 20 years. On the one hand, there exists a significant literature confirming the transfer of knowledge and capabilities through alliances. One the other hand, some studies cast doubt on this alleged positive impact of alliances, highlighting the barriers that firms face in learning from their partners (Hamel, 1991; Crossan and Inkpen, 1994) and underlining the importance of alliance management capabilities (Schilke and Goerzen, 2010). Overall the research reveals that the effectiveness of knowledge flows between partners is highly contingent on the environment and the specific characteristics of firms.

As far as environmental factors are concerned, the central factors which shape knowledge flows can be related with the stability of the industry, the tacitness of the knowledge base and the stage in the industry life cycle. At the same time, alliance management capabilities and strategies (Schilke and Goerzen, 2010), the network positions of partners (Phelps, 2010), prior experience with partners (Gulati *et al.*, 2009) and firms' overall alliance portfolio (Duysters and Lokshin, 2010; Jiang *et al.*, 2010) as well as the balance the firms achieve between exploring and exploiting (Lavie and Rosenkopf, 2006; Hoang and Rothaermel, 2010) are some of the factors which shape the extent to which firms can learn from their partners.

While a variety of factors seem to shape learning processes, the duration of an alliance has been a less studied factor. Most of the studies on alliance duration focus on the determinants of contract duration, rather than the performance effects in the post-alliance phase. The importance of alliance duration mainly stems from the perception of partners in terms of the commitment of their resources. While instability of the industry, rapid technological change and uncertainty can increase the costs of committing resources for long periods (Simonin, 1999), long term relations can also have the benefit of permitting partners to develop shared understandings and a common knowledge base during the execution of the contracts. In addition, the extent to which such long term relations impact the learning from alliances is likely to be moderated by the initial common knowledge base between partners.

The aim of this chapter is to explore these aspects of learning in alliances, through the analysis of 128 technology based alliances which took place between 1984 and 1987, in a variety of industries. In particular, we investigate the extent to which knowledge transfers depend on the duration of the alliance, the scope of the alliance and the initial similarities between partners in terms of their knowledge bases.

The next section presents a theoretical background, as well as our hypotheses. In the third section we explain the data and methodology. And the last section is devoted to some discussions of our results.

Background

Back in the 1980s, a fairly rich literature on strategic alliances focused on the rationales for firms to enter into partnerships in the domestic and global context. The rationales behind alliances were investigated in various ways, both through in depth studies and large scale empirical analysis. In this sense the rationales behind alliances can be explained through a transaction cost perspective, which however falls short of covering strategic behavior of firms, and organizational learning (Kogut, 1989). Organizational learning as a motive behind strategic alliance formation has been since then studied by many researchers, utilizing a wide variety of methodologies and contexts of study. While a large body of literature confirms the knowledge flows between firms through alliances, some cast doubt on the difficulties associated with learning (Hamel, 1991; Crossan and Inkpen, 1994).

During the 1990s, this literature on alliances and organizational learning was to be complemented by the network perspective. Particularly, a leading study is by Powell *et al.* (1996) who emphasized the role of inter firm networks as locus of innovation. They state that firms network with each other not only because they lack resources and need to access others, but because they seek to explore and exploit each other's knowledge bases. In addition, and more importantly, it is the firms' position in this network that has a strong influence on the extent of success. The incorporation of the network approach into the alliance and organizational learning areas proved to yield a very fruitful area of inquiry, resulting in a vast body of work, investigating how network positions relate to learning by firms in the form of exploration and exploitation (Burt, 1992; Ahuja, 2000; Rowley *et al.*, 2000; Gilsing *et al.*, 2008; Phelps, 2010).

According to this literature, network position is related with learning, because it determines firms' access to external knowledge. When relations are embedded in a social context (Granovetter, 1985) where knowledge exchange is frequent and face-to-face, partners can build trust, so that concerns for reputation mitigate possible opportunistic behaviors. In addition, such networks which are rich in social capital facilitate the transfer of tacit knowledge since partners can develop shared meanings and a common understanding. These also can improve efficiency because costs of negotiation are reduced (Uzzi, 1997). Consequently, a clustered network structure facilitates the flow of knowledge (Audretsch and Feldman, 1996; Cowan *et al.*, 2003).

On the other hand, embeddedness can make firms weaker in coping with external shocks and exclude them from knowledge residing outside their own clusters (Uzzi, 1997). Accordingly, network positions which bridge otherwise disconnected clusters can be better for effectiveness of knowledge access (Burt, 1992). Most scholars today agree that the type of network position conducive to performance depends on the context of inquiry, partner specific characteristics and the environment. In this sense, some studies posit that either of the two network positions is beneficial – but not both – or else a hybrid network position is more beneficial (Baum and Rowley, 2008).

Duration of alliances

While network positions can be highly influential in firms' access to knowledge, the duration of these ties also deserves attention. The impact and the determinants of alliance duration on learning have been studied from the lens of different research streams. Transaction cost approaches are predominantly concerned with the design of contracts between firms. In this view, the duration of a contract may depend on the asset specificity (Joskow, 1987), and the extent to which partners are unable to predict future hazards (Simonin, 1999). When the exchange necessitates parties to make relationship specific investments, long term contracting is a more attractive governance mechanism, in the face of unforeseen hazards that may occur during the execution of the contract. In this sense, longer contracts reduce the ambiguity which may be caused by the inability of the partners to foresee the future (Simonin, 1999), and thus protect against opportunistic behaviors. In addition, in high technology industries, where there is rapid innovation, and technologies are complex, long term contracts are better for learning (Pangarkar, 2003).

While these aspects are influential in the initial design phases of a contract, the final impact of duration on the performance of firms has been less studied. On the one hand, long term alliances allow the development of relationship specific skills and capacities which may facilitate knowledge flows between partners. In addition, long term alliances can be better for the transfer of tacit knowledge since partners can have the opportunity to develop shared understandings. In industries where knowledge is highly tacit, a cohesive network structure facilitates learning (Audretsch and Feldman, 1996; Cowan *et al.*, 2003) and that tie age contributes positively to the performance benefits of such closure (Baum *et al.*, 2007). In accordance, our first hypothesis is concerned with the positive impact of duration on learning:

> Hypothesis 1: Alliance duration has a positive impact on learning by partners.

While long term contracts can be preferred when there are risks of opportunistic behaviors, or when there is a high level of asset specificity, or when knowledge is tacit, they can also be costly. Particularly, environmental instability may increase the likelihood of contract termination (Kogut, 1989). Especially in periods of instability, firms value flexibility when committing their resources for long periods. In such environments, deviations from optimal contract incentives significantly raise the cost of long term contracts (Crocker and Masten, 1988). More precisely, longer alliances allow for learning, but the longer the duration, the higher the cost of inflexibilities.

> Hypothesis 2: There is an inverted-U relation between alliance duration and learning.

Technological scope of alliances

In the literature, the scope of collaborations usually refers to the range of different functions that the alliance encompasses (Pisano, 1989; Oxley and Sampson, 2004). There is a fairly rich literature on how the scope of an alliance moderates learning processes. While this definition of scope encompasses different functions, like marketing, R&D, manufacturing and so on, the technological scope of an alliance and its effect on performance has been less studied. Technological scope of an alliance refers to the range of diverse subject matters which the alliance encompasses. The idea here is that, as the width of technological fields expands, so can the potential novelty from the alliance, and also the ambiguity surrounding the subject matter. The relation between technological scope and novelty is based on the Schumpeterian notion of recombination. Recombination refers either to the combination of elements which were previously unconnected, or finding new ways of combining elements which were already associated (Nahapiet and Ghoshal, 1997). One of the building blocks of recombination being variety, it is accepted that the probability of innovation is higher when there is more variety to be recombined (Weitzman, 1998). In addition, radical inventions are more likely to draw upon wider technological fields, than non radical inventions (Schoenmakers and Duysters, 2010).

However, in alliances with a broader scope, partnering firms can face difficulties in specifying the allocation of future property rights. In this sense, broader scope increases the complexity surrounding the alliance, and can result in more frequent changes in the post contract phase (Reuer *et al.*, 2002). In addition, although broader scope can induce firms to invest more in their alliances, it can have lower learning results than alliances with the same function (Amaldoss and Staelin, 2010). As a result, the complexity of alliance scope can augment the motivation to reduce the time span of the contract.

Hypothesis 3a: There is a negative interaction effect between alliance scope and alliance duration as far as their effect on learning is concerned.

However, we argue that the sign of the interaction effect is indeterminate. Broader scope may enhance uncertainties surrounding the contracts (Simonin, 1999; Reuer *et al.*, 2002), since it increases the ambiguity of systems in general (Reed and De Filippi, 1990; Mosakowski, 1997). In this sense when the scope of the alliance is wider, increased duration can yield better learning by permitting firms to have more time in synthesizing complex projects.

Hypothesis 3b: There is a positive interaction effect between alliance scope and alliance duration as far as their effect on learning is concerned.

Knowledge complementarities between partners

The knowledge based theory of the firm is considered to be an extension of the resource based theory, stressing the knowledge of the firm as its key resource. Accordingly, complementarities in knowledge between partners are an important motivation behind strategic alliances (Kogut and Zander, 1992; Grant, 1996). Empirical studies in this tradition usually measure firm complementarities through employing the notion of technological distance between firms. It is possible to argue that initial technological overlap is an important determinant of learning, because it means that firms have a certain initial commonality in their knowledge bases, which increases the extent to which they can learn from each other. Accordingly, we propose that:

> Hypothesis 4: There is a positive relationship between initial technological overlap between partners and their learning from the alliance.

At the same time, this positive effect can diminish as initial overlap increases. Two findings in the literature indicate that, first the likelihood of an alliance between two firms is higher when their distance is at an intermediate level (Mowery *et al.*, 1998). Second, an inverted-U relationship exists between technological distance between firms and their learning (Schoenmakers and Duysters, 2006; Nooteboom *et al.*, 2007; Gilsing *et al.*, 2008). Moreover, this distance diminishes as firms collaborate with each other (Mowery *et al.*, 1998). The underlying logic in this construct is that, when firms are too close in the knowledge space, they have little to add to each other's knowledge; when they are too far, they cannot access the other's knowledge base, and learning is limited. Johnson and Lundvall (1992) get to a very similar conclusion. Accordingly, we propose that:

> Hypothesis 5: There is an inverted-U relationship between initial technological overlap between partners and learning from the alliance.

It is also possible to argue that the initial technological overlap between partners is mediated by the duration of the alliance. First, a high level of initial technological overlap implies that the tacitness of knowledge is less critical in terms of knowledge transfer. The positive impact of duration on learning can be augmented in this case. In addition, with increased overlap between partners, firms can better cope with the risks of future uncertainties.

> Hypothesis 6: There is a positive interaction effect between initial technological overlap between partners and the duration of alliances as far as their impact on learning is concerned.

Figure 11.1 summarizes our hypotheses.

Data and methodology

The primary goal of this chapter is to analyze the knowledge flows between firms through their alliances. For this purpose, we collected the technology based cooperative agreements found in the CATI (Cooperative Agreements and Technology Indicators) database between the years 1984 and 1987. The MERIT CATI relational database covers around 19,000 technology based

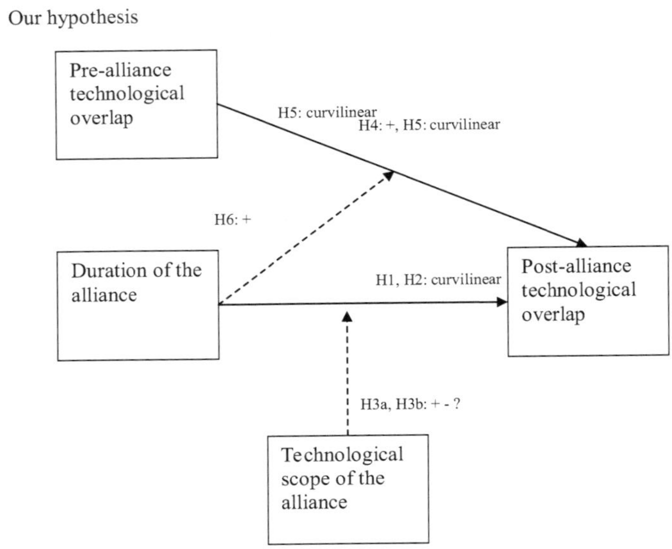

Figure 11.1 Our hypothesis

alliances of nearly 9500 firms. The data have been systematically collected since 1984 (although some of the alliances date back to the end of the 1800s), and it is one of the most widely used databases as far as technology based strategic alliances are concerned. Although data may be incomplete, it is considered as one of the most dependable data sources in this field (Schilling, 2008).

One of the drawbacks of the CATI database is concerned with the duration of alliances, because not all alliances have duration data. Because our main interest is in the longevity of alliances, we included only those alliances in which duration data were available, between 1984 and 1987. Keeping the time period short is important, so as to reduce the impact of time varying factors which might influence learning, capabilities or industry wide determinants of learning. We were further restrained by the availability of data on R&D expenditure for the firms in the sample in the mentioned period, to be used as control variables. We further excluded those firms which had no registered patents either at the beginning of the alliance or at the end. As a result, the total number of dyads in the sample was reduced from 971 to 223. These 223 dyads include a total of 128 technology based alliances,[1] and 143 firms that we include in the regression analysis.

We selected these periods for several reasons. First, these are the years in which alliance activity was very intensive. Second, technological advances in especially Biotechnology and Information Technologies were rather in their early development phases. In this sense, we believe that the *marginal* impact of an additional one year commitment in an alliance could be higher

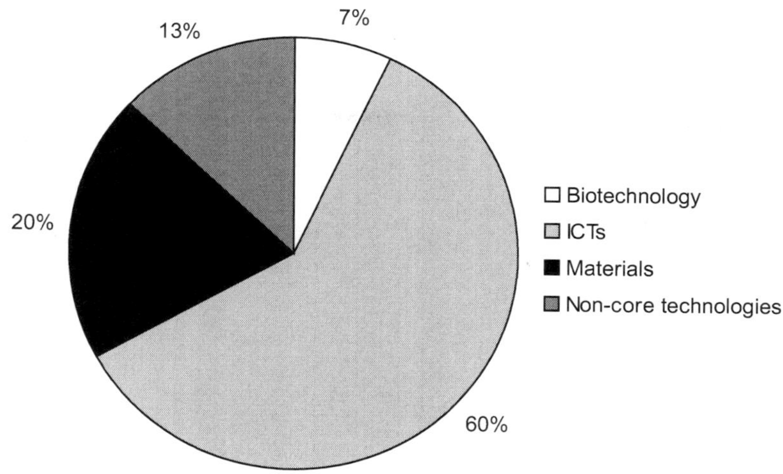

Figure 11.2 Sectoral distribution of alliances in the study

in these early development periods, rather than its effect in more stable phases of the industry. Figures 11.2 and 11.3 shows the breakdown of alliances with respect to sectors and continents.

To explore the learning aspects of alliances, we collected all the patents of the firms in the sample from the EPO PATSTAT database. Thus, we gathered variables that measure the firm's patent portfolio such as number of patents and their citations.

For the regression analysis, we had to keep the period limited (1984–1987), but we also carried out some exploratory analysis covering a broader time horizon, to see the general trends with respect to the variables that we are interested in. For this purpose, Figure 11.4 shows the relationship between learning from the alliance and the duration for all the alliances in which we had the data on duration, between the years 1965 and 1990. Learning is here taken as the difference between initial and final technological overlap between firms (see the section "Dependent variable" below for details on calculation).

As the figure reveals, the extent of negative and zero learning diminishes as the duration of alliance increases. In other words, positive learning is more likely the longer the duration of the alliance. However, it is difficult to draw robust conclusions because the number of observations is not homogeneous with respect to years.

Dependent variable

As a proxy for learning, we used the cross citation index that is commonly found in the literature (Mowery *et al.*, 1998; Schoenmakers and Duysters,

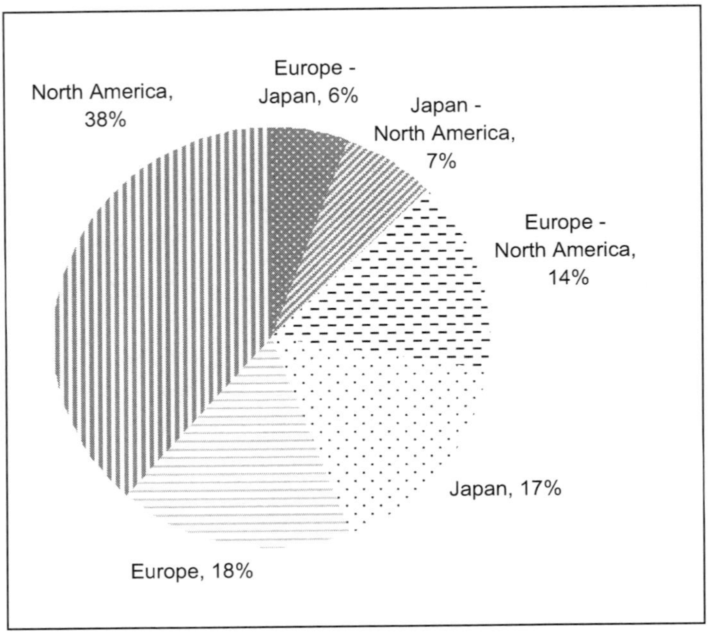

Figure 11.3 Country distribution of the dyads covered in the study

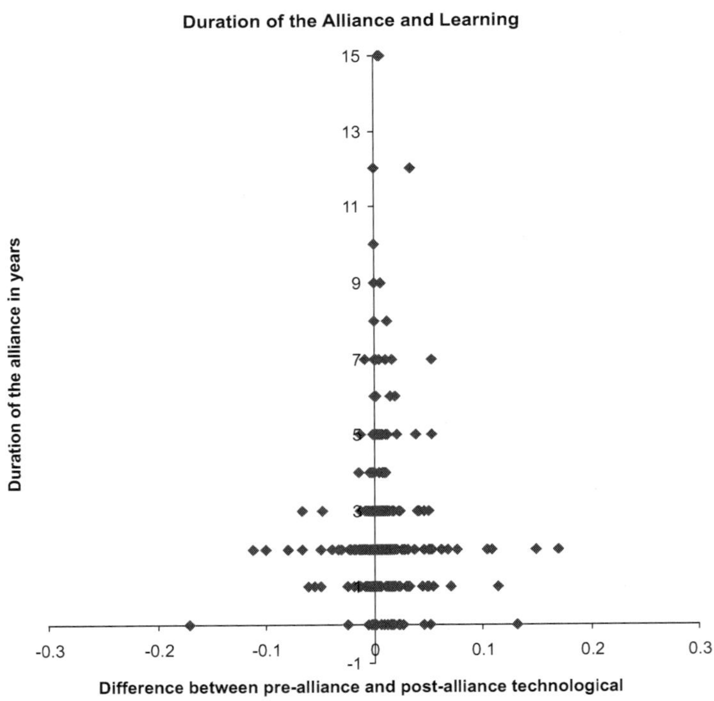

Figure 11.4 Duration and learning

2006; Rothaermel and Boeker, 2008). In particular the cross citation index between firms i and j is given by

$$common\ citation\ index = \frac{citations\ in\ firm\ i\text{'s}\ patents\ to\ firm\ j\text{'s}\ patents}{total\ number\ of\ citations\ in\ firm\ i\text{'s}\ patents} + \frac{citations\ in\ firm\ j\text{'s}\ patents\ to\ firm\ i\text{'s}\ patents}{total\ number\ of\ citations\ in\ firm\ j\text{'s}\ patents}$$

This ratio is calculated taking into account the patents of all the firms in the study. Two sets of patents were collected for each firm in a dyad. First, we collected those patents that the firms applied for and were granted in the five years preceding the alliance beginning (Mowery *et al.*, 1998). In addition, we collected the patents that were applied for in the five years after the end of the alliance. Therefore, we took into account the duration of each alliance separately in data collection. In this way, we obtained approximately 86,000 patents. The difference between the common citation rates will give the relative rise – or fall thereof – of the cross citation rates before and after the alliance. In other words, cross citation rate gives a proxy for the extent to which firms move towards or away from each other in the technology space. The dependent variable used in the regressions is the post cross citation rate, which is the cross citation rate covering the five years after the alliance was ended.

Independent variables

We used a range of independent variables in the models. The duration of the alliance (DUR) is given as the difference between the beginning and the ending of the alliance. The scope of the alliance (SCOPE) is a simple sum of the different technology fields (core and non-core technologies) that the alliance encompasses. The initial technological overlap (OVERLAP_B) is the cross citation rate between the two firms forming the dyad, spanning the five years before the year of the alliance agreement.

In addition we used a range of control variables to account for the total number of patents of the firms in the beginning of the alliance and at the end of the alliance, the average annual sales ratio of the firms in the dyad, as well as the ratio of their average annual R&D intensity. We include dummy variables, for distinguishing between the main technology field of the alliance and to control for whether the two companies are from the same country. Table 11.1 provides some descriptive statistics of the data.

The model utilized is as follows:

OVERLAP_E = f (OVERLAP_B, SCOPE, DUR, Country, Biotech, IT, Materials, R&D, TotalPatents)

Table 11.1 Descriptive statistics

Variable	*Description*	*Mean*	*SD*	*Max*	*Min*
OVERLAP_E	*Technological overlap at the end of the alliance*	0.0097	0.0188	0.109	0
OVERLAP_B	*Technological overlap at the beginning of the alliance*	0.0050	0.0138	0.0952	0
OVERLAP_B SQ	*Overlap at beginning squared*	0.0002	0.0009	0.0091	0
DUR	*Duration of the alliance*	2.0090	0.8718	5	0
DURSQ	*Duration squared*	4.7928	4.3326	25	0
SCOPE	*Scope of the alliance*	2.8198	0.6753	6	2
Total Patents Beg	*Total number of patents of two firms in the period of five years before the alliance*	444.937	573.999	2448	0
Total Patents End	*Total number of patents of two firms in the period of five years after the alliance*	897.599	1083.217	4991	6
Country	*Dummy indicating whether country of origin of two companies are the same*	0.5946	0.4921	1	0
BIOTECH	*Dummy indicating biotechnology field*	0.0691	0.2436	1	0
IT	*Dummy indicating information technology field*	0.5991	0.4912	1	0
MATERIALS	*Dummy indicating materials technology field*	0.199	0.400	1	0
R&D	*R&D intensity ratio between two firms*	0.529	0.278	0.999	0
Sales	*Sales ratio between two firms*	0.275	0.287	0.983	0

Results

The regression results of models 2, 3, 4 and 5 reveal first that there is an inverted-U relationship between alliance duration and final technological overlap between partners. This corroborates our hypotheses 1 and 2. In other words, while the duration of the alliance can have a positive impact on learning, this positive effect diminishes for longer terms.

The regression models highlight no significant interaction effect between alliance duration and scope, rejecting hypotheses 3a and 3b. Hypothesis 6 is also rejected, which tests for the positive interaction effect between initial technological overlap and duration of the alliance.

As far as the effect of initial technological overlap on final overlap is concerned, we do not detect a significant curvilinear relation. Rather, the models 1–5 show that initial technological overlaps between partners have a positive effect on learning, which corroborates Hypothesis 4. There is no diminishing effect on learning as partners are closer to each other at the beginning of an alliance. As a result, Hypothesis 5 is not confirmed.

All the models show that the R&D ratio between firms has a negative and significant impact on learning. In other words, the more asymmetric firms are in terms of their R&D investments, the less they learn from each other as a result of the alliance. In addition, the models detect a significant effect neither for sector nor country variables. One of the interesting results is concerned with the significant positive impact of initial number of patents, and the final learning levels. This result reflects the importance of initial technological capabilities of firms in learning processes.

Discussion and prospects for future research

Learning from alliances depends on range factors, as a very rich literature reveals. In this chapter, we focused on the impact of duration, technological scope and the extent of overlap in the partners' knowledge profiles. One of our results reveals an inverted-U relationship between the duration of alliances and the final technological overlap. This can partly explain the discussions in the literature with respect to the impact of duration on learning. While longer term alliances can be preferable in terms of partners' opportunities to build shared meanings as far as complex tasks are concerned, they are also disadvantageous when technology evolves fast and along uncertain paths. In the latter case, firms may prefer not to commit their resources for long periods. Indeed, these factors seem to suggest an inverted-U shaped function between duration and performance, which our data also confirms. According to our results, there is no significant interaction effect between duration and initial technological overlap, and neither is there an interaction effect between alliance scope and duration. However, this might also be related with the measurement of technological scope. Coverage of a wide range of fields in the alliance may not imply that the alliance is more

Table 11.2 Some results of the OLS regressions

	Dependent variable: OVERLAPE				
	Model 1	*Model 2*	*Model 3*	*Model 4*	*Model 5*
Intercept	−0.007	−0.012	−0.013	−0.013	−0.020
	−0.392	*−1.026*	*−1.103*	*−1.118*	*−1.156*
Overlap B	0.446	0.428	0.453	0.234	0.234
	*1.95**	*1.91***	*1.86***	*2.76***	*2.74***
Overlap B SQ	−3.185	−2.994	−2.808		
	−0.974	*−0.934*	*−0.882*		
Scope	0.003				0.002
	0.282				*0.558*
Scope SQ	0.000				
	−0.108				
Duration		0.011	0.012	0.012	0.013
		*2.35***	*2.37***	*2.37***	*2.12***
Duration SQ		−0.003	−0.003	−0.003	−0.003
		*−2.57***	*−2.60***	*−2.59***	*−2.54***
Duration * Overlap B			−0.022		
			−0.193		
Duration * Scope					−0.0001
					−0.387
Country	0.002	0.001	0.002	0.002	0.002
	0.835	*0.489*	*0.637*	*0.658*	*0.818*
R&D	−0.007	−0.006	−0.006	−0.006	−0.006
	−1.627	*−1.71**	*−1.78**	*−1.83**	*−1.69**
Total Patents B	0.000	0.000	0.000	0.000	0.000
	*1.98**	*1.79**	*1.76**	*1.92**	*1.99**
Total Patents E	0.000	0.000	0.000	0.000	0.000
	1.224	*1.568*	*1.610*	*1.69**	*1.570*
Biotechnology	0.011	0.012	0.012	0.012	0.012
	1.010	*1.163*	*1.135*	*1.180*	*1.157*
Information Tech	0.007	0.007	0.007	0.007	0.007
	0.780	*0.734*	*0.713*	*0.729*	*0.717*
Materials	0.005	0.003	0.003	0.003	0.003
	0.525	*0.275*	*0.265*	*0.295*	*0.328*
n	209	209	209	211	209
F	7.44	8.15	8.12	9.72	7.07

t-*tests in italics*

complex in terms of its knowledge content. One of the questions that can be addressed in the future is whether these results are valid also in relatively mature industries.

Last but not least, a brief note on data and methodology. Although utilizing alliance and patent data can give insight into broadly observed regularities, and they are commonly used in the innovation literature, we should be careful in interpreting the results. Hamel (1991: 84), in exploring the measurement problems related to learning from alliances states that:

> Because patterns of causality are extremely complex in most real-world administrative systems, traditional deductive-analytic methodologies force the researcher to declutter the phenomenon by: (1) substituting crude proxies for difficult-to-measure determinants or outcomes; (2) assuming away some of the multidimensionality in causal relationships; and/or (3) narrowing the scope of research. In doing so, much of the potential value of the research is lost. The problem is not that the resulting theories are under-tested (i.e. they fail a test of rigor), but that they are under-developed (i.e. they are so partial in coverage that they illuminate only a fragment of the path between choice, action and outcome).

In this sense, our study should be taken as exploratory, in the sense of highlighting a possible impact of duration, which seems intuitively plausible. However, generalizations are always difficult. Case studies can be a useful complementary approach to validate findings from large databases, and our study suggests that duration and technological scope of alliances can play a role in the learning processes, a relationship which deserves investigating with individual case studies as well.

Notes

1 CATI includes a variety of alliance types. In this study we took into account only the technology based alliances, and we excluded equity based alliances.

References

Ahuja, G. (2000) "Collaboration Networks, Structural Holes and Innovation: A Longitudinal Study," *Administrative Science Quarterly*, **45**(3): 425–53.

Amaldoss, W. and Staelin, R. (2010) "Cross-Function and Same-Function Alliances: How Does Alliance Structure Affect the Behavior of Partnering Firms?" *Management Science*, **56**(2): 302–17.

Audretsch, D. and Feldman, M. (1996) "R&D Spillovers and the Geography of Innovation and Production," *American Economic Review*, **86**(3): 630–40.

Baum, J. and Rowley, T. (2008) *Network Strategy*, Greenwich, CT: JAI Press.

Baum, J., McEvily, B. and Rowley, T. (2007) "Better With Age? Tie Longevity and the Performance Implications of Bridging and Closure," *Organization Science*, **23**(2): 529–46.

Burt, R.S. (1992) *Structural Holes: The Structure of Competition*, New York: Academic Press.

Cowan, R., Jonard, N. and Ozman, M. (2003) "Knowledge Dynamics in a Network Industry," *Technological Forecasting and Social Change*, **71**(5): 469–84.

Crocker, K.J. and Masten, S.E. (1988) "Mitigating Contractual Hazards: Unilateral Options and Contract Length," *RAND Journal of Economics*, **19**(3): 327–43.

Crossan, M. and Inkpen, A. (1994) "Promise and Reality of Learning Through Alliances," *Thunderbird International Business Review*, **36**(3): 263–73.

Duysters, G. and Lokshin, B. (2010) "Determinants of Alliance Portfolio Complexity and Its Effect on Innovative Performance of Companies," *The Journal of Product Innovation Management*, **28**(4): 570–85.

Gilsing, V., Nooteboom, B., Vanhaverbeke, W., Duysters, G. and van de Oord, A. (2008) "Network Embeddedness and the Exploration of Novel Technologies: Technological Distance, Betweenness Centrality and Density," *Research Policy*, **37**: 1717–31.

Granovetter, M. (1985) "Economic Action and Social Structure: The Problem of Embeddedness," *American Journal of Sociology*, **91**: 481–93.

Grant, R.M. (1996) "Towards a Knowledge-based Theory of the Firm," *Strategic Management Journal*, **17**(7): 109–22.

Gulati, R., Lavie, D. and Singh, H. (2009) "The Nature of Partnering Experience and Gains from Alliances," *Strategic Management Journal*, **30**: 1213–33.

Hamel, G. (1991) "Competition for Competence and Inter-Partner Learning within International Strategic Alliances," *Strategic Management Journal* **12**: 83–103.

Hoang, H. and Rothaermel, F.T. (2010) "Leveraging Internal and External Experience: Exploration, Exploitation, and R&D Project Performance," *Strategic Management Journal*, **31**(7): 734–58.

Jiang, R., Tao, T. and Santoro, M. (2010) "Alliance Portfolio Diversity and Firm Performance," *Strategic Management Journal*, **31**: 1136–44.

Johnson, B. and Lundvall, B.-Å. (1992) "Closing the Institutional Gap," *Revue d'Economie Industrielle*, **59**: 111–23.

Joskow, P. (1987) "Contract Duration and Relationship-Specific Investments: Empirical Evidence from Coal Markets," *The American Economic Review*, **77**(1): 168–85.

Kogut, B. (1989) "The Stability of Joint Ventures: Reciprocity and Competitive Rivalry," *The Journal of Industrial Economics*, **38**: 183–98.

Kogut, B. and Zander, U. (1992) "Knowledge of the Firm and the Evolutionary Theory of the Multinational Corporation," *Journal of International Business Studies*, **24**(4): 625–45.

Lavie, D. and Rosenkopf, L. (2006) "Balancing Exploration and Exploitation in Alliance Formation." *Academy of Management Journal*, **49**: 797–818.

Mosakowski, E. (1997) "Strategy Making under Causal Ambiguity: Conceptual Issues and Empirical Evidence," *Organization Science*, **8**(4): 414–42.

Mowery, D.C., Oxley, J.E. and Silverman, B.S. (1998) "Technological Overlap and Interfirm Cooperation: Implications for the Resource Based View of the Firm," *Research Policy*, **27**: 507–23.

Nahapiet, J. and Ghoshal, S. (1997) "Social Capital, Intellectual Capital and Creation of Value in Firms," *Academic of Management Best Paper Proceedings*, **11**(3): 35–9.

Nooteboom, B., Vanhaverbeke, W., Duysters, G.M., Gilsing, V.A. and van den Oord, A. (2007) "Optimal Cognitive Distance and Absorptive Capacity," *Research Policy*, **367**: 1016–34.

Oxley, J.E. and Sampson, R.C. (2004) "The Scope and Governance of International R&D Alliances," *Strategic Management Journal*, **25**: 723–49.

Pangarkar, N. (2003) "Determinants of Alliance Duration in Uncertain Environments: The Case of the Biotechnology Sector," *Long Range Planning*, **36**: 269–84.

Phelps, C. (2010) "A Longitudinal Study of the Influence of Alliance Network Structure and Composition on Firm Exploratory Innovation," *Academy of Management Journal*, **53**: 890–913.

Pisano, G.P. (1989) "Using Equity Participation to Support Exchange: Evidence from the Biotechnology Industry," *Journal of Law, Economics and Organization*, **5**(1): 109–26.

Powell, W.W., Koput, K.W. and Smith-Doerr, L. (1996) "Inter-organizational Collaboration and the Locus of Innovation: Networks of Learning in Biotechnology," *Administrative Science Quarterly*, **41**: 116–45.

Reed, R. and DeFillippi, R.J. (1990) "Causal Ambiguity, Barriers to Imitation, and Sustainable Competitive Advantage," *Academy of Management Review*, **15**: 88–102.

Reuer, J.J., Zollo, M. and Singh, H. (2002) "Post-Formation Dynamics in Strategic Alliances," *Strategic Management Journal*, **23**: 135–51.

Rothaermel, F.T. and Boeker, W. (2008) "Old Technology Meets New Technology: Complementarities, Similarities, and Alliance Formation," *Strategic Management Journal*, **29**(1): 47–77.

Rowley, T., Behrens, D. and Krackhardt, D. (2000) "Redundant Governance Structures: An Analysis of Structural and Relational Embeddedness in the Steel and Semiconductor Industries," *Strategic Management Journal*, **21**: 369–86.

Schilke, O. and Goerzen, A. (2010) "Alliance Management Capability: An Investigation of the Construct and Its Measurement," *Journal of Management*, **36**(5): 1192–1219.

Schilling, M.A. (2008) "Understanding the Alliance Data," *Strategic Management Journal*, **30**(3): 233–60.

Schoenmakers, W. and Duysters, G. (2006) "Learning in Strategic Technology Alliances," *Technology Analysis and Strategic Management*, **18**(2): 245–64.

Schoenmakers, W. and Duysters, G. (2010) "The Technological Origins of Radical Inventions," *Research Policy*, **39**(8): 1051–59.

Simonin, B. (1999) "Ambiguity and the Process of Knowledge Transfer in Strategic Alliances," *Strategic Management Journal*, **20**: 595–623.

Uzzi, B. (1997) "Social Structure and Competition in Interfirm Networks: The Paradox of Embeddedness," *Administrative Science Quarterly*, **42**(1): 35–67.

Weitzman, M.L. (1998) "Recombinant Growth," *Quarterly Journal of Economics*, **113**: 331–60.

12 Deal values for acquisitions by larger incumbents in high-technology industries

The influence of non-linearities and moderation effects

Marcus Wagner

Introduction

Innovation activities in high-technology industries such as semiconductors, biotechnology or electronic design automation (EDA) are diverse and driven by a variety of determinants. This, together with the cumulative and rapid nature of innovation in those industries, makes innovation management a very challenging task (Yoshikawa, 2003). For example, in the semiconductor industry, levels of research and development (R&D) input are strongly affected by the highly cyclical nature of the industry, which experienced its most severe downturn in 2000 and 2001. R&D expenditure dropped significantly in this period and has not recovered so far. In parallel, there is also evidence of innovation-related acquisition activity in the industry (Bloningen and Taylor, 2000; Sanchanta, 2007).

Another example of a challenging environment for innovation management is the EDA industry, a particular sub-segment of the semiconductor industry focused on chip design and covering a number of complex processes ranging from logical layout to device testing. As far as the supply side is concerned, EDA has a simple market structure with three large firms dominating and a number of small firms being active, the latter being frequently acquired by larger firms in the industry. Increasingly, the products of the electronic design automation industry also get integrated into manufacturing processes in order to enable direct feedback from the production to the design stage in turn making innovation processes and their management even more challenging.

Finally, looking at biotechnology as the third sector mentioned above, it can also be characterised by strengthened patent rights, especially in the United States (Hall and Ziedonis, 2001), and rapid technological change for cumulative technologies. In this sector again acquisitions are a frequent phenomenon as are intensive collaboration and cooperation activities (Pangarkar, 2003; Rothaermel and Deeds, 2004; Jack, 2006, 2007; Hofmann, 2007) which are also relevant for the trend towards open innovation and the knowledge-based view in general (Cohendet *et al*., 2000; Amin and Cohendet, 2004; Cohendet and Simon, 2007). The purpose of this chapter is to analyse these three industries in more detail and the determinants of acquisition value in terms of R&D characteristics of acquirers with a specific focus on non-linear relationships, interactions and variations across sectors. The remainder of the chapter first derives hypotheses and then tests these empirically for a set of larger firms in the three high-technology industries described earlier to better understand what determines acquisition values of innovative or entrepreneurial start-ups from an acquirer's perspective. Finally, the results are discussed and more general managerial implications are derived.

Literature and hypotheses

Classical themes of technology and innovation management such as the concepts of architectural and competence-destroying innovation and the distinction between radical and incremental innovation have particular relevance in high-technology industries (Cohendet *et al*., 2000). Issues at stake for example are whether small start-ups are largely responsible for incremental innovation or whether radical innovation requires external technology sourcing (as was observed in the past in high-technology industries). This chapter analyses these aspects empirically for acquisitions as one specific form of external technology sourcing and tests hypotheses derived from theories rooted in new institutional economics and transaction cost economics (Williamson, 1975). For example, Sangiovanni-Vincentelli (2003) argues that in the EDA industry, in-sourcing of outside technology may be needed to bring about radical innovation there and Pangarkar and Klein (2001) and Zhang *et al*. (2007) point to the relevance of external technology sourcing in the biotechnology industry. Absorptive capacity can increase the value that an acquirer can derive from buying a target firm to source technology externally (Cohen and Levinthal, 1990), because it supports learning from the target (Levinthal and March, 1993). Therefore, individual deal values and ultimately the average annual value of acquisitions made should increase with increasing absorptive capacity (usually operationalised as R&D intensity, i.e. the ratio of R&D expenditure to sales). These considerations lead to the following hypothesis:

H1: The average annual monetary value of acquisitions of acquired firms is positively associated with the R&D intensity of acquiring firms.

Another stream of scholarly work of relevance, namely the economic theory of acquisitions and the numerous empirical studies related to it (e.g. Nesta and Saviotti, 2006) suggests a qualification of the effect of R&D intensity (Milgrom and Roberts, 1992). Economic theory has proposed a number of reasons for acquisitions (Morris and Hay, 1991). For example, under the assumption that the stock market is efficient, motives for acquisitions can be increased market power, reduced advertising and other promotional expenditure or efficiency gains which could not be realised without the acquisition. Other explanations proposed for acquisitions are managerial or allocational acquisitions or conglomerate mergers aimed at risk reduction (Morris and Hay, 1991). Here, 'allocational' refers to acquisitions which are motivated by a reallocation of assets to managers who make more efficient uses of them. This situation is possible if the assumption of fully efficient firms is no longer made. To address this effect, Belderbos *et al.* (2004) include the square of R&D intensity in a similar research context. Absorptive capacity in their theoretical conceptualisation can have a negative effect on deal value because at very high levels, acquirers derive most of the benefits from an acquisition through their own activities. Therefore, the following hypothesis is proposed:

> *H2: The average annual monetary value of acquisitions of acquired firms is negatively associated with the square of the R&D intensity of acquiring firms.*

The previous considerations on acquisitions show that these can be one efficient means for firms to keep the level of their innovative activities high by acquiring successful start-ups in the industry. However, innovative output (e.g. measured in terms of patents granted) is another important characteristic of acquiring firms that potentially determines individual deal values and hence ultimately the average annual monetary value of acquisitions (Griliches, 1981; Bloom and van Reenen, 2002). Desyllas and Hughes (2008) analyse the association of R&D and patenting with acquisitions in a sample of broadly defined high-technology industries. They find that decreasing returns from exploiting a firm's existing knowledge base and the choice between doing or buying R&D are main drivers for the acquisition of innovative firms. If a firm has more patents, relative to its size or economic activity (i.e. a higher patenting intensity), then it is in a better position to appropriate the benefits from an acquisition (Teece, 1986; Hurmelinna-Laukkanen, 2009). A further theoretical justification for this is that if firms have very high innovation outputs, then this in itself can result in a significantly heightened potential from product or other complementarities that should have a positive effect on individual deal values and hence also on the average annual monetary value of acquisitions. This leads to the following hypothesis:

> *H3: The average annual monetary value of acquisitions of acquired firms is positively associated with the patenting intensity of acquiring firms.*

Finally, the interaction between innovation efforts (as proxied by R&D intensity) and innovation output (as measured by means of patenting intensity) needs to be considered in its effect on the value (or price) and cost of acquisition. This can be related to the literature on the obstacles to innovation in larger firms. Henderson (1993) and Hauschildt (1999) provide examples for situations in which firms may not be able to carry out specific types of innovation. One response of firms to not being able to carry out an innovation at acceptable cost can be the acquisition of start-ups in order to make up for their missing capabilities (e.g. Markides and Geroski, 2005). The better the conversion of innovation efforts into output works in an acquiring firm, the less likely it is to derive value from an acquisition. It is mainly those firms that are either weak in inputs or outputs or both, that derive value from an acquisition. Therefore the interaction of R&D intensity and patenting intensity matters as an additional explanatory variable for the average annual value of acquisitions and hence the following hypothesis can be proposed:

H4: The average annual monetary value of acquisitions of acquired firms is negatively associated with the interaction of patenting intensity and R&D intensity of acquiring firms.

Data and method

To address the hypotheses developed in the previous section, a combination of qualitative and quantitative methods was pursued. Semi-structured qualitative interviews on the hypotheses were carried out with industry experts and staff from small and large firms in the three industries analysed. This approach, which involved interview guidelines, is advocated methodologically (e.g. Miles and Huberman, 1994). In addition to the aspects addressed in the last section, questions also related to the reasons of acquisitions and the relevance of innovation aspects for acquisitions. Interview content was transcribed and the written material was then coded inductively in order to derive categories using the Nvivo software to provide qualitative insights on acquisitions in the industries studied. This qualitative analysis was complemented by a quantitative analysis to formally test the above hypotheses.

Data for this quantitative analysis were collected at the SFB (Sonderforschungsbereich) 541 of Humboldt University Berlin (Germany) from the SDC Platinum and Worldscope Disclosure databases. They comprise the largest firms in the three industries under study during the period 1982–2004. In order to compare them, all firms making up 80 per cent of the industry by sales value were included in the analysis, resulting in 14 EDA, 50 semiconductor and 21 biotechnology firms being analysed. Data were collected on a number of variables concerning various firm characteristics (leverage, sales, R&D expenditure, liquidity, patents granted and firm location). Patents of a large firm as measure of innovation output take time to issue. Since this chapter follows the convention to assign patents granted to the application year, only data up to and including 2002 can be used. Furthermore, it was not possible to use operating margin and cash flow as measures for profitability, since these were empirically highly correlated with R&D intensity. Therefore, sales growth was used as a joint proxy for profitability and industry potential. Due to missing data on individual variables, the final sample used in the econometric estimation was further reduced. Especially, values were not disclosed for all deals. Table 12.A1 in the Appendix summarises descriptive statistics and correlations of all explanatory variables. Time dummies are included in all models, which partly accounts also for supply-related factors (to the degree these are time-related, e.g. recession phases).

Random and fixed effects panel models are used to estimate effects on the dependent variable. The difference between the fixed effects and the random effects model is based on whether the time-invariant effects are correlated with the regressors (which is the case for the fixed effects) or (in case of the random effects model) not (Johnston and DiNardo, 1997). For these models, the specification is:

$$u_{it} = c_i + \varepsilon_{it}. \tag{1}$$

$$s_{it} = \alpha + \beta' \mathrm{X}_{it} + c_i + e_{it}, \tag{2}$$

where $i = 1, \ldots,$ N units under observation and $t = 1, \ldots, T$ time periods for which data were collected. s_{it} denotes the dependent variable, that is, the average annual monetary value of the acquisitions for firm i in period t, $\mathbf{X}_{it}$ represents a set of independent variables, β' a vector of coefficients, c_i the unobserved heterogeneity of firm i and e_{it} an idiosyncratic error that satisfies $E[e_{it}|\mathbf{X}_{it}, c_i] = 0$. The model is estimated through GLS assuming no correlation between e_{it} and c_i. For the fixed effects model, other than the random effects model, the assumption is that the individual effect c_i is correlated with the time-variant independent variables $\mathbf{X}_{it}$. This means that although the basic specification given in (1) and (2) remains, the interpretation differs, in that the disturbance c_i is a constant (and thus represented by a dummy variable) for each unit of analysis, i.e. here for each specific firm. The fact that the disturbance is a constant in the fixed effects model implies that all time-invariant variables are dropped during the estimation. To decide which of the variants is better fitting, the Hausman test is involved. If the Hausman test statistic is significant, then the fixed effects model is more appropriate.

Results

Results are first stated for a pooled sample of firms in all three industries. In the following, the models are then estimated for each industry separately to better gauge variation between industries and to at least partly account for supply-demand interactions, i.e. the notion of an acquisition 'equilibrium'. In each case, a base model without and an extended model including non-linear and interaction effects are estimated.

Results for the pooled sample

As can be seen from Table 12.1 reporting the results of the base model, the average value of acquisitions in general is comparatively more strongly related to the size of the firm than to innovation-related variables. The joint effect of the industry dummies included in the model reported in the right column of Table 12.1 is found to be non-significant ($Chi^2 = 2.73$, p-value > $Chi^2 = 0.2551$), which suggests that as far as the variation of average acquisition is concerned value pooling is acceptable given the relative homogeneity of the three sectors included in the analysis with regard to this variable. The joint effect of year dummies is significant which also justifies their inclusion *ex post* and advocates that they are also retained in the industry-specific analyses.

As can be seen from Table 12.2 reporting the extended model, the average value of acquisitions in general is comparatively more strongly connected to innovation-related variables when accounting for interaction effects and non-linearity. More specifically, patenting intensity is positively and the interaction of patenting with R&D intensity negatively associated with acquisition value, which is consistent with both H3 and H4. The next three sections summarise the results for each industry separately.

Results for the EDA industry

Tables 12.3 and 12.4 summarise the results in the EDA industry. Concerning the factors associated with the average annual value of the acquisitions made by firms in the EDA industry, Table 12.3 shows that sales have a significant positive association with the value of acquisitions, whereas if a firm is headquartered in Asia they have a significant negative association. The year dummies are found to be jointly significant.

Next to the base model, the results for the extended model in the EDA industry are reported in Table 12.4. As can be seen from Table 12.4, while the model is overall significant, only the year dummies are significant, suggesting heterogeneity within the EDA industry, that makes identification of individual predictors in the extended model difficult.

Table 12.1 Base model for all sectors jointly, dependent variable: average annual value of acquisitions (mn €)

Variables	*Estimates*	*Estimates*
Financial leverage (total assets to total equity)	−1.2449 (0.9357)	−1.1944 (0.9392)
Current ratio (current assets to current liabilities)	1.7046 (1.3939)	0.7515 (1.5746)
Sales growth (% over previous year)	0.0221 (0.0201)	0.0216 (0.0205)
Sales (natural logarithm of net sales in 1000 €)	16.8670 (7.3102)**	13.7522 (5.7863)**
R&D intensity (R&D expenditure to net sales in %)	0.0099 (0.0161)	−0.0022) (0.0181)
Patenting intensity (Patents granted by application year to net sales)	0.1367 (0.1914)	0.1422 (0.1938)
Company headquartered in Europe (dummy; 1 = yes; base category: United States)	−23.9458 (40.1074)	−33.4676 (35.6073)
Company headquartered in Asia (dummy; 1 = yes; base category: United States)	146.341 (139.0906)*	147.0505 (139.2865)
DJS industry Technology (dummy; 1 = yes; base category: Other)	-	−53.0470 (49.3511)
DJS industry Healthcare (dummy; 1 = yes; base category: Other)	-	−20.6551 (56.5757)
Constant	−228.6962 (83.1652)***	−160.6726 (101.4076)
R^2 within	0.0336	0.0332 0.0890
R^2 between	0.0793	0.0482
R^2 overall	0.0470	
No. of observations (No. of groups)	610 (81)	610 (81)
F	34.13***	32.69***
Hausman specification test Chi²	4.42	6.43

Notes: Significance levels: * $p < 0.1$; ** $p < 0.05$; *** $p < 0.01$; heteroskedasticity-robust standard errors in parentheses; unbalanced panel data with observations per group: min = 1; max = 15; average = 7.5; random firm and year effects included; test for joint significance of year dummies: Chi² = 10.79, p-value > Chi² = 0.7671 (model without industry dummies); Chi² = 32.84, p-value > Chi² = 0.0049 (model with industry dummies).

Results for the biotechnology industry

As concerns the value of acquisitions, Table 12.5 shows that for the biotechnology industry it is significantly positive associated with the size of the acquiring firm, as measured by sales and it is significantly negative associated with a firm being located in Europe, i.e. European firms acquire at significantly lower prices than firms in the US, *ceteris paribus*.

Table 12.2 Extended model for all sectors jointly, dependent variable: average annual value of acquisitions (mn €)

Variables	*Estimates*
Financial leverage (total assets to total equity)	1.0280 (2.9085)
Current ratio (current assets to current liabilities)	4.5508 (5.7254)
Sales growth (% over previous year)	0.0147 (0.0816)
Sales (natural logarithm of net sales in 1000 €)	42.3637 (36.7317)
R&D intensity (R&D expenditure to net sales in %)	0.4849 (0.3035)
Square of R&D intensity (R&D expenditure to net sales in %)^2	−0.0001 (0.0001)
Patenting intensity (Patents granted by application year to net sales)	1.0743 (0.4193)**
Interaction of Patenting intensity and R&D intensity (R&D expenditure to net sales in % * Patents granted)	−0.0486 *** (0.0184)
Company headquartered in Europe (dummy; 1 = yes; base category: United States)	52.2877 (73.7181)
Company headquartered in Asia (dummy; 1 = yes; base category: United States)	−27.1067 (16.0881)*
Constant	−652.4111 (274.3312)**
R^2 within R^2 between R^2 overall	0.0486 0.0609 0.0284
No. of observations (individuals)	610 (81)
F	2.95***
Hausman specification test Chi^2	13.39*

Notes: Significance levels: * $p < 0.1$; ** $p < 0.05$; *** $p < 0.01$; heteroskedasticity-robust standard errors in parentheses; unbalanced panel data with observations per group: min = 1; max = 15; average = 7.5; fixed firm and year effects included; test for joint significance of year dummies: $Chi^2 = 21.61$, p-value > $Chi^2 = 0.1184$.

Next to the base model, the results of estimating the extended model for the biotechnology industry alone are reported in Table 12.6. Table 12.6 reveals strong heterogeneity within the biotechnology sector, which makes any significant predictions from the extended model difficult. However, this is the only case in which sectoral variability is so strong that it is problematic for the overall model quality (p=0.22).

Table 12.3 Base model for the EDA industry, dependent variable: average annual value of acquisitions (mn €)

Variables	*Estimates*
Financial leverage (total assets to total equity)	−6.8481 (11.7189)
Current ratio (current assets to current liabilities)	0.2594 (5.1601)
Sales growth (% over previous year)	−0.2083 (0.1917)
Sales (natural logarithm of net sales in 1000 €)	24.5336 (7.1548)***
R&D intensity (R&D expenditure to net sales in %)	1.2023 (1.0475)
Patenting intensity (Patents granted by application year to net sales)	−231065.9 (212743)
Company headquartered in Europe (dummy; 1 = yes; base category: United States)	70.52536 (87.1526)
Company headquartered in Asia (dummy; 1 = yes; base category: United States)	−23.7969 (14.1068)*
Constant	−248.9254 (86.3693)***
R^2 within	0.0776
R^2 between	0.3867
R^2 overall	0.1543
No. of observations (individuals)	95 (14)
F	1063.75***
Hausman specification test Chi^2	3.17

Notes: Significance levels: * $p < 0.1$; ** $p < 0.05$; *** $p < 0.01$; heteroskedasticity-robust standard errors in parentheses; unbalanced panel data with observations per group: min = 3; max = 11; average = 6.9; random firm and year effects included; test for joint significance of year dummies: Chi^2 =18.56, p-value > Chi^2 =0.0462.

Results for the semiconductor industry

For the case of the semiconductor industry, Table 12.7 shows that firm size is significantly positively associated with the average annual value of a firm's acquisitions. Furthermore, patenting intensity is almost significantly positively associated with the latter, which confirms H3.

Next to the base model, the results of the extended model in the semiconductor industry are reported in Table 12.8. As can be seen from Table 12.8 concerning the factors associated with the average annual value of the acquisitions made by firms in the semiconductor industry, it is found that again patenting intensity has a significant positive association with the value

Table 12.4 Extended model for the EDA industry, dependent variable: average annual value of acquisitions (mn €)

Variables	*Estimates*
Financial leverage (total assets to total equity)	8.6097 (12.2604)
Current ratio (current assets to current liabilities)	3.8432 (6.8115)
Sales growth (% over previous year)	0.0169 (0.1957)
Sales (natural logarithm of net sales in 1000 €)	18.4234 (20.5455)
R&D intensity (R&D expenditure to net sales in %)	0.8396 (2.4668)
Square of R&D intensity (R&D expenditure to net sales in %)^2	−0.0035 (0.0137)
Patenting intensity (Patents granted by application year to net sales)	−24.3195 (66.5972)
Interaction of Patenting intensity and R&D intensity (R&D expenditure to net sales in % * Patents granted)	0.6427 (2.7093)
Company headquartered in Europe (dummy; 1 = yes; base category: United States)	−33.0207 (33.2492)
Company headquartered in Asia (dummy; 1 = yes; base category: United States)	37.8325 (64.4686)
Constant	−225.5175 (263.9582)
R^2 within	0.3275
R^2 between	0.5238
R^2 overall	0.3882
No. of observations (individuals)	95 (14)
F	34.27***
Hausman specification test Chi^2	6.19

Notes: Significance levels: * $p < 0.1$; ** $p < 0.05$; *** $p < 0.01$; heteroskedasticity-robust standard errors in parentheses; unbalanced panel data with observations per group: min = 3; max = 11; average = 6.8; random firm and year effects included; test for joint significance of year dummies: Chi^2 =17.75, p-value > Chi^2 =0.0593.

of acquisitions, whereas the interaction between a firm's R&D and patenting intensities has a significant negative association.

Discussion and future research

The objective of the analysis reported here is to test four hypotheses relating acquirers' characteristics to average deal values. It basically finds support at the aggregate level for H3 and H4. A disaggregated analysis of the individual sectors

Table 12.5 Base model for the biotechnology industry, dependent variable: average annual value of acquisitions (mn €)

Variables	*Estimates*
Financial leverage (total assets to total equity)	−0.9459 (4.3625)
Current ratio (current assets to current liabilities)	4.4728* (2.5697)
Sales growth (% over previous year)	−0.0040 (0.0353)
Sales (natural logarithm of net sales in 1000 €)	68.5278 (32.0973) **
R&D intensity (R&D expenditure to net sales in %)	0.0875 (0.0585)
Patenting intensity (Patents granted by application year to net sales)	19758.9 (13497.21)
Company headquartered in Europe (dummy; 1 = yes; base category: United States)	−163.5689 (74.9174) **
Company headquartered in Asia (dummy; 1 = yes; base category: United States)	−16.7617 (121.5723)
Constant	−728.7524 (360.3179)**
R^2 within	0.0199
R^2 between	0.3584
R^2 overall	0.0755
No. of observations (individuals)	242 (21)
F	23.70***
Hausman specification test Chi^2	14.21

Notes: Significance levels: * $p < 0.1$; ** $p < 0.05$; *** $p < 0.01$; heteroskedasticity-robust standard errors in parentheses; unbalanced panel data with observations per group: min = 5; max = 15; average = 11.5; random firm and year effects included; test for joint significance of year dummies: Chi^2 =11.08, p-value > Chi^2 =0.6798.

involved in the analysis reveals that there is however considerable variation across sectors and that the effects found are strongest in the semiconductor industry. This suggests a need for a more detailed analysis of the main sectoral characteristics to better understand the drivers of variation. Analysing initially the average sales across all years it is found that these are slightly larger for EDA (mean US$629,000, minimum US$226,000 and maximum US$9 million) than for the biotechnology industry (mean US$562,000, minimum US$60,000 and maximum US$10 million). However, both industries are relatively similar by this measure when compared with the semiconductor industry, where average sales are considerably larger (mean US$3.5 million, minimum US$95,000 and maximum US$31 million). Hence one reason for the stronger effects in the semiconductor industry could be economies of scale which appear to be significantly stronger there than in the other two industries.

Table 12.6 Extended model for the biotechnology industry, dependent variable: average annual value of acquisitions (mn €)

Variables	*Estimates*
Financial leverage (total assets to total equity)	−5.4141 (16.0088)
Current ratio (current assets to current liabilities)	2.1838 (5.3211)
Sales growth (% over previous year)	0.0330 (0.0917)
Sales (natural logarithm of net sales in 1000 €)	50.6489 (33.0275)
R&D intensity (R&D expenditure to net sales in %)	0.1976 (0.2786)
Square of R&D intensity (R&D expenditure to net sales in %)^2	−0.0001 (0.0001)
Patenting intensity (Patents granted by application year to net sales)	29.7006 (67.2037)
Interaction of Patenting intensity and R&D intensity (R&D expenditure to net sales in % * Patents granted)	−0.0128 (0.0536)
Company headquartered in Europe (dummy; 1 = yes; base category: United States)	−125.5757 (110.1007)
Company headquartered in Asia (dummy; 1 = yes; base category: United States)	78.1623 (129.7858)
Constant	−1487.401 (488.8261)***
R^2 within	0.0960
R^2 between	0.3271
R^2 overall	0.1364
No. of observations (individuals)	242 (21)
F	24.33
Hausman specification test Chi^2	9.86

Notes: Significance levels: * $p < 0.1$; ** $p < 0.05$; *** $p < 0.01$; heteroskedasticity-robust standard errors in parentheses; unbalanced panel data with observations per group: min = 5; max = 15; average = 11.5; random firm and year effects included; test for joint significance of year dummies: Chi^2 =11.43, p-value > Chi^2 =0.6522.

Concerning the variation in sales growth, this is found (based on pooled data per sector across all years) to be stronger in biotechnology (mean 114 per cent, minimum −100 per cent and maximum 5500 per cent) than in the EDA industry (mean 33 per cent, minimum −31 per cent and maximum 717 per cent). In the semiconductor industry, it averages 164 per cent and the variation (minimum −99 per cent, maximum 9868 per cent) is stronger than for both the EDA and biotechnology industry, which is consistent with the high revenue fluctuation well established in the history of the semiconductor industry (Levy, 1994). However, the differences across the three industries

Table 12.7 Base model for the semiconductor industry, dependent variable: average annual value of acquisitions (mn €)

Variables	*Estimates*
Financial leverage (total assets to total equity)	−0.3282 (3.1085)
Current ratio (current assets to current liabilities)	16.2925 (31.9372)
Sales growth (% over previous year)	−0.0043 (0.0377)
Sales (natural logarithm of net sales in 1000 €)	0.01767 (0.0063)***
R&D intensity (R&D expenditure to net sales in %)	-1.8720 (5.7030)
Patenting intensity (Patents granted by application year to net sales)	0.3419 (0.2976)
Company headquartered in Europe (dummy; 1 = yes; base category: United States)	-
Company headquartered in Asia (dummy; 1 = yes; base category: United States)	-
Constant	−103.1219 (113.026)
R^2 within	0.0283
R^2 between	0.0128
R^2 overall	0.0042
No. of observations (individuals)	273 (46)
F	2.27**
Hausman specification test Chi^2	10.19*

Notes: Significance levels: * $p < 0.1$; ** $p < 0.05$; *** $p < 0.01$; heteroskedasticity-robust standard errors in parentheses; unbalanced panel data with observations per group: min = 1; max = 15; average = 5.9; fixed firm and year effects included; test for joint significance of year dummies: Chi^2 =19.01, p-value > Chi^2 =0.2132.

are not as clear-cut as for sales levels with the semiconductor industry being more similar to biotechnology than both are to EDA. Hence sales growth does not seem a crucial factor explaining differences in the effects across industries.

Looking at R&D intensity and patenting intensity, the pooled average for the biotechnology industry is 164 per cent for R&D intensity. The average patenting intensity for the biotechnology industry (patents per thousand US$ sales) is 0.0003 (patents per thousand US$ sales). Compared to this, in the EDA industry, the average R&D intensity is 47 per cent. The average R&D intensity is pushed upwards by very high outlier values for one firm (Magma Design Automation, which is said to have been significantly supported by the semiconductor firm Intel to become an opponent

Table 12.8 Extended model for the semiconductor industry, dependent variable: average annual value of acquisitions (mn €)

Variables	*Estimates*
Financial leverage (total assets to total equity)	0.9295 (3.0402)
Current ratio (current assets to current liabilities)	−16.5058 (33.3014)
Sales growth (% over previous year)	0.0582 (0.2044)
Sales (natural logarithm of net sales in 1000 €)	80.2686 (66.4789)
R&D intensity (R&D expenditure to net sales in %)	13.8259 (15.2774)
Square of R&D intensity (R&D expenditure to net sales in %)^2	−0.0026 (0.2409)
Patenting intensity (Patents granted by application year to net sales)	1.7526 (0.5882)*
Interaction of Patenting intensity and R&D intensity (R&D expenditure to net sales in % * Patents granted)	−0.0941 (0.0315)***
Company headquartered in Europe (dummy; 1 = yes; base category: United States)	-
Company headquartered in Asia (dummy; 1 = yes; base category: United States)	-
Constant	−947.8316 (385.1286)**
R^2 within	0.0659
R^2 between	0.0312
R^2 overall	0.0231
No. of observations (individuals)	273 (46)
F	1.85***
Hausman specification test Chi^2	30.71***

Notes: Significance levels: * $p < 0.1$; ** $p < 0.05$; *** $p < 0.01$; heteroskedasticity-robust standard errors in parentheses; unbalanced panel data with observations per group: min = 1; max = 15; average = 5.9; fixed firm and year effects included; test for joint significance of year dummies: Chi^2 = 18.91, p-value > Chi^2 = 0.2180.

to the established incumbents in the EDA industry). For most other firms in the EDA industry, the R&D intensity is in the range of 10–15 per cent. Concerning patenting intensity, the mean value is of a similar order of magnitude in EDA and in biotechnology, namely 0.0002 patents per thousand US$ sales. Comparing biotechnology and EDA with the semiconductor industry it becomes however clear that for the latter patenting intensity (with a mean of 0.00007 patents per thousand US$ sales) is lower than for EDA and biotechnology (which have relatively similar averages, and also minimum and maximum values for the number of patents per thousand US$ sales). This result is however consistent with the low effectiveness of patents as an appropriation mechanism in the semiconductor industry: the semiconductor industry patents less frequently in relation to sales (Cohen and Nelson, 2001). Concerning the R&D intensity of the semiconductor industry (mean approx. 14 per cent), it is found that it is similar to that of the EDA industry (after accounting for outlier values in the latter) which is again consistent with earlier findings (Sangiovanni-Vincentelli, 2003). Also the R&D intensity in both the EDA and semiconductor industries is considerably lower than for biotechnology, which mirrors the higher uncertainty of research-related investments in the latter.

Therefore, given that H3 and H4 mainly concern patenting and partly R&D intensity, the differences with regard to patenting intensity again suggest that the semiconductor industry is further apart from both EDA and biotechnology than the latter two are from one another, which is likely one important explanation for the differing effect strength identified between the industries. Furthermore, the higher similarity of semiconductors to EDA in terms of R&D intensity can explain why H1 and H2 are not supported in the econometric analysis.

In terms of future research, it must be acknowledged that especially for the EDA and biotechnology industries, it cannot be ruled out that the non-significant findings concerning the hypotheses are due to the relatively low number of observations. While this is somewhat inherent to the industry structure, future research should attempt a more comprehensive analysis of individual sectors involving large numbers of observations.

Unfortunately pooling data from the EDA and biotechnology industries is not available as an easy remedy to this situation since, as far as the main dimensions are concerned, only sales levels and patenting intensity of these two industries are similar, but not sales growth (where biotechnology resembles more the semiconductor industry) and R&D intensity (where EDA resembles more the semiconductor industry). Furthermore, it was not possible to include target data because this is more difficult to source since records are partly deleted after acquisition and due to the fact that targets are usually smaller and rarely stock listed. This implies more limited reporting requirements and hence lower data availability. However, this is not considered too much of a limitation because acquirers' characteristics 'fit' frequently with target characteristics so that one can derive one from the other, i.e. the more acquirer characteristics are available, the less information on targets is needed.

Overall, the results show that the patenting and R&D intensities of firms are only rarely associated significantly with the average annual value of acquisitions. The most consistent predictor found is sales and hence firm size, suggesting that deep pockets associate with high deal values. This is notable, since earlier research found R&D and patenting intensities being more strongly associated with the number of (both total and technology-related) acquisitions (Desyllas and Hughes, 2008; Wagner, 2011), but not

Table 12.9 Summary of results for average annual value

All sectors	*EDA*	*Biotechnology*	*Semiconductors*
Sales (+) Asia (+/−) Patenting (+) Interaction (−)	Sales (+) Asia (−)	Sales (+) Liquidity (+) Europe (−)	Sales (+) Patenting (+) Interaction (−)

so much with more value-oriented indicators such as (in this chapter) deal value or the number of target patents in the five years prior to acquisition (Wagner, 2011). Beyond that, geographical influences could also be identified, as summarised in Table 12.9.

The findings do not generally support the notion that acquisition value benefits from lack of absorptive capacity indicated by lower R&D intensity or that acquisition value is negatively affected by jointly very high absorptive capacity and output efficiency. This seems to apply across all sectors. In this respect, in all industries acquisition values seem to be more strongly related to the size of firms and to a lesser extent their geographical locations, the latter of which is consistent with research on the global variation of R&D investment (Martinez-Torres and Toral, 2009).

Appendix

Table 12.A1 Summary statistics and correlations[a]

Variables	*Mn.*	*SD*	*Min.*	*Max.*	*Correlations* 1	2	3	4	5	6
1 Financial leverage	2.35	6.56	0.72	166.11						
2 Current ratio	3.62	4.36	0.33	45.67	−0.08**					
3 Sales growth	46.39	217.85	−96.06	3462.91	−0.05	0.2***				
4 Sales	9.31	2.55	3.38	15.52	−0.01	−0.02	−0.03			
5 R&D intensity	54.95	220.98	0	4069.5	−0.01	0.22***	0.04	−0.08**		
6 Patenting intensity	312.95	588.89	0	4331.0	0.01	0.04	−0.01	−0.06**	0.87***	
7 European firm	0.08	-	0	1	−0.003	−0.09**	−0.02	0.17***	−0.05	
8 Asian firm	0.09	-	0	1	0.14***	−0.12***	0.02	−0.1***	−0.07*	−0.01**

Notes: [a] Mn.: sample mean; SD: standard deviation; Min.: minimum value; Max.: maximum value; *: p < 0.1; **: p < .05; ***: < 0.01.

References

Amin, A. and Cohendet, P. (2004) *Architectures of Knowledge: Firms, Capabilities and Communities*, Oxford: Oxford University Press.

Belderbos, R., Carree, M. and Lokshin, B. (2004) 'Cooperative R&D and firm performance', *Research Policy*, Vol. **33**: 1477–92.

Bloningen, B.A. and Taylor, C.T. (2000) 'R&D intensity and acquisitions in high-technology industries: Evidence from the US electronic and electrical equipment industries', *Journal of Industrial Economics*, Vol. **XLVIII**: 47–70.

Bloom, N. and van Reenen, J. (2002) 'Patents, real options and firm performance', *Economic Journal*, Vol. **112**: C97–C116.

Cohen, W.M. and Levinthal, D.A. (1990) 'Absorptive capacity: A new perspective on learning and innovation', *Administrative Science Quarterly*, Vol. **35**: 128–52.

Cohen, W.M. and Nelson, R.R. (2001) *Protecting Their Intellectual Assets: Appropriability Conditions and Why U.S. Manufacturing Firms Patent (or Not)*, Cambridge: NBER (NBER Working Paper 7552).

Cohendet, P. and Simon, L. (2007) 'Playing across the playground: Paradoxes of knowledge creation in the videogame firm', *Journal of Organizational Behavior*, Vol. **28**: 587–605.

Cohendet, P., Llerena, P. and Marengo, L. (2000) 'Is there a pilot in the evolutionary firm?', in N. Foss and V. Mahnke (eds), *New Directions in Economic Strategy Research*, Oxford: Oxford University Press, 95–115.

Desyllas, P. and Hughes, A. (2008) 'Sourcing technological knowledge through corporate acquisition: Evidence from an international sample of high technology firms', *Journal of High Technology Management Research*, Vol. **18**, No. 2: 157–72.

Griliches, Z. (1981) 'Market value, R&D, and patents', *Economics Letters*, Vol. **7**: 183–7.

Hall, B.H. and Ziedonis, R.H. (2001) 'The patent paradox revisited: An empirical study of patenting in the U.S. semiconductor industry, 1979–1995', *RAND Journal of Economics*, Vol. **32**, No. 1: 101–28.

Hauschildt, J. (1999) 'Opposition to innovations – destructive or constructive?' in K. Brockhoff, A. Chakrabarti and J. Hauschildt (eds), *The Dynamics of Innovation*, Berlin: Springer, 217–40.

Henderson, R. (1993) 'Underinvestment and incompetence as responses to radical innovation: Evidence from the photolithographic alignment equipment industry', *RAND Journal of Economics*, Vol. **24**, No. 2: 248–70.

Hofmann, S. (2007) 'Biotechnology-Riese Amgen kauft weiter ein', *Handelsblatt*, 6 June, p. 19.

Hurmelinna-Laukkanen, P. (2009) 'The availability, strength and efficiency of appropriability mechanisms – protecting investments in knowledge creation', *International Journal of Technology Management*, Vol. **45**, No. 3/4: 282–90.

Jack, A. (2006) 'M&A may be best medicine for mid-sized pharma groups', *Financial Times*, 26 September, p. 20.

Jack, A. (2007) 'AZ to move into vaccines with $16bn US deal', *Financial Times*, 24 April, p. 15.

Johnston, J. and DiNardo, J. (1997) *Econometric Methods*, New York: McGraw-Hill.

Levinthal, D.A. and March, J.G. (1993) 'The myopia of learning', *Strategic Management Journal*, Vol. **14**: 95–112.

Levy, D. (1994) 'Chaos theory and strategy: Theory, application and managerial implications', *Strategic Management Journal*, Vol. **15**: 167–78.

Markides, C. and Geroski, P. (2005) *Fast Second: How Smart Companies Bypass Radical Innovation to Enter and Dominate New Markets*, San Francisco: Jossey-Bass.

Martinez-Torres, M.R. and Toral, S.L. (2009) 'International comparison of R&D investment by European, US and Japanese companies', *International Journal of Technology Management*, Vol. **49**, No. 1/2/3: 107–22.

Miles, M.B. and Huberman, A.M. (1994) *Qualitative Data Analysis: An Expanded Sourcebook*, Thousand Oaks/London: Sage.

Milgrom, P.R. and Roberts, J. (1992) *Economics, Organizations and Management*, Upper Saddle River: Prentice-Hall.

Morris, D.J. and Hay, D.A. (1991) *Industrial Economics & Organisation – Theory & Evidence*, Oxford: Oxford University Press.

Nesta, L. and Saviotti, P. (2006) 'Firm knowledge and market value in biotechnology', *Industrial and Corporate Change*, Vol. **15**: 625–52.

Pangarkar, N. (2003) 'Determinants of alliance duration in uncertain environments: The case of the biotechnology sector', *Long Range Planning*, Vol. **36**: 269–84.

Pangarkar, N. and Klein, S. (2001) 'The impacts of alliance purpose and partner similarity on alliance governance', *British Journal of Management*, Vol. **12**, No. 4: 341–53.

Rothaermel, F.T. and Deeds, D.L. (2004) 'Exploration and exploitation alliances in biotechnology: A system of new product development', *Strategic Management Journal*, Vol. **25**: 201–21.

Sanchanta, M. (2007) 'Shortlist drawn up for Sanyo chip unit sale', *Financial Times*, 23 May, p. 19.

Sangiovanni-Vincentelli, A. (2003) 'The tides of EDA', *IEEE Design and Test of Computers*, November–December: 59–74.

Teece, D.J. (1986) 'Profiting from technological innovation: Implications for integration, collaboration, licensing and public policy', *Research Policy*, Vol. **15**: 285–305.

Wagner, M. (2011) 'Acquisition as a means for external technology sourcing: Complementary, substitutive or both?', *Journal of Engineering and Technology Management*, Vol. **28**: 283–99.

Williamson, O.E. (1975) *Markets and Hierarchies: Analysis and Antitrust Implications*, New York: Free Press.

Yoshikawa, T. (2003) 'Technology development and acquisition strategy', *International Journal of Technology Management*, Vol. **25**, No. 6/7: 666–74.

Zhang, J., Baden-Fuller, C. and Mangematin, V. (2007) 'Technological knowledge base, R&D organization structure and alliance formation: Evidence from the biopharmaceutical industry', *Research Policy*, Vol. **36**: 515–28.

13 Marshalling resources and dynamic capabilities to overcome innovative barriers and foster creativity in the service sector

Caroline Hussler, Laurence Frank, Matthieu Bunel and Patrick Eparvier

Introduction

"Capturing the ways in which companies are innovating services" is presented by Ostrom *et al.* (2010, p. 12) as one of the priorities for academic research on services. This echoes Menor and Roth (2007)'s argument according to which "current theory and understanding of the strategies and tactics for developing new services is inadequate" (p. 825). Although many papers have addressed the determinants of innovative activity (see for instance DeJong and Vermeulen (2003) for a review of the service sector) only a scarce number of them examine the flip side of the coin incarnating the causes pertaining to a lack of innovative dynamism within firms. Very little is known about barriers to innovation, their perception, their intensity and their consequences on the innovative activities of the service sector. Yet, the state of the art of research on innovation in the service sector reveals the existence of obstacles and roadblocks to the marketing and the delivery of new services and activities. To understand what really hampers innovation in the service sector, we examine the difficulties firms reported in their attempts to design new customer value propositions. As the literature agrees on the peculiar nature of innovation undertaken in different service sectors (Kvalshaugen *et al.*, 2008) we are inclined to think that innovative obstacles vary across the sub-fields of the service industry. We deliberately put manufacturing activities aside and concentrate on the service sector as it accounts for two thirds of the added value produced by the developing world and is considered the backbone of its economy. Concentrating the analysis on barriers encountered by service industries will provide some insights into the perceived lack of innovative capacity compared to the manufacturing sector (Wölfl, 2005; Miles, 2008; Cordellier, 2009).

Our main assumption is that limited innovativeness stems from one or several paralyzing obstacles. The Resource Based-View (RBV) provided several explanations on restrained or unused innovation capacity. Penrose (1995) came to observe that innovation happens when firms mobilize their

Box 13.1 Innovation styles in service industries

If service industries and firms share two common features: intangibility and interactivity (Miles, 2008), the service sector remains heterogeneous. As pointed out by Tether (2003) comparing services to manufacturing fosters a broad conception of services and to view all service firms as adopters or supply dominated. However, diversity in services has been shown to be far greater within the manufacturing sector (Hughes and Wood, 2000; Miles, 2008): service firms range from personal services (often considered as being organized on a small scale and being based on highly specialized skills) to financial and insurance services (mostly organized at a national or an international scale and using high tech networking technologies) and going through business services (targeted at providing their activities to intermediate businesses rather than final consumers and becoming global as their customers develop abroad). This motivates us to distinguish four sub-categories of services.

The peculiarities of services innovation have been often neglected; "services have been considered as non-innovative or supplier-dominated recipients of technology" (Rubalcaba *et al.*, 2010). However, the pioneering taxonomy on innovation in services developed by Soete and Miozzo (1990) distinguishes between three sets of innovation styles in the service sector: supplier dominated, scale-intensive and science-based. More recently den Hertog (2000) identified four different innovation modes in service sectors: technological, client-interface, conceptual and delivery innovation. Gallouj and Savona (2009) provided a typology for six innovative models:

> radical innovation (the creation of a totally new product), improvement innovation (improvement in certain characteristics), incremental innovation (substitution or addition of characteristics to a certain product), *ad hoc* innovation (interactive construction of a solution to a particular problem), recombinative (new combinations of technological characteristics) and formalization innovation (putting service characteristics "into order")
>
> (Rodriguez and Camacho, 2010, p. 45)

All those theoretical contributions suggest that innovation might have both a hard (technology-based) and soft (organizational) side in service sectors, the proportion of each side being probably a function of the category of services under analysis.

asset base, capitalize on knowledge and skills built over time and trust the audacity of managers. Nevertheless, changes in the productive opportunity will be restricted to the extent to which the firm does not see opportunities for expansion, is unwilling to act upon them or is unable to respond to them (p. 32). RBV advocates "competing on capabilities" rather than choosing a product market position or making traditional resource investments as the appropriate task for strategic management (Collis, 1994). The theory of dynamic capabilities brought the RBV to another level when additional factors had been taken into account to explain growth potential and the ability to innovate. A dynamic capability is "the capacity of an organization to purposefully create, extend or modify its resource base in order to survive and prosper under conditions of change" (p. 1). Teece *et al.* introduced in 1997 the concept of "organizational capabilities" to characterize the possibility of exploiting existing internal and external firm-specific competences to address changing environments and build a competitive advantage: "end products are the final goods and services produced by the firm based on utilizing the competences that it possesses" (p. 516). According to scholars, the most recurrent problem faced by innovative firms appears to be the lack of (financial) resources. However, our assumption is that the reported difficulty in accessing resources underpinning innovation is subordinated to the difficulty faced by firms in taking advantage of the assets they possess and in their capability to design new products and services with existing means. In a nutshell the lack of innovation might be explained by a lack of dynamic capabilities and not necessarily by resource shortages.

As the main objective of this chapter is to identify and characterize obstacles faced by service innovators, we resorted to an empirical study based on the content of two databases: the French part of the fourth edition of the Community Innovation Survey (CIS 4) which provided a detailed list of potential obstacles to innovation, and the EAE[1] files storing administrative information on effective characteristics of firms. Our first objective is to confront these two sources of information in order to identify barriers faced by firms of the service sector on the path to innovation. In a second stage, as our ambition is to enrich the debate on the origins of those barriers, we consider the RBV and the Theory of Dynamic Capabilities to be appropriate theoretical frameworks to explain service innovation dynamics.

The content of this chapter is organized as follows: the first section reviews the main results drawn from contemporary literature on barriers to innovation and their origins. We also explain how the service innovation perspective interacts with RBV and dynamic capabilities and we provide a preliminary analysis of service innovation impediments. Methodology and empirical choices are detailed in the following section. Finally, we present our empirical results and comment on our main findings and their managerial implications.

Barriers to innovation and their origins: literature review

Reported obstacles

Managerial literature on innovation in the service sector delivers a recurrent observation about new and small firms facing different types of obstacles to innovation, the most frequent ones being the lack of resources or the intensity of market rivalry (Katila and Shane, 2005). As service sectors typically involve a greater share of small firms in comparison to the manufacturing industries (Tether *et al.*, 2002), obstacles related to insufficient levels of resources systematically appear. Being active in large markets is presented as the leading barrier because most firms are unable to manage all the complementary assets required to be competitive in all areas of a vast market (Schoonhoven *et al.*, 1990). In addition, services are generally delivered where they are produced (Hauknes, 2001; Segal-Horn and Dean, 2007) and they remain less globalized (see Brock and Alon, 2009 for an in-depth study on this phenomenon). Due to the significant impact of customer interaction and intangibility factors, services are less standardized, less focused on products and less centralized/dispersed (den Hertog *et al.*, 2010) and all these factors are of potential influence on innovation capacity.

At the same time, the related economic literature draws on evidence provided by innovation surveys (Mohnen and Rosa, 2000; Baldwin and Lin, 2002; Galia and Legros, 2004) and rather unanimously demonstrates that the larger the firm, the greater the importance attached to barriers to innovation. Consequently, the characteristics of the firm seem to have an effect on perceived obstacles to innovation and on the difficulty in accessing the required resources. Surprisingly, only a few papers actually provide an industry-specific analysis of innovation barriers. Landry and Amara (2005) are a noteworthy exception and their fine-grained analysis enabled them to conclude that the intensity of perceived obstacles significantly differs depending on the industrial sector which is studied rather than on the firm's size. A second valuable observation stems from this Canadian study that highlights that the main barrier perceived by surveyed firms lies in a shortage of qualified manpower. Astonishingly, most of them were not even aware of the existence and the nature of training programs available on their territory, all this strongly suggesting a discrepancy between the perceived obstacles and their accuracy. The literature also reveals a repeated conclusion (on the fact) that being involved in R&D and innovative activities influences the probability and the intensity of the perceived obstacles to innovation. Indeed, most of the existing papers based on declarative databases show that being innovative increases the firm's consciousness of innovative obstacles. Last but not least, innovation in services appears to have its own features, behaviors and needs (Gallouj, 2002; Tether, 2003). The specificities of the process of a new service development – apparently less sensitive to technological push than to market pull – encompass other peculiarities inherent to the service sector (such as size and age) which might generate specific related obstacles in those industries.

Declared impediments

Until now, only a few scholars tested whether the awareness of obstacles really prevents firms from being successful in their innovative activities. An exception is to be found in Tourigny and Le's (2004) study which indicates the opposite: none of the revealed impediments actually stops innovation, showing that firms are actually able to overcome the obstacles featured in innovation surveys. In other words, one has to be cautious and differentiate "barriers as revealed" that are perceived as factors potentially constraining innovative activities, from "barriers as deterrents" which prevent firms from being successful in their innovative activities, the gap between declared and actual barriers impeding innovation being potentially substantial (d'Este *et al.*, 2008, p. 4). This evidence echoes Iammarino *et al.*'s (2007) skepticism about the accuracy of the informative content of declarative data on obstacles to innovation in designing appropriate strategies and policies to support firms in their innovative activities. In short, it means that one tries to account for an objective problem that is measured in a subjective way (through the perception of respondents). It thus seems important to distinguish obstacles as declared by firms from those really paralyzing innovation. To resolve this, d'Este *et al.* (2008) suggested analyzing barriers perceived by innovators on the one hand and the impediments declared by firms which have abandoned their innovative projects on the other.

It also seems interesting to analyze the origin of perceived barriers and specify their nature. For instance, Loewe and Dominiquini (2006) explain that many CEOs do not know how to overcome innovation barriers because their company suffers from a distorted approach to innovation: they consider innovation their top priority but do not allocate the required resources to it, most of them being absorbed by quarterly earnings. Whether resources are central to innovative capabilities is thus examined in the next section.

As Drucker (1998) points out: "systematic innovation begins with the analysis of the sources of opportunities and depending on the context, sources will have a different importance … Because innovation is both conceptual and perceptual would-be innovators must go out and look" (p. 8). In 1956, Penrose already pointed out that dynamic capabilities include the capacity with which managers identify the need for change, formulate a response to such a need and implement a course of action. Teece (2007) recalled the significance of managerial capability to sense opportunities. Accordingly, innovation pulled by human resources and organizational capabilities (rather than technologically pushed) leans on internal managerial competences which are even more critical in dynamic settings.

Resources and capabilities as underlying root causes

The existing literature tends to focus on the reasons explaining why some firms outperform others in the competitive market. Instead, we decided to learn from their drawbacks and to investigate the reasons underpinning difficulties faced throughout the innovation process. We hypothesize that a lack of specific resources (tangible and intangible) hampers the ability to innovate. The RBV identified key-resources which allow firms to create value, a patrimony Ness and Brechin (1988) named the firm's "technology" that includes not only the machines and other hardware organizations used to achieve theirs ends, but also the skills, knowledge, training of employees, approaches, strategies and procedures, and even the characteristics of the objects (inputs or outputs) on which work is performed (p. 256). Miller (1998, p. 509) added "configuration" to technology as a distinctive feature to explain the firm's performance through "the degree to which an organization's elements are orchestrated and connected by a single theme."

Applying this literature to the service sector, den Hertog *et al.* (2010, p. 498) identified six core dynamic service innovation capabilities:

- signaling user needs (as firms depend on their users to co-develop and co-produce new services, Teece, 2007) and technological options;
- the ability to translate a fuzzy idea into a new service;
- an unbundling capability allowing the later recombination of existing elements;
- co-producing and orchestrating service innovation outside the boundaries of the firm;
- scaling up and stretching the service already provided;
- learning and adapting.

This taxonomy acts as a starting point and analytical grid to investigate innovation capabilities further and how firms marshal, enlarge their resource-base to think new services and become creative in their attempts to offer unprecedented services. Creativity here refers to the ability to rearrange existing resources to build a distinctive or an atypical organization model by assembling particular constellations of assets inside an enterprise, because by employing such assets these organizations can produce highly differentiated and innovative goods and services that consumers want (Helfat *et al.*, 2007, p. 23).

The main originality of this chapter resides in its exploratory character and in the areas of freedom we deliberately left in the research design so as to allow unexpected information to arise. More details on the methodological framework are provided in the following section.

Methodology and data sources

To tackle barriers and difficulties experienced by firms in the innovation process, we resorted to an in-depth statistical analysis of difficulties reported by French firms. The sample of firms comprises those which are active in the service sector but distinguishes firms which have effectively innovated from those which have abandoned their innovation project. We deemed this distinction necessary to perform a refined analysis of obstacles delaying or killing innovation. Declared obstacles were subsequently compared to resources mentioned in accounting or administrative documents. This comparison allowed us to test whether obstacles as perceived by firms reflect effective limited access to resources or demonstrate a difficulty in taking advantage of owned assets, hence the assumption that a lack of managerial skills and weak dynamic capabilities foster barriers to innovation. Data were extracted from two databases. The French part of the CIS 4 database covering the period from 2002 to 2004 is administered by the French National Institute for Statistics (INSEE) which stores data on the origin, characteristics and difficulties experienced in the innovation process by firms of five employees and more. We created sub-samples out of the categories provided by the summary economic classification (*nomenclature économique de synthèse*) used by INSEE and this enabled us to distinguish four types of services namely consultancy, real estate, finance, personal services and intermediate services. The CIS also supplied information about factors slowing down or interrupting firms' innovative activities and highlighted seven obstacles to innovation:

1. Uncertain demand for innovative goods and services
2. Difficulty in finding a business partner
3. Lack of qualified manpower
4. Limited availability of financial means
5. Market dominated by established firms
6. Lack of information on markets and technologies
7. Unaffordable cost of direct innovation

We noted that some of these obstacles implicitly refer to a lack of resources (3, 4, 7) whereas others can be attributed to a lack of dynamic capabilities. We interpreted a difficulty in finding a business partner as a lack of "co-production capabilities." The uncertainty related to the demand of innovative goods and services or the lack of information about markets and technologies implied deficient "signaling capabilities." We considered these seven obstacles underlying root causes challenging either the resource-base or dynamic capabilities of these firms.

We matched these CIS survey data with the information supplied by the national inquiry on firms (EAE) which collects data on characteristics of firms such as geographic location, legal status, financial performance, qualification of human resources (from unqualified to very qualified manpower) and also the size of their main market. Matching this information enabled us to compare the effective resources firms possessed with the problems they complained about. Although our analysis concentrates on the service sector, comparisons are made with the results provided by studies on manufacturing industries.

Box 13.2 Using CIS data to study innovation in the service sector

CIS surveys provide information on the input and output of a firm's innovative activities, as well as on the modalities of these activities. On the output side, the CIS provides a definition of innovation to help managers fill in the questionnaire. More precisely,

> Innovation is defined as major changes aimed at enhancing your competitive position, your performance, your know-how or your capabilities for future enhancements. These can be new or significantly improved goods, services or processes for making or providing them. It includes spending on innovation activities, for example on machinery and equipment, R&D, training, goods and service design or marketing.

Using CIS data thus leads us to consider innovation as the result of a purposive action, a deliberate project and to neglect the potentially spontaneous and unanticipated nature the innovative process might be of. Nevertheless, the CIS covers all facets of innovation (hard and soft ones) at stake in service firms, as it distinguishes product innovation, process innovation, organizational innovation and marketing innovation.

The use of CIS data was also questioned by Drejer (2004) because of the neglected idiosyncrasies observed within the service sector. However, the CIS survey is representative of the French industrial sectors and gathers data on more than 25,000 firms out of which 12,000 assert that they have been engaged in an innovative project, 30 percent of these being active in the service sector. Moreover, we choose to use it as we believe that differences between the behaviors of different service sectors might be even greater than those between services and manufacturing (Hughes and Wood, 2000).

The next section presents the results of our empirical studies and provides insights into the reported difficulties experienced in the attempts to innovate.

Empirical results

The originality of this study resides in the identification of effective and perceived deterrent barriers to innovation in the service sector. The results of our empirical study are presented in three steps: first, we describe the perceived obstacles and see if they are evidence of a lack of resources. Second, we analyze the consequences of these reported obstacles on innovation. Finally, we focus on the effective means available to firms in order to establish the link with perceived obstacles.

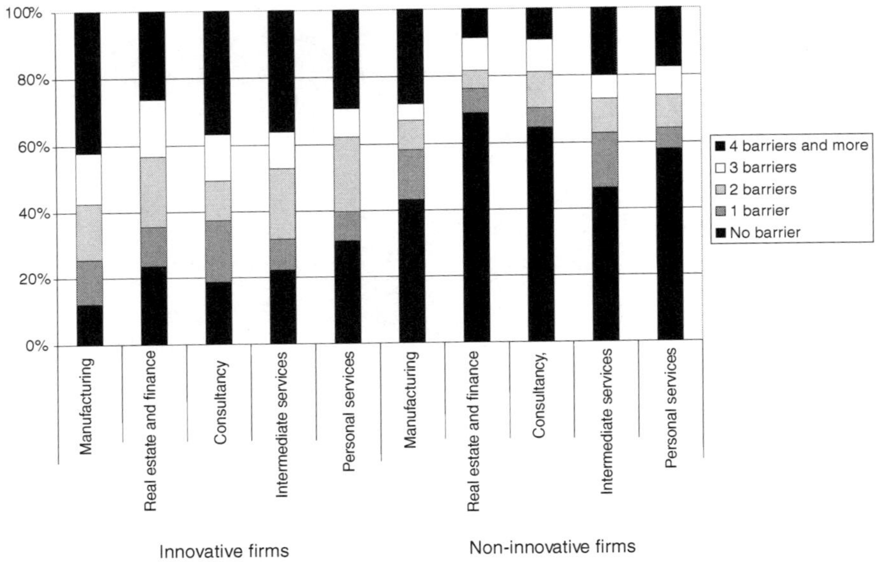

Figure 13.1 Total number of perceived barriers by sector

Perceived obstacles: distribution and predominance by sector

The first diagram (Figure 13.1) presents the perceived barriers to innovation by sector and shows significant variations in the respective distributions of their dominance.

Significant variations can be observed between innovative firms and non-innovative firms. These results confirm the literature indicating that innovative firms tend to overestimate obstacles to innovation compared to the non-innovative ones, regardless of the type of industry under study. The non-innovative firms belonging to the consultancy industry which did not declare any barrier are four times more numerous than their counterparts in the innovative group. The main reason is that non-innovative firms (having no innovation project in the pipeline) care less about the potential difficulties encountered on the innovation path. Another valuable finding resides in the systematic tendency of the service sector to underestimate difficulties compared to the manufacturing industry, whatever the service sector. In personal services, 30 percent of the surveyed population of innovative firms does not witness any barriers. As for consultancy services, behaviors are similar to those of the manufacturing industry: a low number of firms do not see any barriers within the large number of situations in which firms suffer from the accumulation of concomitant obstacles. This suggests that differences between services and goods are a matter of "degree rather than of substance" (Coombs and Miles, 2000, p. 99) and that new service and new product development are not systematically of a different kind. In terms of commonalities, the diagram exhibits variations among service sectors when the number of perceived obstacles is of lesser concern. Personal services seem less exposed (as they complain less about innovative difficulties) whereas consultancy services mimic barriers reported by the manufacturing firms. This is partly attributable to the nature of personal services which can be deployed with basic skills and technologies, a small scale of operations (Kvalshaugen *et al.*, 2008) and thus require less human and financial resources to be undertaken and improved.

At this stage, our results put two explanations forward with regard to difficulties reported in the service sector:

1. Innovation in the service industries is less difficult to undertake than in other industrial fields; or
2. Firms are unable to identify the difficulties hampering their innovation activities. However, this statement should be qualified as it might draw from an inappropriate list of obstacles proposed by the CIS survey. It is striking to note that the list of obstacles proposed to surveyed firms omits internal managerial problems or conceptualizing capacity (den Hertog *et al.*, 2010), the seven obstacles mainly dealing with external contingencies.

The next section refines these preliminary results and focuses on the nature of the perceived obstacles.

Variety of perceived obstacles among the service sectors

The most critical impediments encountered by firms (as declared in CIS) are presented in Figures 13.2, 13.3 and 13.4. The sum of percentages associated with obstacles exceeds 100 percent reflecting a plural choice of sources of difficulties.

There is no change in the ranking of barriers whether innovative or non-innovation firms are concerned. Figure 13.2 shows the importance of limited financial capacity and illustrates the main impediment to the innovative process as it impacts 41 percent of surveyed firms.

The high cost of innovation ranks second and affects 37 percent of the firms while the lack of qualified manpower is considered the third most important problem, for 34 percent of them. Similar findings can be found in the sub-population of innovative firms (Figure 13.3). The obstacle recurrently disparaged by manufacturing firms is the lack of financial means along with the excessive cost of innovation which suggests that innovativeness in this sector is constrained and driven by money.

For firms having abandoned their innovation projects (Figure 13.4), slightly different results appear on their perception of barriers to innovation as they rank the uncertainty of demand as the third difficulty they deal

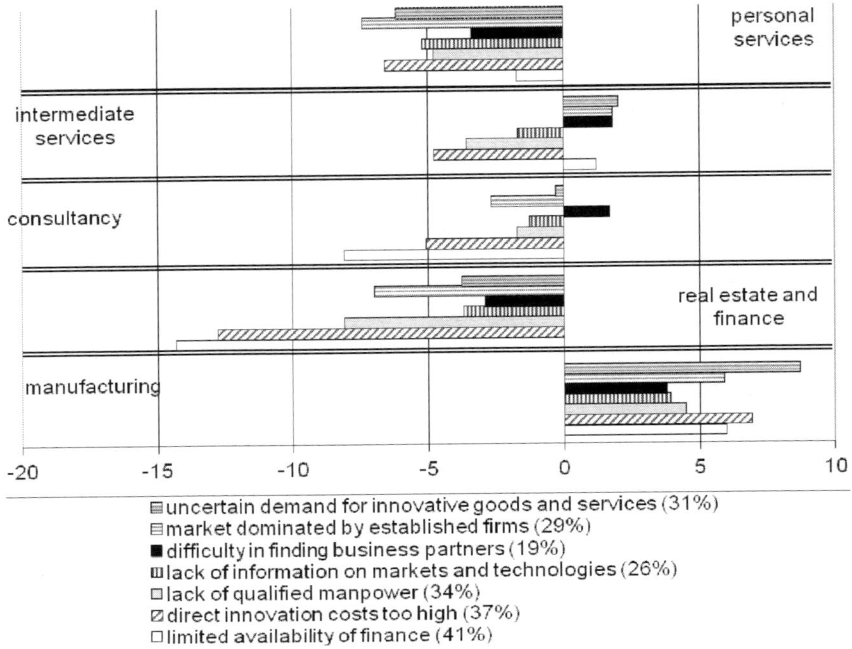

Figure 13.2 Barriers to innovation: whole sample

How to read the figure: within the whole sample of firms, 41 percent state that the lack of financial means slows down their innovative activity. This share is six points more (thus 47 percent) for firms from the manufacturing sectors exclusively.

Source: CIS 4 and EAE surveys.

with in their attempts to innovate. In addition, firms which have abandoned an innovation project on the go have a tendency to report more obstacles than those of the innovative sub-group. With regard to industrial differences, the manufacturing sector systematically declares more obstacles to innovation than the total group the group as whole (3.8 additional points for the difficulty in finding partners and 8.7 additional points for market uncertainty). Firms which have effectively innovated acknowledge more frequent difficulties than the whole group although the discrepancy between the two subsets is less significant (between 0.3 and 6.8 points). Firms having abandoned their innovation project show a perception of risk quite similar to those of the whole group.

Digging into the different services sectors revealed discrepancies depending on the type of service provided:

- Real estate and financial services gather firms which rather underestimate obstacles to innovation (3 to 14 fewer points than the whole

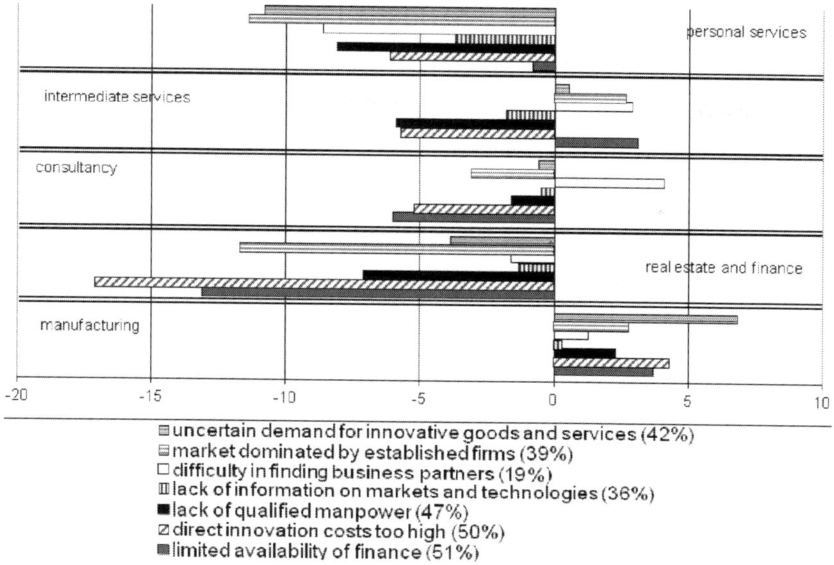

Figure 13.3 Barriers to innovation: innovative firms only

population). Firms with a successful innovation track record put the lack of qualified manpower forward as the main obstacle, most of the value stemming from the competences of human resources involved. Firms which have abandoned an ongoing innovation project blame the difficulty in assessing an uncertain demand for innovative services. Although the required knowledge is formal it is applied to unknown challenges (Kvalshaugen *et al.*, 2008). The main difficulty for those intending to innovate lies in the capacity to understand customers' needs so as to provide the appropriate solution. The ability to design a good customer value proposition is conditional to a "signaling capability" that enables the firm to take advantage of development opportunities and consequently of innovation.

- Firms evolving in intermediate services and personal services show similar profiles and echo the whole population's opinion when it comes to hierarchy of obstacles. Innovative firms in these two sectors rank limited finances and excessive costs first when explaining their difficulties in innovating. Non-innovative firms in intermediate services (respectively personal services) consider an uncertain demand (respectively a lack of qualified manpower coupled with a lack of information) as a major cause of failure in their path to innovation.

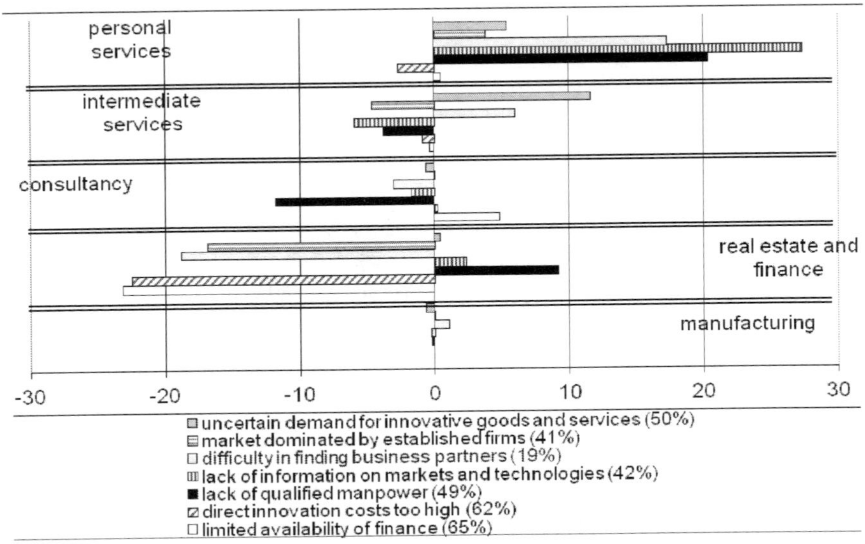

Figure 13.4 Barriers to innovation: firms with abandoned innovation project only

- The surveyed firms report again fewer obstacles especially in terms of financial means as the need for expensive infrastructure is low. Moreover, firms in this sector have difficulties in ordinating the obstacles: four of them score similarly. Firms which have abandoned their innovation projects consider limited availability of funds to be the first difficulty in importance.

The nature of obstacles faced by innovative firms and firms which have abandoned their innovation projects often differ. If all sectors do not suffer similar problems, the main and most recurrent obstacle cited by surveyed firms in the service sector revolves around the lack of (human/financial) resources. Whether a lack of resources induces weak dynamic capabilities will be discussed once all findings have been presented. However, a distinction is to be made between obstacles slowing down innovation and obstacles effectively hampering innovation depending on sectors. The next step of this empirical study investigates the relative impact of those disabilities on innovation further.

Hindrance of perceived obstacles on innovation

Figure 13.5 illustrates the impact of constrained perceptions on the final decision made by firms in terms of innovation.

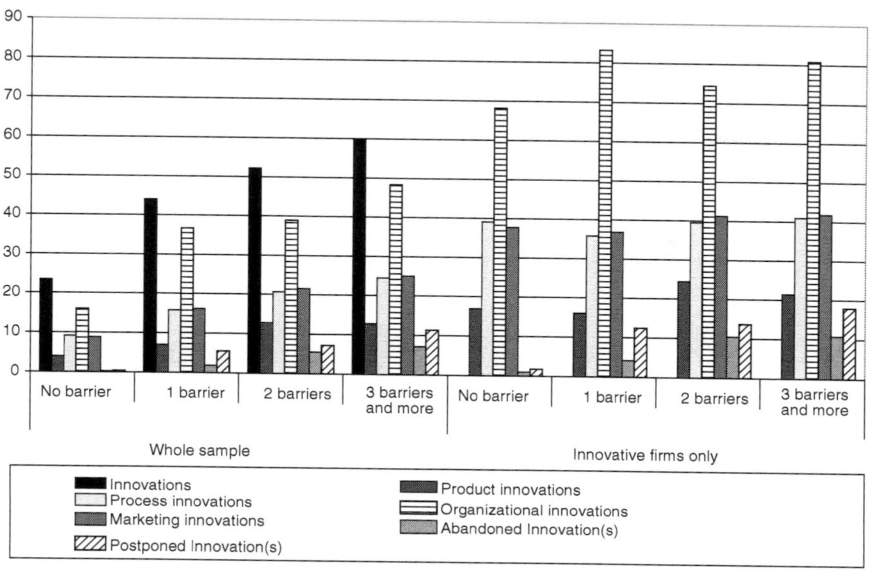

Figure 13.5 Constraints faced and innovation introduced in the service sector

How to read the figure: in the whole sample of firms, 23 percent of the firms which do not bear any barrier to innovation did effectively innovate, whereas this proportion climbs to 60 percent for the firms which bear three barriers or more.

Source: CIS 4 and EAE surveys.

Among the set of firms which did not report any obstacles 23 percent innovate and so do 60 percent of those reporting three different obstacles and more. Surprisingly, the total number of undertaken innovations increases with the number of perceived obstacles when all firms of the French service sector are being considered. The analysis refined to the different types of innovation undertaken (product, process, organizational, marketing) confirms this trend and stresses that despite reported barriers to innovation, firms engage, persevere and finally succeed in their efforts. This observation actually jeopardizes the assumption that perceived obstacles derive from effective problems. The positive and quasi-linear correlation on the whole population is less salient when only innovative firms are considered. We also observe that the proportion of firms which report process or marketing innovation is relatively stable regardless of the number of perceived constraints. However, as these two obstacles are of a strategic nature, they have to be overcome no matter the difficulty. With regard to the percentage of organizational innovations, it first rises and then drops before rising again. All in all, no significant and durable decrease appears when the number of obstacles increases. Predictably, failed innovative projects (postponed or abandoned) are positively correlated with the number of barriers faced by firms.

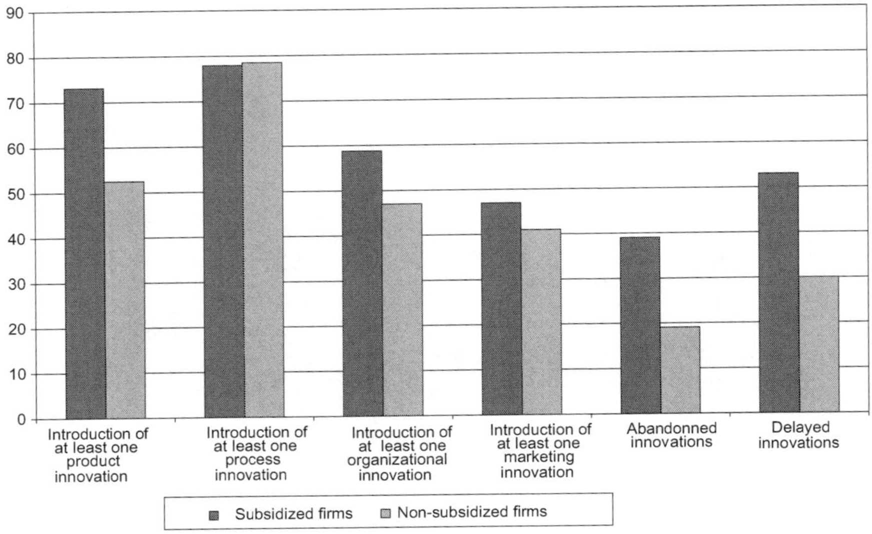

Figure 13.6 Impact of public support on the service sector

Two hypotheses can be drawn from these results. First, firms report plenty of barriers in obtaining additional public support although this strategy is highly questionable. The CIS survey indicates (Figure 13.6) that support supplied by public actors mostly benefits product innovation rather than marketing and organizational innovations. However, the latter encompass most of the innovative outcomes in the service sectors (Miles, 2008; Cordellier, 2009) and sponsored firms abandon their innovation projects more frequently and bear more delays than non-sponsored ones. All those findings suggest that public support does not necessarily catalyze an innovative activity in the service sector as government programs may not be tailored for services industries.

Second, obstacles faced by service sector firms are trivial and more easily overcome than those of the manufacturing industries. While some of the obstacles seem to hinder the innovative activity, none of them succeeded in deterring firms in their attempts, as if nothing really can hamper innovation in these industries.

We came to believe that reported obstacles do not necessarily reflect and account for effective difficulties. The next section intends to test this hypothesis with empirical statistics.

Effective characteristics and perceived obstacles

In this part, we focus on effective means available to firms and on their relation with perceived obstacles. We first investigate whether service firms are less innovative because they are less competent (compared to manufacturing firms) in leading an innovative project to its success. We then examine whether perceived obstacles underpin an effective lack of resources or reflect feeble dynamic capabilities.

Risk aversion or incapacity to run an innovative project?

If the list of seven obstacles provided in the CIS 4 survey does not match the six dynamic capabilities identified by den Hertog *et al.* (2010), we believe that a weak "conceptualizing capability" can be proxied by data calculated on other items of the survey. We consider a lack of conceptual capability highly detrimental to the chance to turn an idea into a completed project. Thanks to data on completed, abandoned or delayed projects, we could establish a failure rate as a proxy of "conceptualizing capability." Table 13.1 summarizes these results.

In the absolute, the first row of the table reflects low rates of innovation and the second a low rate of abandonment in the service sector in comparison with manufacturing firms. It is worth noticing that personal services firms have three times fewer abandoned projects and that consultancy and real estate exhibit the highest percentage of abandoned projects. Looking at the failure rate, manufacturing sectors appear more risk-taking in the sense that half of their innovation projects are abandoned. Seemingly, service firms tend to be more risk-adverse and decide to invest time and money in an innovative project when the likelihood to succeed runs high (which might explain the lower failure rate). Results on delayed information seem to emphasize this finding: as soon as a service firm engages in an innovative process, the uncertainty relates to the end date of the project rather than to the success of it. Again, consulting firms behave quite similarly to manufacturing industries while personal services exhibit the lowest innovative and failure rates. Personal services which engage in innovative activities do lead their projects to success: they seem to possess the "conceptualizing capability."

Our findings suggest that service sector firms suffer from elementary cultural flaws namely "fear of failure" and "lack of interest for innovation" which explains the sector's relative passivity when it comes to innovation. Manufacturing industries appear more risk-taking and abandon many projects on the run while the service firms venture into new projects only when the chance to succeed is high. We have paid attention to success factors until now and deemed it relevant to also investigate whether firms may suffer other disabilities in the next section.

Perceived obstacles not derived from effective resources

Table 13.2 presents whether perceived impediments vary according to the firm's effective resources.

Table 13.1 Innovative projects by sector: success and failure

Nature of innovation	*Manufacturing*	*Real estate and finance*	*Consultancy*	*Intermediate services*	*Personal services*	*Total*
Product or process innovation(s) completed, in progress or abandoned	35.6%	27.0%	25.7%	21.1%	17.2%	28.4%
Abandoned innovation(s)	11.2%	6.6%	5.9%	4.7%	3.9%	7.9%
Delayed innovation(s)	13.1%	9.3%	9.9%	5.7%	5.3%	10.1%
Abandoned or delayed innovation(s)	18.9%	12.4%	12.2%	9.2%	6.4%	14.0%
Failure rate*	53.0%	46.0%	47.5%	43.5%	37.2%	49.3%

* A given firm might declare it has abandoned one innovation project AND has endured a delay on another innovative project during the period under analysis. But what we are interested in is the probability for a given firm to see its innovative project either abandoned or delayed, but not necessarily both. Thus, the failure rate ratio accounts for the proportion of firms that declare being involved in innovative activities and which face either an abandonment or a delay exclusively, divided by the proportion of firms that have been engaged in innovative activities.

We first examine the potential relation between the actual qualification level of human resources in firms and the claim of these same firms about the lack of qualified workforce. We can see that firms who reported a lack of qualified manpower do not have different characteristics in terms of human resources as the total of surveyed firms. Of these firms, 24 percent have manpower without any qualification. Regarding firms reporting a shortage of qualified personnel, the distribution of qualifications among the manpower is quite similar (24.4 percent – 42.9 percent – 32.7 percent). In this case perceptions of obstacles and quality of workforce are not correlated. The perceived lack of qualified human resources is not supported by an effective shortage of qualified personnel. This result expresses a difficulty in acknowledging the value of internal competences and strengths and in transforming them into innovation potential and accordingly a deficit in "learning capabilities."

Concerning innovative firms, we note that those who declare enduring a qualified manpower shortage exhibit an underrepresentation of highly qualified employees while having more middle qualified employees than the whole population of innovative firms. Even if they do not have significantly fewer unqualified employees, they lack highly qualified staff although this

Table 13.2 Innovation impediments as perceived by service sector firms depending on manpower qualification

		Distribution of firms with reported lack of qualified manpower	*Distribution of firms regardless of perceptions*
Whole sample	% of employees with no qualification	24.4	24.0
	% of employees with a middle or high qualification	42.9	41.7
	% of employees with a high qualification	32.7	34.3
	Total	100	100
Innovative firms only	% of employees with no qualification	20.4	19.9
	% of employees with a middle or high qualification	42.5	39.6
	% of employees with a high qualification	37.1	40.5
	Total	100	100
Firms that have abandoned innovation only	% of employees with no qualification	21.4	16.8
	% of employees with a middle or high qualification	33.4	33.3
	% of employees with a high qualification	45.3	49.9
	Total	100	100

situation does not seem to impede their innovation potential. A lack of highly qualified manpower tends to be a perceived barrier rather than a deterrent one. Finally, firms which simultaneously reported a lack of manpower and a drop of an innovative project exhibit a higher proportion of unqualified manpower than the group of firms who have abandoned an innovation. A minimal qualification of staff may then act as a deterring factor for innovation and underpins failure. At the same time, however, they host more than 45 percent of qualified employees within their staff. A qualified manpower is not the panacea for innovation. Actually, the peculiarity of this sector is that services are produced and consumed almost simultaneously, implying an ad hoc production and almost no possible standardization. As soon as an activity is less "routinizable," creativity and learning capabilities are required along with a minimum level of qualifications to answer the customers' needs and be inventive.

There is no significant link between effective levels of qualification within firms, perceived obstacles and innovative performance. The difficulty in hiring qualified manpower does not systematically constitute an effective impediment preventing firms from being innovative. In addition, a lack of highly qualified manpower and the absence of qualification lead to different consequences in terms of innovation. This suggests that the number of employees is insufficient to guarantee innovation: "learning capabilities" matter more to harness creativity of individuals and to catalyze organizational learning.

Impact of market size

Table 13.3 summarizes the relation between market size and perceived obstacles in the service sector.

We note that when the geographical scope of the market increases the probability for a given firm to report barriers also rises. In addition, firms which declare some innovative barrier(s) belong less to the category of firms operating in a local market and more to those with larger markets. More precisely, the innovative barrier for underrepresented firms with local markets relates to a lack of information on technology or on the market. This actually confirms that a firm's capability to succeed in foreign markets leans on the capability to learn from new (foreign) sources of information on clients, competition or technologies (Brock and Alon, 2009). Such a double competence is not necessarily accessible to all firms and thus reinforces the decisive role played by the "signaling capability" in the innovation process. Teece *et al.* (1997) pointed out that firms own unequal capabilities when it comes to seizing opportunities in an impetuous environment.

Concentrating on innovative firms allows us to observe that those with a market of national scale adopt a different behavior than those of the whole group with the same market. Indeed, innovative ones show a tendency to minimize risk. Moreover, firms which simultaneously suffer from uncertain demand and difficulty in innovating are more frequently involved in local markets than the entire population of innovative firms. Conversely, as soon as they are involved in international markets, innovative firms no longer seem to endure great uncertainty (14.3 percent and 12.3 percent).

Our results suggest that firms with international development readiness seem better equipped to overcome any innovative barrier. This result can be attributable to upper risk-taking behavior among international service providers. An alternative explanation lies in potential variations in size as international service providers usually are established and large firms. However,

Table 13.3 Innovation impediments as perceived by firms depending on the scope of their market

		Lack of information on markets and technologies	*Difficulty in finding business partners*	*Market dominated by established firms*	*Uncertain demand for innovative goods and services*	*Distribution of the firms whatever their perception*
Whole sample	Local market	57.2	57.7	58.9	59.1	61.1
	National market	22.0	21.1	22.6	21.7	21.0
	European market	9.5	9.0	7.8	8.5	8.1
	Extra-European market	11.3	12.3	10.7	10.6	9.8
	Total	100.00	100.00	100.00	100.00	100.00
Innovative firms only	Local market	51.3	52.9	53.6	54.1	51.4
	National market	23.6	21.5	24.1	24.0	24.6
	European market	11.2	10.7	9.0	9.7	9.7
	Extra-European market	13.8	14.9	13.3	12.3	14.3
	Total	100.00	100.00	100.00	100.00	100.00

recent conclusions drawn from CIS data by Rubalcaba *et al.* (2010, p. 28) point out that

> the position of services in the variables does not seem to be conditioned by size (these are independent effects). Even if they are well endowed for innovation, international service firms more frequently mention cost and the difficulty to find a collaborative partner as barriers.

In short, these firms lack a "co-producing capability" which slows down their innovative capacity.

We presented a detailed analysis of obstacles to innovation faced by firms in the service sector with the objective to identify hindering factors. These barriers being identified, we suppose that appropriate (preventive and curative) strategies can be developed by the actors (private and public) to boost innovation. The next section summarizes our main findings and initiates a discussion on derived managerial implications.

Main findings and conclusion

Theoretical contribution

Our results indicate that limited access to resources is detrimental to innovation capacities and that these resources are of a different nature:

- *Tangible resources* such as money, technologies, patents, equipment and manpower. We found that limited availability of funds and manpower is presented as the most critical problem in the service sector. These resources are generally available outside the firm's boundaries and involve negotiations with third parties. Access to some of these means can partially be solved by public policies via empowering programs in a given territory. However, because most hard resources cannot be produced with idiosyncratic capabilities, they have to be negotiated with "suppliers." Access to these resources then depends on the quality of relational skills and networks.
- *Intangible resources* such as managerial skills, expertise or knowledge. As Drucker (1998) remarked, knowledge-based innovation usually demands not one kind of knowledge but many (p. 7). Insufficient capacity to take advantage of existing resources questions the quality of internal management capabilities.

We highlighted the consequences of weak or non-existent "signaling" and "conceptualizing" capabilities hindering innovation. As these capabilities are in essence related to human skills and competences, we deduce that most of the reported barriers relate to intangible resources. Additionally, when it comes to obstacles faced by innovative firms and those willing-to-innovate, not all obstacles have the same impact on innovation. Despite all barriers and roadblocks, firms in the service sector continue to innovate which suggests that obstacles rather slow down innovation than completely stop it. An alternative explanation can be drawn from the service sector firms' capabilities to succeed in overcoming these barriers and in pursuing their innovative projects to their completion. There is strong evidence that the service sector firms developed a significant "conceptualizing capability" which allows them to efficiently turn their ideas into concrete innovations.

Our empirical study also reflects the peculiarities of service innovation (as compared to innovation in goods). For instance, information on new technologies and funding seems to be of higher concern in the sector of goods than in services. Some service firms show a weak "signaling capability" reflecting a difficulty in effectively foreseeing and reading market changes. The picture on obstacles is uneven and depends on the studied service sector. Business services which represent the biggest share of innovative firms perceive more obstacles (even more than manufacturing firms) while consulting firms have no reported financial difficulties, although they seem hindered by a weaker "conceptualizing capacity."

Another objective of this study was to investigate whether the difficulty in transforming existing resources into innovation expresses a lack of competences and weak dynamic capabilities. We therefore confronted the reported obstacles with the owned resources and found that: 1) there is no direct correlation between firms' declared shortage of qualified human resources and the effective qualification levels in those firms and 2) an increase of the market size of innovative firms is not correlated with the perceived market-related obstacles. What really seems to slow down the firms' willingness to innovate is again an insufficient "signaling capability." Some firms show a weak capacity to make teams more innovative, a flaw we consider more critical than resource scarcity for innovative activities. A shortage of resources may slow innovation down whereas managerial incompetence can block the innovation process from the start.

Dynamic capabilities as a shield against obstacles to creativity and innovation

One of the dynamic "incapabilities" indentified by Teece *et al.* (1997) is the inability of managers to accurately decipher the economic environment so as to identify new business opportunities and competition. Our findings reveal that some service firms lack "signaling capabilities" which translates into a difficulty in effectively foreseeing and reading market changes. We deem these inabilities as highly critical because they are supposed to act as prerequisites for innovative actions and further conceptualization. How well a dynamic capability enables an organization to make a living by creating, extending or modifying its resource base refers to "evolutionary fitness" (Helfat *et al.*, 2007). Higgins (1999) argues that several trends have caused a renewed interest for market research including increased customer focus, more demands for resource productivity and heightened domestic and international competition. The continuous collection of information requires innovative thinking and more creativity about how things are done. The difficulty in envisioning the market in a non-traditional way and in ignoring a legitimate fear of trend effects may also moderate entrepreneurial enthusiasm. Sutton (2001) approaches innovation and creativity as a mindset shift which, once overcome, causes people to see old problems in new ways and help companies break from the past. Managing for innovation often means shifting traditional, rational approaches to hiring, management and risk 180 degrees. The study did not allow us to test managerial audacity and propensity to take new pathways and the picture on obstacles is uneven and depends on the studied service sector. Business services perceive more obstacles while consulting firms have no reported financial difficulties, although they seem hindered by a weaker "conceptualizing capacity." Devising innovative services and new ways to please customers brings the difficulty in prototyping a service in a 3D picture.

The next section addresses managerial implications of this study and delves into the ability of managers to take up opportunities arising in the business environment.

Managerial implications and conclusion

This chapter provides a theoretical contribution on barriers to innovation and examines what role resources and dynamic capabilities play in the innovation process. The refined typology elaborated by den Hertog *et al.* (2010) enabled us to identify obstacles and capability shortages for each service sector. However, one limit of this study resides in considering resources and capabilities as independent root causes of innovative behavior, whereas some of them are evidently correlated[2] (for instance a highly educated manpower often presented with higher learning and adaptation skills or a lack of financial resources ruining the chance to hire highly qualified staff). Data and methodology used in this research only allowed us to observe co-occurrence between obstacles and innovation regardless of causality and/or time lag. Future studies will be performed to control these effects in order to supplement this preliminary quantitative study with qualitative analysis.

We deem the results of this empirical investigation of utmost interest for managers as it may provide them with "food for thought" to develop competences skills within their teams to stimulate and speed up the pace of innovation. Do weak dynamic capabilities put managerial competences in question? Adner and Helfat (2003) use the term "dynamic managerial capabilities" to refer to the capacity to create, extend or modify the resource base of an organization. Helfat *et al.* (2007) (re)confirm that the concept of dynamic capabilities challenges organizational and managerial competences. Our results tend to suggest that firms suffering from a lack of innovation capacity should invest in strengthening or acquiring negotiation skills to access the missing financial or technological resources they need to launch an innovation project. As far as collaborative capabilities are concerned, Dyer and Kale (in Helfat *et al.*, 2009) explain why some firms are better in partnering than others and identified drivers of relational advantage: careful selection of partners, complementary capabilities, substantial knowledge flow between partners and the creation of relationship-specific assets. The reported difficulty in finding a business partner is an innovation barrier in a nutshell as partnering owns more complex features.

A priori, creativity in the service sector should be easier and inexpensive compared to product development as it might require less financial means and R&D efforts to design new services and to market them. As creativity in services seems less constrained by innovation stages, patenting, engineering skills and technical challenges, the service sector should be more versatile and develop dynamic capabilities of a peculiar nature. At the end of this chapter, we more precisely conclude that the key to innovation in the service sector rather lies in finding a good orchestra conductor, able to integrate and reconfigure internal and external resources, than in hiring a virtuoso and buying expensive instruments.

Notes

1 EAE: Enquête Annuelle d'Entreprise – annual report on French firms.
2 We also consider (the lack of) each resource as an independent obstacle, neglecting to account for the (cumulative?) effect of the co-occurrence and interdependence of different obstacles on innovation.

References

Adner, R. and Helfat, C. (2003) "Corporate effects and dynamic managerial capabilities," *Strategic Management Journal*, Special Issue **24**(10): 1011–25.

Baldwin, J. and Lin, X. (2002) "Impediments to advanced technology adoption by Canadian manufacturers," *Research Policy*, **31**: 1–18.

Brock, D. and Alon, I. (2009) "Internationalization of professional service firms," *International Business: Research, Teaching and Practice*, **3**(1): 52–70.

Collis, D.J. (1994) "Research note: How valuable are organizational capabilities?" *Strategic Management Journal*, **1**: 143–52.

Coombs, R. and Miles, I., (2000) "Innovation measurement and services: The new problematique," in S. Metcalfe and I. Miles (eds), *Innovation systems in the service economy: measurement and case study analysis*. Kluwer academic publishers: Dordrecht, the Netherlands: 83–102.

Cordellier, C. (2009) *Innovations et performance des sociétés de services et de l'industrie manufacturière, comparaisons*. INSEE série de documents de travail de la direction des statistiques d'entreprises. E2009:6.

DeJong, J. and Vermeulen, P. (2003) *Organizing successful new service development: a literature review*. Scales-Paper N200307. EIM Business and Policy Research.

Den Hertog, P. (2000) "Knowledge intensive business services as co-producers of innovation," *International Journal of Innovation Management*, **4**(4): 491–528.

Den Hertog, P., Van Der Aa, W. and De Jong, M.W. (2010) "Capabilities for managing service innovation: Towards a conceptual framework," *Journal of Service Management*, **21**(4): 490–514.

D'Este, P.P., Iammarino, S., Savona, M. and Von Tuzelmann, N. (2008) *What hampers innovation?* Evidence from the UK CIS4. SPRU Electronic.

Drejer, I. (2004) "Identifying innovation in surveys of services: A Schumpeterian perspective," *Research Policy*, **33**(3): 551–62.

Drucker, P. (1998) "The discipline of innovation," *Harvard Business Review*. HBR Classic November–December.

Galia, F. and Legros, D. (2004) "Complementarities between obstacles to innovation: Evidence from France," *Research Policy*, **33**: 1185–99.

Gallouj, F. (2002) *Innovation in the service economy: the new wealth of nations*, Edward Elgar Publishing: Cheltenham.

Gallouj, F. and Savona, M. (2009) "Innovation in services: A review of the debate and a research agenda," *Journal of Evolutionary Economics*, **19**: 149–72.

Hauknes, J. (2001) "Service economy in international encyclopedia of business and management," In *Handbook of economics*. International Thompson Press: London.

Helfat, C., Finkelstein, S., Mitchell, W. et al. (2007) *Dynamic capabilities*. Blackwell Publishing: Oxford.

Higgins, L.F. (1999) "Applying principles of creativity management to marketing research efforts in high-technology markets," *Industrial Marketing Management*, **28**(3): 305–17.

Hughes, A. and Wood, E. (2000) "Rethinking innovation comparisons between manufacturing and services: The experience of the CBR SME survey in the UK," in S. Metcalfe and I. Miles (eds), *Innovation systems in the service economy: measurement and case study analysis*. Kluwer academic publishers: Dordrecht, the Netherlands.

Iammarino, S., Sanna-Randaccio, F. and Savona, M. (2007) *The perception of obstacles to innovation. Multinational and domestic firms in Italy*. Working Paper Beta No. 2007–11.

Katila, R. and Shane, S. (2005) "When does lack of new resources make new firms innovate?" *Academy of Management Journal*, **48**(5): 814–29.

Kvalshaugen, R., Hydle, K.M. and Bloch, C. (2008) "Towards a taxonomy for business service innovation: Literature review," *Servinno*, April.

Landry, R. and Amara, N. (2005) *Les obstacles à l'innovation: perspectives pour le développement cohérent des entreprises innovantes de la Montérégie*. Valortech.org, Québec.

Loewe, P. and Dominiquini, J. (2006) "Overcoming the barriers to effective innovation," *Strategy and Leadership*, **34**(1): 24–31.

Menor, L.J. and Roth, A.V. (2007) "New service development competence in retail banking: Construct development and measurement validation," *Journal of Operations Management*, **25**(4): 825–46.

Miles, I. (2008) "Patterns of innovation in service industry," *IBM System Journal*, **47**(1): 115–28.

Miller, D. (1998) "Configurations revisited," *Strategic Management Journal*, **17**(7), 505–12.

Mohnen, P. and Rosa, J. (2000) *Les obstacles à l'innovation dans les industries de service au Canada*. Working Paper Cirano, 2004–14.

Ness, G. and Brechin, S. (1988) "Bridging the gap: International organizations as organizations," *International Organization*, **42**(2): 245–73.

Ostrom, A.L., Bitner, M.J., Brown, S.W. *et al.* (2010) "Moving forward and making a difference: Research priorities in the science of service," *Journal of Services Research*, **13**(1): 4–36.

Penrose, E. (1995) *The theory of the growth of the firm*. 3rd edition. Oxford University Press: Oxford.

Rodriguez, M. and Camacho, J. (2010), "Are knowledge-intensive business services so 'hard' innovators? Some insights using Spanish microdata," *Journal of Innovation Economics*, **1**(5): 41–65.

Rubalcaba, L., Gago, D. and Gallego, J. (2010) "On the differences between goods and service innovation," *Journal of Innovation Economics*, **5**(1): 17–38.

Schoonhoven, C., Eisenhardt, K. and Lyman, K. (1990) "Speeding products to market: Waiting time to first product introduction in new firms," *Administrative Science Quarterly*, **35**: 177–207.

Segal-Horn, S. and Dean, A. (2007) "The globalization of law firms: Managerial issues," *International Journal of Service Industry Management*, **18**(2): 206–19.

Soete, L. and Miozzo, M. (1990) *Trade and development in services: a technological perspective*. MERIT: Maastricht.

Sutton, R. (2001) "The weird rules of creativity," *Harvard Business Review*, Sept.

Teece, D. (2007) "Explicating dynamic capabilities: The nature and microfoundations of (sustainable) enterprise performance," *Strategic Management Journal*, **28**: 1319–50.

Teece, D., Pisano, G. and Shuen, A. (1997) "Dynamic capabilities and strategic management," *Strategic Management Journal*, **18**(7): 509–33.

Tether, B. (2003) "The sources and aims of innovation in services: Variety between and within sectors," *Economics of Innovation and New Technology*, **12**(6): 481–505.

Tether, B., Miles, I., Blind, K., Hipp, C., De Liso, N. and Cainelli, G. (2002) *Innovation in the service sector: analysis of data collected under community innovation survey*. CIS2, CRIC Working Paper No. 11. University of Manchester, UK.

Tourigny, D. and Le, C.D. (2004) "Impediments to innovation faced by Canadian manufacturing firms," *Economics of Innovation and New Technology*, **13**(3): 217–50.

Wölff, A. (2005) *The service economy in OECD countries*. OECD-STI Working Paper 2005/3.

14 Research, teaching and commercial activities

The behaviour of laboratories in an entrepreneurial university

Mireille Matt and Véronique Schaeffer

Introduction

The traditional role of academic research in society is to contribute to the accumulation of scientific knowledge in every field. One of the engines of a creative economy being knowledge sharing, academic researchers and universities have an important role to play in fostering creativity. At the end of the 1990s the entrepreneurial model of university emerged (Clark, 1998; Etzkowitz, 1998) and became the implicit model underlying science and technology policies discourses. Beside their traditional missions of education and research, universities have to develop the commercialisation of the results of research activities, to set up partnerships with economic and political actors, to define patenting policies and to encourage the entrepreneurial behaviour of researchers (Etzkowitz, 2003; Rasmussen *et al.*, 2006). The generalisation of the entrepreneurial model generates debates questioning the pertinence of a unique paradigm guiding the evolution of a diversified academic system, the role of universities in society and the impact of the commercialisation of the results of research on scientific communities (Conceiçao and Heitor, 1999; Florida and Cohen, 1999; Geuna *et al.*, 2004; Mailhot and Schaeffer, 2005, Cohendet *et al.*, 2006). The diversification of the missions of universities raises new challenges for universities. They have to favour the development of entrepreneurial and commercial activities without being detrimental to teaching and research activities (Etzkowitz, 1996, 2003).

The aim of our contribution is to explore the compatibility between teaching, research and commercial activities at the level of laboratories. The economy of science provides many quantitative studies about the impact of the commercialisation of science on research productivity (Breschi *et al.*, 2007; Stephan *et al.*, 2007; Azoulay *et al.*, 2009). Others studies explored the relation between research and teaching performance (Fox, 1992; Gander, 1995; Marsh and Hatie, 2002). An originality of this contribution is to explore simultaneously the compatibility between teaching, research and commercial activities.

We use the case of a French scientific university having an old tradition of research, highly engaged in the development of the commercialisation of the results of research and presenting the features of an entrepreneurial university. Another originality of our contribution stems from the richness of our database. In quantitative approaches, commercial activities are often reduced to patenting ones. We seize their diversity through the number of patents, the share of contractual research in the total funding of the laboratories and the number of publications with industrial co-authors.

While many previous studies on the complementarity between the three main missions of universities focus on the individual level, we focus on the laboratory level. Carayol and Matt (2004), using the same database as we do to study the determinants of productivity at the laboratory level, propose a typology of labs that highlights different styles of research organisation and productivity. Stephan (1996) underlines that research is a collective activity, so that factors explaining the productivity of research activities have to be considered at the collective level of production and not only at the individual one.

In the first part of the chapter we present a literature review about the complementarity between teaching, research and commercial activities in universities. The second part describes University Louis Pasteur and the type of data collected. In the third part we develop a typology of labs. The last part stresses the benefits of diversity within universities and the challenges it induces to university managers.

A review of the literature

The literature deals with the complementarity between teaching, research and commercial activities regarding on the one hand the effect of commercial activities on the productivity of research and on the other hand the relation between teaching and research.

Commercialisation of research and research productivity

Many authors addressed the question of the impact of the commercialisation of research on academic productivity and found a positive link between the two activities. Azoulay *et al.* (2007, 2009) examined the patenting behaviour of 3862 scientists in biomedical fields from 1967 to 1999. They showed that patents and publications, encoding similar pieces of knowledge, are two outputs of a same process. Azoulay *et al.* (2009) examined the relation between patenting and publication in terms of quality and found a positive weak effect on the quality of publication. Fabrizio and Di Minin (2008) studied the relation between publishing and patenting in several institutions and academic areas, Stephan *et al.* (2007) used the Survey of Doctorate Recipients, Thursby and Thursby (2011a) considered the research and invention disclosure of faculty at 8 US universities from 1983 to 1999. Their results confirmed that publication and patenting are complementary and not substitute activities for faculty members. Research with a commercial potential did not divert faculty from their role in basic research. Fabrizio and Di Minin (2008) also showed that the positive relation between publications and patents declines with an increasing number of patents held by the faculty researcher. Thursby and Thursby (2011b) used data on research and invention disclosures of science and engineering faculty at 11 US universities and showed that recent divulgation of an invention has a positive impact on both government and industrial funding. However, when divulgation is repeated several times the effect on government funding becomes negative.

European researches confirmed this positive relation between patenting and publishing. Breschi *et al.* (2007) investigated the scientific productivity of Italian academic inventors from 1978 to 1999 and used a longitudinal data set comprising 299 academic inventors. They matched this population with an equal number of non-patenting researchers on the basis of the number and the quality of publications. They found a strong and positive relationship between patenting and publishing, even in basic science. This result is also shown by Carayol and Matt (2004) in the case of a large French scientific university (ULP), Calderini *et al.* (2007) with a longitudinal analysis of patent and publication performances of Italian researchers in Materials Science and Van Looy *et al.* (2006) using the population of the 32 inventors of the University of Leuven. Calderini *et al.* (2007) distinguished in a same field of research between applied and fundamental research. For scientists engaged in applied research there is a positive link between publication and patent performance. For scientists engaged in more fundamental research, it is more difficult to find the time to produce industrial applications. Moreover, Breschi *et al.* (2007) found that the correlation between patenting and publishing is strongest when patents are owned by business partners and not by scientists or universities. They suggested that the positive correlation between patenting and scientific productivity relies on the advantages derived from solid links with industry. Van Looy *et al.* (2006) showed that inventors publish more than non-inventors, even before the first patenting activity is observed. As such, this observation suggests the presence of the 'star scientist' phenomenon. At the same time, it is apparent that involvement in patenting activities increases the publication difference in favour of inventors. Inventors tend to publish more in scientifically oriented journals than non-inventors. Involvement in both activities may produce mutually beneficial effects. For some inventors the insights they obtain from a detailed scrutiny and awareness of the patent literature contribute to improving the quality of their fundamental research.

Research and teaching

The connection between research and teaching is a key feature of higher education. Teaching and research activities are considered as mutually reinforcing. Marsh and Hattie (2002) proposed a synthesis of the arguments supporting this positive connection. Research contributes to the quality of teaching because teachers involved in research activities are more likely to be at the forefront of their discipline, to use their research to update a teaching topic, to transmit to students the ability to develop a critical approach rather than passively accepting facts. Inversely, teaching can contribute to research. Teaching leads academics to have a clear vision of the whole knowledge in a field. They can identify missing elements of knowledge and research conducted by teachers can benefit from interactions with students when they present the results of their research.

But several quantitative studies about the influence of teaching on research productivity support a reverse position for which teaching and research are antagonistic activities. Gander (1995) studied 31 academic departments of the University of Utah over the period 1991–1994 and explored the relationships between faculty research and teaching productivity. He showed that the teaching productivity increases and the research productivity decreases when the number of students increases. He also showed that the teaching productivity decreases and the research productivity increases when the faculty size increases. Fox (1992) used a US survey of almost 4000 social science faculty on work attitude and practices. She showed that those who have a high productivity in research have a weak investment in teaching. They spend fewer hours preparing their courses and consider teaching a less important task than research. Teaching can generate a conflict of engagement with research activities (Etzkowitz, 1996).

The aim of our contribution is to analyse the complementarity between teaching, research and commercial activities at the level of laboratory. For Callon *et al.* (1997) the evaluation of the production of laboratories has to take into consideration five dimensions of the activities of laboratories: knowledge sharing with peers, training for research through research, participation in the creation of collective goods, participation in the economic dynamics through technology transfer and participation in public debates. Laboratories are research units but their members are also involved in time consuming teaching or administrative activities that might influence their research efficiency. According to us the teaching dimension has to be added to measure the production of the members of a laboratory, because researchers are also involved in teaching activities.

Presentation of the data

After a presentation of the main features of University Louis Pasteur, we present the data we used and the way we collected them. Then we define the variables we used to measure the involvement in teaching, research and commercialisation of the results of research at the level of laboratories.

University Louis Pasteur (ULP) as an entrepreneurial university

In 2009 ULP merged with the two other universities of Strasbourg and is now a part of the University of Strasbourg, which is one of the biggest universities in France with 42,000 students and 4600 employees. ULP was rather large and diversified in the French context, enrolling 18,500 students plus 1260 doctoral ones. It had an old tradition of fundamental research and its researchers received numerous national and international scientific prizes, including Nobel Prizes. Overall, ULP was one of the largest French universities in terms of research. The Third European Report on Science & Technology Indicators 2003 ranked ULP 1st among French universities in terms of impact and 11th among European universities. Its research activity was structured around 13 Research Federations pooling 74 research labs. Research and teaching activities covered a wide range of subjects: Medical Sciences, Mathematics, Computer Science, Physics, Chemistry, Life Sciences, Geology, Geophysics, Astronomy and Engineering Sciences. Human and social sciences were also present with Economics, Management, Geography, Psychology and Educational Sciences. One peculiarity of the French system is the association of university research facilities with those of public research organisations (PROs) such as the National Centre for Scientific Research (CNRS) and the National Institute for Health and Medical Research (INSERM). At ULP a majority of university labs were associated either with CNRS or with INSERM. They received public funds and positions from the Ministry of Research and from the associated PROs. Half of the 1650 researchers of ULP were from CNRS or INSERM. ULP was highly engaged in the valuation of research activities. In 1987, it established its own technology transfer office (TTO) called 'ULP-Industrie', one of the first TTOs created in France. In 2000, it created an incubator and about 40 start-ups have been created since 2000. The increasing number of patents, contractual activity, co-publications with industrial scientists, start-ups and professional teaching programmes undoubtedly provide evidence of an increasing and important entrepreneurial behaviour of ULP.

Type of data

Information related to the laboratories and to the personnel was extracted from administrative reports. Each laboratory issues them every four years and they are evaluated at the level of the Ministry of Higher Education, CNRS and INSERM. They account for the scientific activity of the laboratory and the project for the next four years. To collect data about the publications of the permanent researchers of the laboratories we used the ISI Web of Science. The French patent office provided data about patents of researchers of ULP laboratories.

The laboratories

The laboratory database provides information about the disciplinary orientation, the annual budget applied for at the Ministry of Higher Education and at the associated PROs, the number of researchers, their institutions of origin, the number of PhDs and Post-docs, the administrative and technical staff. We recorded 84 labs, but for our analysis we kept only 59 labs. We excluded all the medical labs, as some of their information is rather unclear.

The personnel

The personnel database includes information about all the permanent researchers or professors: their laboratory of origin, the type of position, date of birth, employer and discipline of research. In France, the permanent scientists may occupy either a full-time research position or a 'teaching-research' position. We will use the share of 'teaching-research' positions as an indicator of the degree of involvement of the laboratories in teaching activities.

The publications

For each permanent scientist present in our database in 2000, we collected their published papers using SCI, SSCI and Arts and Humanities in the ISI Web of Science database. We found 19,850 publications over the period 1997–2001 covered in this chapter. This amount includes some double counting as some ULP researchers have co-authored papers. To solve this problem, we divided each paper by the number of co-authors and we obtained around 4000 normalised publications for the considered period. We also gathered information about the external co-authors and especially the co-authorship with industrial partners.

The patents

Concerning the French patents and their international extensions, which were invented by at least one ULP permanent researcher, we used the patent data provided by the French Patent Office (INPI). Over the period 1997–2001, the database contains around 300 patents. After a cleaning process, which eliminated all the extensions, we ended up with 152 patents.

The contracts

'ULP-industrie' and the CNRS TTO provided information concerning all the contractual activities of the permanent researchers, with a private company, a French ministry, the European Commission and the local public authorities. Our database contains different types of contracts: services, research contracts, co-development, grants. Between 1997 and 2001, ULP signed around 2100 contracts for a total income of 100 million euros.

The variables

We analysed the activity of each laboratory between 1997 and 2001 in order to capture the importance of teaching, research and commercial activities. We measured the following dimensions of the activity of laboratories:

Table 14.1 Definition of the variables

Variables	*Description of the variables*
pubmean	Number of normalized publications per permanent researchers
pubindmean	Number of publications per permanent researchers co-authored with industrial partners
patmean	Number of patents per permanent researchers
indcontshar	Amount of contracts signed with industrial partners divided by the total budget (all types of contracts + budget from PROs and MHER)
pubcontshar	Amount of public contracts (signed with public actors or within a public programme) divided by the total budget (all types of contracts + budget from PROs and Ministry of research)
profshare	% of permanent researchers that occupy a professor type of position

commercial outputs (patents, industrial co-authorship, contractual activity), research production (publications) and teaching orientation (teaching-research positions) of the members of the laboratories. The variables we used are presented in Table 14.1.

Table 14.2 summarises some descriptive statistics concerning our variables. On average, the share of budget derived from industrial contracts is 13.5 per cent and from public contracts 23 per cent, the maximum being respectively 86 per cent and 79 per cent. The share of professor type positions is on average 51 per cent, some labs involving only researchers and others only teaching-research positions. The average number of normalised publications per permanent researcher is 2.6 and the average number of publications per permanent researcher co-authored with industry equals 0.5. The average number of patents per permanent scientists is 0.1 with an important variation between labs.

Table 14.3 presents the correlation between the variables. Publications are positively correlated with patents and with co-authorship with industry; and patents and industrial co-authorship are also positively correlated. The share of professors is negatively correlated with publications per permanent researcher. Industrial contracts are slightly positively correlated with industrial co-publications and patents.

Results

We used quantitative variables (Table 14.1) to conduct a Principal Component Analysis (PCA) and an Ascendant Hierarchical Classification (AHC). The aim is to identify which dimensions of the activities of the laboratories are

Table 14.2 Descriptive statistics of the variables

	Mean	*Std deviation*	*Min*	*Max*
pubmean	2.5912	1.8379	0.0588	8.9352
profshar	0.5153	0.3010	0.0000	1.0000
pubindmean	0.5056	0.5609	0.0000	2.3636
patmean	0.1234	0.2330	0.0000	1.2857
indcontshar	0.1353	0.1864	0.0000	0.8629
pubcontshar	0.2286	0.1944	0.0000	0.7910

Table 14.3 Correlation coefficients between our variables

	indcontshar	*pubcontshar*	*profshar*	*pubindmean*	*pubmean*	*patmean*
indcontshar	1.0000					
pubcontshar	−0.1686	1.0000				
profshar	−0.1032	0.1643	1.0000			
pubindmean	0.2164	−0.1697	−0.2803	1.0000		
pubmean	0.0459	0.0454	−0.4711	0.5508	1.0000	
patmean	0.3095	−0.0174	−0.3596	0.4400	0.5317	1.0000

associated in order to identify specific behaviours of labs corresponding to different realities of the complementarity between the various activities.

Principal Component Analysis

The result of the PCA may be first summarised by the eigenvalues. Figure 14.1 shows the six eigenvalues and the percentage of variance explained by each of them. Taking account of the eigenvalue and the cumulated percentage of variance explained, we select three axes that represent 75 per cent of the total variance. Figure 14.2 is a projection of the six active variables on the first factorial plan.

Table 14.4 describes the most significant variables associated with the first three factorial axes.

The first component, which represents 40.5 per cent of the total variance, corresponds to the number of normalised publications per permanent researcher, the number of patents per permanent researcher and the number of co-publications with industry per permanent researcher. It distinguishes between labs highly involved in research and commercial activities through industrial co-publications and patents on the right hand side of the axis and labs that are not, on the left hand side (Figure 14.3). Even if the share of professors (profshar) does not contribute highly to the construction of the first

#	Eigenvalue	Per cent	Cumul
1	2.4324	40.54	40.54
2	1.1482	19.14	59.68
3	0.9200	15.33	75.01
4	0.7006	11.68	86.69
5	0.4760	7.93	94.62
6	0.3227	5.38	100.00

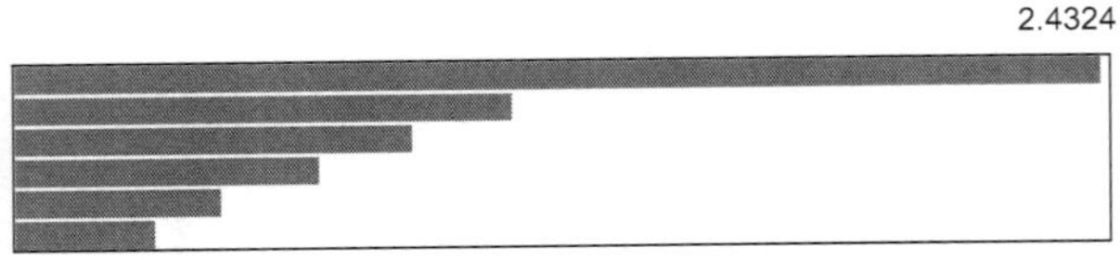

Figure 14.1 Eigenvalues and percentage of explained variability

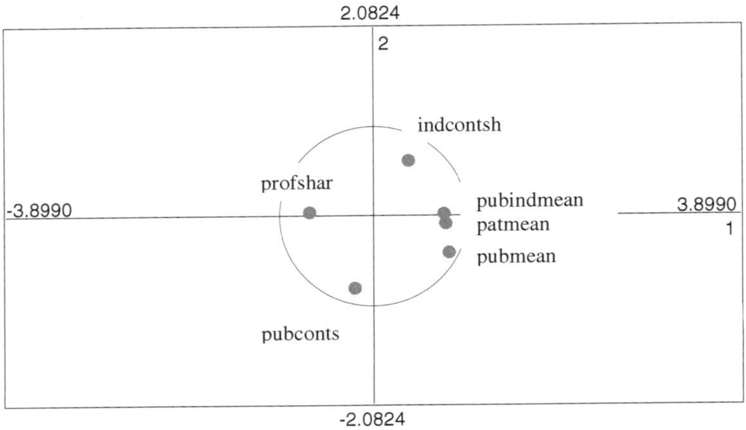

Figure 14.2 Projection of the variables on the plan 1×2

component, its left hand side position would suggest that the labs highly involved in commercial activities consist mainly of full-time researchers.

The second component (19 per cent of the variance) represents a measure of the share of industrial contracts (positive side) by opposition to the share of public contracts (negative side). It sets apart labs contracting with private companies and labs that use public actions to get external money. These public actions involve projects from the European Framework Programme, but also regional funds, subsidies from French ministries, PROs and public associations.

The third component (15 per cent of the variance) corresponds to the involvement in contractual activities (positive side). It distinguishes between labs looking for external extra money and labs relying heavily on annual funds provided by the supporting institutions (MHER, CNRS and/or INSERM).

The dispersion of labs through the first factorial plan distinguishes clearly between the labs highly involved in research and in commercial activities (the three labs on the right border of Figure 14.3) and labs weakly involved in research and commercial activities (those outside the circle on the left

Table 14.4 Variables defining the three components

Variables defining the three first axes	
First axis	
Negative side	Positive side Publications/permanent researcher Industrial co-authorship/permanent researcher Patent/permanent researcher
Second axis	
Negative side Share of public contracts	Positive side Share of industrial contracts
Third axis	
Negative side	Positive side Share of public contracts Share of industrial contracts

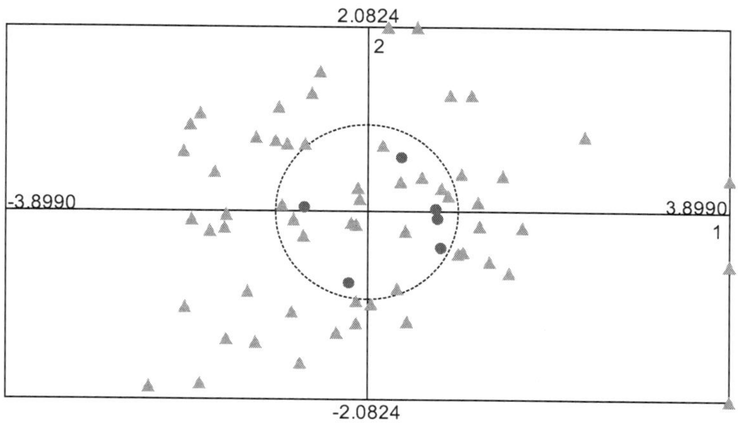

Figure 14.3 Projection of labs in the first factorial plan

hand side of the graph are weakly involved). This first factorial plan also allows us to separate the labs working intensively with industry via industrial contracts (two labs on the upper border of Figure 14.3) and those getting involved in public contracts (a few labs located outside the circle in the lower part of the graph).

The Hierarchical Classification

The Hierarchical Classification allows us to identify seven homogeneous classes. In retaining these seven classes, the within-classes variance is 33

Table 14.5 Means of the variables for each class

	Class 1 (13)	*Class 2 (2)*	*Class 3 (3)*	*Class 4 (11)*	*Class 5 (10)*	*Class 6 (9)*	*Class 7 (11)*
pubmean	3.5314	0.6619	6.7802	3.8584	1.5362	1.2155	1.5059
profshar	0.5213	0.5000	0.1057	0.2633	0.3473	0.8610	0.7444
pubindmean	1.1766	0.2404	1.6966	0.4327	0.1709	0.1460	0.1075
patmean	0.0956	0.1923	0.9185	0.1335	0.1098	0.0280	0.0070
indcontshar	0.1549	0.7966	0.3090	0.1120	0.0378	0.1402	0.0524
pubcontshar	0.1312	0.0417	0.2947	0.3114	0.0760	0.1049	0.5167

Table 14.6 Characteristics of each class

	(−) Characteristics (+)	
Class 1		Industrial co-authorship publication
Class 2		Share of industrial contracts
Class 3		Patents Publication Industrial co-authorship
Class 4	Share of professors	Publication
Class 5	Share of public contracts Publication Industrial co-authorship Share of industrial contracts	
Class 6	Publication Industrial co-authorship	Share of professors
Class 7	Share of industrial contracts	

per cent of the total variance. Table 14.5 exhibits for each class the means of the different variables.

Table 14.6 underlines the variables contributing to the formation of the classes.

We could pool these seven classes into three groups according to the types of activities they are conducting. A first group (Classes 1, 2, 3) pools labs with commercial activities. One class (Class 4) represents research-oriented labs. The last group involves the labs weakly involved in commercial activities (Classes 5, 6 7).

First group: Laboratories with commercial activity

Classes 1, 2 and 3 include labs highly involved in commercial activities. They co-publish with industry or establish contracts with firms or patent their results or combine these types of activities. Except for Class 3, which exhibits the highest scores in terms of publication, patents and industrial co-authorship, the other labs are more specialised in one or the other activity. We may also underline that Class 1 and Class 3 combine research and commercial activities whereas Class 2 concentrates mainly on commercial activities.

CLASS 1: LABS PUBLISHING WITH INDUSTRY

The common behaviour of the 13 laboratories of this class is to publish with industry 2.3 times more than the average. Considering the overall publications they also publish much more than the average. These labs do not patent and do not use public contracts as a source of funding, but obtain as much industrial contracts as the average of our sample. They are rather small in terms of number of researchers and employ an equal share of full-time researchers and teaching-research positions. The main disciplines of these labs are very diverse: biotechnology, fundamental neurosciences, information technologies, chemistry and material science.

CLASS 2: LABS SIGNING CONTRACTS WITH INDUSTRY

The two small labs of this class are characterised by an important share of industrial contracts in their budget (around 80 per cent). One of them is active in information technologies, employs only professors, who have created start-ups and have patented. Their overall publishing activity is rather weak and they have almost no public contracts. The second lab is very small, active in polymers and new materials. It involves only full-time researchers. They have no start-up, no patents and almost no public contracts and exhibit weak publishing performances.

CLASS 3: THE ENTREPRENEURIAL LABS

In Class 3, we find three labs highly engaged in commercial activities. Patents, publications and industrial co-authorship constitute the main contributors of this class. Through interviews we conducted in labs we also know that at least two of them have created start-ups. Moreover, these labs publish much more than the average (2.6 times more than average). They are small and the scientists mainly occupy full-time research positions (90 per cent). These labs belong to different disciplines: biotechnologies, cellular and molecular biology and physics.

Second group: Laboratories with a strong involvement in research activities

The second group of labs corresponds to the Class 4 and the labs are characterised by their important involvement in research activities.

CLASS 4: PUBLISHING FULL-TIME RESEARCHERS

The 11 labs are characterised by a high number of publications per permanent researcher and a small share of teaching-research positions (26 per cent). This class exhibits the second score in terms of publication (1.5 times more than the average; i.e. 3.85 papers per permanent researcher). In terms of commercial activities, this group behaves like the average of our sample. In other words the labs do have some commercial activities, but they are not specialised in a particular action. Their goal is probably to first concentrate on the scientific mission. The main disciplines are biotechnology, genetics, cellular and molecular biology, physics and chemistry of condensed matter, new materials and chemistry.

Third group: Laboratories with weak links with industry

Finally, the last group of classes are not characterised positively by their commercial activities. Table 14.6 underlines that these classes are formed due to the absence of industrial contracts or industrial co-authorship. This does not mean a total absence of links with industrial partners or no contribution to the socio-economic sphere, as we will show.

CLASS 5: NON-ENTREPRENEUR FULL-TIME RESEARCHERS

The low share of public contracts, the weak publication performance, the low industrial co-authorship and the high share of full-time researchers influence the formation of this class. They patent less and get fewer public contracts than the average. The main disciplines are: agronomy, science and information technologies, cellular and molecular biology, physics, fundamental neurosciences, biotechnologies.

CLASS 6: TEACHING-ORIENTED LABS

Class 6 groups nine labs characterised by an important share of teaching-research positions (86 per cent on average) and low publication performances. They exhibit low performances in terms of contractual activities and patenting. The labs are on average rather small. The reality summarised in this class is rather diverse in terms of disciplines. There are two labs in social sciences (management and psychology), one lab in new materials and one in physics that offer professional teaching programmes. The other labs are active in engineering, biotechnology, genetics and neurophysiology.

CLASS 7: TEACHING-ORIENTED LABS WITH PUBLIC CONTRACTS AND PROFESSIONAL CURRICULA

These 11 labs have almost no industrial contracts. They are characterised by a high proportion of professors (80 per cent) and by the highest score in terms of public contracts. Concerning patents, publications and industrial co-authorships, they behave almost like the labs of Class 6 (i.e. the other teaching-oriented labs). Among the 11 labs, 9 propose professional teaching programmes. The disciplines are the following: economics, geography, education sciences, earth sciences, maths, mechanics, information technologies and biology.

Discussion and conclusion

The results show that none of the 64 laboratories of the university is simultaneously active on research, teaching and commercial activities. In the first group 18 laboratories are more involved than the average in commercial and research activities. In these labs, at least half of the researchers are full-time researchers not involved in teaching activities. In the third group, laboratories have a high share of teaching researchers. They are less involved in commercial activities and in research than the average. So research and commercial activities are mutually reinforcing but seem to be conflicting with teaching activities. We find at a collective level the same result as at the individual level. This result is obvious. As researchers spend more time teaching, they have less time for research. However, to be able to conclude that research and teaching are conflicting activities we should have indicators about the impact of teaching. If the involvement of researchers in teaching activities has a positive effect on the quality of teaching we cannot conclude that research and teaching are conflicting. It would be interesting to compare the quality of training programmes in which teachers are connected to research activities to others with no connection with research. One of the arguments for the connection of research and teaching activities is that researchers pass on to students the ability to develop a critical approach rather than passively accepting facts or knowledge. This kind of positive effect is difficult to measure but it could be approached by the professional integration of students. This kind of positive effect on the quality of education is difficult to measure but it could be approached by the professional integration of students, through variables such as the time they spend looking for employment and the salaries they are paid for their first jobs.

Among the 18 laboratories of the first group, which are more involved than the average in commercial activities, only 3 laboratories are highly active on the full spectrum of commercial activities (industrial contracts, co-publication with industrial partners, patenting and public contracts). The others are either specialised in industrial contracts or in industrial co-publishing, and not in the full spectrum of commercial activities. They are also more involved in teaching activities than the other labs of this first group. We interviewed the directors of laboratories having an important share of industrial contracts in their budgets, and they stressed the strong link existing between their teaching activities and their industrial contracts.

They are involved in engineers' education, and their relations with industrial firms are not limited to their research activities.

Almost half of our labs (30) are not involved in commercial activities in the sense that they have rather weaker performance than the average in terms of patent, co-authorship with industry and industrial contracts. Nevertheless, among these labs, a majority of teaching labs have socio-industrial relationships via professional teaching programmes and public contracts. Laboratories are thus specialised in one or two missions, but none has the ability to conduct the three missions at the same time. Research, teaching and commercialisation seem to be complementary activities at the university level, but not at the laboratory level.

The results suggest that the entrepreneurial character of a university does not come from the addition of laboratories acting uniformly for the development of commercial activities. The internal diversity between laboratories allows the university to benefit from the complementarity between the various activities of a university. This diversity between labs is an answer to the diversity of links existing between science and technology and to the diversification of roles played by universities: research, training, contribution to the economic and social dynamic. Knowledge sharing between science and industry constitutes a key factor in the knowledge creation process. Thus to stimulate creativity public authorities and universities should favour the development of knowledge sharing environments. These environments are not only fostered by research collaborations between firms and research labs, but they are also achieved through forums organised by scientists for industrialists, through lifelong training, through placements of students and through academic presentations of research results to industrialists.

However, internal diversity can be a source of conflict. The coexistence in the same university of research teams able to promote the results of their research and others having no short-term commercial applications raises challenges for managers. The first concerns the allocation of resources and the criteria for allocating research budgets to laboratories. The non-entrepreneurial labs can benefit from commercial activities through a redistribution of resources in order to maintain fields of research irrespective of their economic interest. But the risk of redistributing contractual research funds is to create a disincentive effect to developing contractual research within the university. The university managers have thus to find a balance between neglecting fields of research having no short-term socio-techno-economic interests and a mutualisation of benefits that could be detrimental in terms of incentives.

The second challenge concerns the strategic choice of the disciplines the university wants to support and the criteria used to guide this choice. The more frequent indicators of research results, especially at the individual level, measures the impact of research on the scientific community. Their predominance is not adapted to a diversified system, even if their importance is evident in a system where it is necessary to encourage scientific excellence. Internal diversity may be characterised by differences in terms of performances. It thus becomes unavoidable to develop a set of appropriate evaluation criteria, despite the difficulty in comparing the scientific production of different fields, the economic valuation of various disciplines and the teaching performance of different curricula. Conflicts related to the evaluation criteria may obviously occur: they should take account of teaching, research and commercial activities and not only of scientific excellence. A mismanagement of these conflicts could harm the organisational coherence and performance of the university.

The previous discussion induces us to conclude that the diversity of labs within universities allows them to meet a variety of goals. The laboratories more oriented towards commercial activities undoubtedly favour direct diffusion of knowledge towards industry and foster industrial innovation. The variety of disciplines present on the different campuses generates the possibility of answering interdisciplinary questions. The development of professional teaching programmes and the overall high-level teaching activities of some laboratories play social and spatial roles. Lifelong training programmes, internships of students and the frequent organisation of academic conferences stimulate the creativity at a regional level. To stabilise these conclusions, we should refine our analysis by considering the disciplinary differences, in terms of research practices, profitability and performance. However, the variables taken into account in this analysis do not allow us to conclude about the legitimacy of all the labs.

References

Azoulay, P., Ding, W. and Stuart, T. (2007) 'The determinants of faculty patenting behavior: Demographics or opportunities?' *Journal of Economic Behavior and Organization*, **63**: 599–623.

Azoulay, P., Ding, W. and Stuart, T. (2009) 'The impact of academic patenting on the rate, quality, and direction of (public) research output', *Journal of Industrial Economics*, **57**: 637–76.

Breschi, S., Francesco, L. and Montobbio, F. (2007) 'The scientific productivity of academic inventors: New evidence from Italian data', *Economics of Innovation and New Technology*, **16**: 101–18.

Calderini, M., Franzoni, C. and Vezzulli, A. (2007) 'If star scientists do not patent: The effect of productivity, basicness and impact on the decision to patent in the academic world', *Research Policy*, **36**: 303–19.

Callon, M., Larédo, P. and Mustar, P., (1997) *The Strategic Management of Research and Technology*, Economica International, Paris.

Carayol, N. and Matt, M. (2004) 'Does research organization influence academic production? Laboratory level evidence from a large European university', *Research Policy*, **33**: 1081–1102.

Clark, R. (1998) *Creating Entrepreneurial Universities: Organisational Pathways of Transformation*, Pergamon/Elsevier Science, Oxford.

Cohendet, P., Mailhot, C. and Schaeffer, V. (2006) 'European universities under the pressure of globalisation', in E. Lorenz and B.-A., Lundvall (eds), *How Europe's Economies Learn: Coordinating Competing Models*, Oxford University Press, Oxford.

Concei ç ao, P. and Heitor, M. (1999) 'On the role of the university in the knowledge economy', *Science and Public Policy*, **26**(1): 37–51.

Etzkowitz, H. (1996) 'Conflicts of interest and commitment in academic science in the United States', *Minerva*, **34**(3): 258–77.

Etzkowitz, H. (1998) 'The norms of entrepreneurial science: Cognitive affects of the new university-industry linkages', *Research Policy*, **27**(8): 823–34.

Etzkowitz, H. (2003) 'Research groups as "quasi-firms": The invention of the entrepreneurial university', *Research Policy*, **32**: 109–21.

Fabrizio, K.R. and Di Minin, A. (2008) 'Commercializing the laboratory: Faculty patenting and the open science environment', *Research Policy*, **37**(5): 914–31.

Florida, R. and Cohen, W.M. (1999) 'Engine or infrastructure? The university role in economic development', in L.M Branscomb, F. Kodama and R. Florida (eds), *Industrializing Knowledge: University–Industry Linkages in Japan and the United States*, MIT Press, London: 589–610.

Fox, M.F. (1992) 'Research, teaching and publication productivity: Mutuality versus competition in academia', *Sociology of Education*, **65**: 293–305.

Gander, J.P. (1995) 'Academic research and teaching: A case study', *Technological Forecasting and Social Change*, **49**: 311–19.

Geuna, A., Salter, A.J., Steinmueller, W.E. and Hoffman, Y.E. (eds) (2004) *Science and Innovation: Rethinking the Rationales for Funding and Governance*, Edward Elgar Publishers, Cheltenham.

Mailhot, C. and Schaeffer, V. (2005) 'Universities specificities and the emergence of a global model of university: How to manage these contradictory realities', in P. Llerena and M. Matt (eds), *Innovation Policy in a Knowledge-based Economy*, Springer, Berlin: 339–59.

Marsh, H.W. and Hattie, J. (2002) 'The relation between research productivity and teaching effectiveness', *Journal of Higher Education*, **73**(5): 603–41.

Rasmussen, E., Moen, O. and Gulbrandsen, M. (2006) 'Initiatives to promote commercialization of university knowledge', *Technovation*, **26**: 518–33.

Stephan, P.E. (1996) 'The economics of science', *Journal of Economic Literature*, **34**: 1199–1235.

Stephan, P.E., Gurmu, S., Sumell, A. and Black, G. (2007) 'Who's patenting in the university? Evidence from the survey of doctorate recipients', *Economics of Innovation and New Technology*, **16**(2): 71–99.

Thursby, J.G. and Thursby, M.C. (2011a) 'Has the Bayh-Dole act compromised basic research?' *Research Policy*, **40**: 1077–83.

Thursby, J.G., and Thursby, M.C. (2011b) 'Faculty participation in licensing: Implications for research', *Research Policy*, **40**: 20–9.

Van Looy, B., Callaert, J. and Debackere, K. (2006) 'Publication and patent behaviour of academic researchers: Conflicting, reinforcing or merely co-existing?' *Research Policy*, **35**(4): 596–608.

Part III

Economics of creativity, networks and markets

15 The public economics of creativity

Economies of scope in technological infrastructure[1]

Moshe Justman

Introduction

The Schumpeterian perspective that informed so much of Ehud Zuscovitch's thinking and writing on economics stressed the role of creative innovation as an antidote to the meager profit margins of conventional production under neoclassical competitive conditions. Creative innovation allows one to be a price-maker rather than a price-taker, a principle he applied in his personal and professional life as often as he could, preferring as a rule the intellectual adventure of the new and untried over routine, quotidian academic production. This chapter reflects a pattern of mutual influence between Ehud and myself that colored much of our joint work – and some of the work we produced separately – the economics of innovation and technological change, reflecting Ehud's well-articulated Schumpeterian perspective and my own neoclassical training.[2]

Specifically, the present chapter is an attempt to embed a heterodox perspective on technological infrastructure (TI), developed in Tassey (1991), Justman *et al.* (1993), Justman and Teubal (1995), and Teubal *et al.* (1996) within a fully specified neoclassical framework. It is an approach that emphasizes the special characteristics of TI that distinguish it from other forms of industrial infrastructure, viewing it as a network of differentiated intermediate inputs that supports the competitiveness of domestic manufacturing in global markets and attracts international capital for investment in local industry.

The need for TI support is considered here first in the context of a small open economy operating unilaterally in an integrated world economy, where the scope of TI supplied by the private sector, without intervention, is shown to be less than the scope that maximizes net domestic output, indicating a specific role for public intervention. The analysis is then set in a multinational context in which many countries pursue similarly conceived TI policies as instruments of international fiscal competition. When such policies proliferate, a concerted scaling down of unilaterally optimal TI policies may benefit all concerned. This suggests a role for supra-national organizations such as the European Union in coordinating TI policies among its member countries (Haaland and Kind, 2006) as well as for national governments coordinating regional TI policies.[3]

The structure of the paper is as follows. The next section characterizes TI in more detail and offers a brief overview of the literature on strategic public support for industrial infrastructure. The third section provides a formal analysis of TI in a small open economy, indicating a role for public intervention. The fourth section extends the analysis to a multinational setting, and considers the possible role of international coordination. The final section concludes.

Technological infrastructure

Following Justman and Teubal (1995), technological infrastructure (TI) should be thought of as a network of specialized providers of technological services that support domestic manufacturing firms and enhance their ability to compete effectively in global markets. Typical services provided by TI might include: product design; identification, screening and testing of new production technologies; quality control and application of international standards; solution of ecological problems facing firms in a particular sector or region; and so on.[4] They are characterized by economies of scope but initially there is neither supply nor demand for such services – they are neither available nor missing – and so a process of market building must be triggered. In some cases, government will provide these services directly, but more often it will act in support of cooperative private sector initiatives, e.g., through subsidies or special taxes earmarked for use by an industry association.[5] Over time, many of these services will spin off as private enterprises operating on a commercial basis.

Here we focus on the following common elements of TI:

- Its economic value is indirect (or "pre-competitive"); the goods and services it offers are intermediate inputs.
- It has a generic quality, geared to serving many different users, not aimed exclusively at individual innovations or firm-specific R&D projects.
- It exhibits economies of scope deriving from the need to provide a broad spectrum of differentiated capabilities.
- It cannot be imported: it must be available domestically to be effective.
- Initially, there is neither the capability to supply these goods and services, nor are there well-articulated needs for them.

The differentiation of TI distinguishes it from conventional infrastructure – roads, power, water – where economies of scale rather than scope predominate. TI is more specific and more varied, therefore more difficult to plan or direct; because of its generic quality it may be difficult to identify its potential users *ex ante*. Poorly defined demand for TI implies a need to build new markets and create new capabilities.

The present analysis builds on several strands in the literature. Early work on industrial infrastructure saw it essentially as a problem of timing, stemming from the indivisibility of most infrastructure investments (Weitzman, 1970). Optimal timing is readily characterized, but in implementing it a problem of simultaneity arises: a critical mass of directly productive activity (DPA) is necessary before it is worth establishing the infrastructure, but DPA may not be feasible or profitable without the infrastructure (Chenery, 1959). This suggests a role for government in building infrastructure in advance of economic conditions that would render it profitable, where this is a necessary condition for economic development (Justman, 1995). In a world of imperfect information this clearly has its perils (Hirschman, 1958), though the risk involved may be manageable where the infrastructure needed is conventional and multi-purpose, as in the case of roads, electricity and so on.

A second important strand of the literature on which this chapter builds is fiscal competition, surveyed by Wildasin (2006). Studies by Spencer and Brander (1983) and Neary and Leahy (2000) analyze the use of subsidies for research and development in international fiscal competition; Haaland and Kind (2006) focus on the related issue of centralization or decentralization of technology policy within an economic union; and Justman *et al.* (2002, 2005) examine the role of national investment in differentiated infrastructure in fiscal competition.

The present paper focuses on fiscal competition through the development of TI services, building on earlier analyses of the characteristics that distinguish TI from conventional infrastructure, cited above, notably the role of economies of scope and the information asymmetries that preclude governments directly providing these services in an effective manner.[6] The contribution of the present paper is in its analysis of the implications of these characteristics for the design of technological infrastructure policy as an instrument of fiscal competition.

Technological infrastructure in a small open economy

We begin by considering a small country in which there are two types of firms: manufacturing firms that produce final output, an internationally tradable commodity, which we use as our numeraire, and differentiated expert consultancies that provide intermediate technological services. We assume that production of final output requires capital, available in perfectly elastic supply on an international financial market, and locally available technological services. We index the manufacturing firms by i, assume that they are competitive, and set their sum measure equal to one. Firm i's output is denoted y_i and its capital k_i. The cost of capital is r. The different types of technological services available domestically are indexed by s. We assume that s varies continuously in the range $0 \leq s \leq n$, where n measures the scope of the country's technological infrastructure, and denote the quantity of service s used by firm i by x_{is}. To fix ideas, let firm i's production be:[7]

$$y_i = (\int_0^n x_{is}^{\theta} ds)^{\alpha/\theta} k_i^{\beta} \tag{1}$$

where $0 < \theta < 1$, $\alpha + \beta < 1$, and $\alpha/\theta < 1$. We abstract from other immobile inputs in fixed supply that induce decreasing returns to scale in capital and intermediate TI services. Each type of technological service is provided by at most one consultancy (some may not be available at all). To establish itself the consultancy must first expend a units of final product to acquire its specialized capability, and can then provide its expert services in unconstrained quantity at a constant marginal cost of v.

Equilibrium without intervention

Denote the price of service s by p_s. Producing firms seek to maximize their profits:

$$\Pi_i = (\int_0^n x_{is}{}^{\theta} ds)^{\alpha/\theta} k_i^{\beta} - \int_0^n p_s x_{is} ds - rk_i \tag{2}$$

And consultancies providing technological services seek to maximize their profits:

$$\psi_s = (p_s - v)\int_0^1 x_{is} di - a \tag{3}$$

Setting the value of the marginal product of x_{is} equal to its price, equation (2) implies that the elasticity of demand faced by each provider of TI services is $1/(1-\theta)$. Equating marginal revenue to marginal cost then yields

$$p_s = v / \theta \tag{4}$$

As this is the same for all s, it implies that firm i demands the same quantity of each type of available expert service, $x_{is} = x_i$. Next, setting the marginal product of each factor of production equal to its price gives:

$$\partial \Pi_i / \partial k_i = \beta n^{\alpha/\theta} x_i^{\alpha} k_i^{\beta-1} = r \tag{5}$$

$$\partial \Pi_i / \partial x_{is} = \alpha n^{\alpha/\theta-1} x_i^{\alpha-1} k_i^{\beta} = v / \theta \tag{6}$$

This implies that all x_i are equal, $x_i = x$. And if providers of TI services are in zero-profit equilibrium then, from (4)

$$\psi_s = v[(1-\theta)/\theta]\int_0^1 x_i di - a = v[(1-\theta)/\theta]x - a = 0 \tag{7}$$

Equations (5), (6) and (7) determine k, x and n in the profit-maximizing equilibrium.

The profit-maximizing equilibrium does not maximize net output

We take net output – output less the cost of capital and cost of producing expert services – as our measure of performance: $q = y - rk - na - nvx$. The symmetry of production implies that net output will be maximized when $x_{is} = x$ and $k_i = k$ for all manufacturing firms i and TI services s, so that each firm produces

$$y = n^{\alpha/\theta}x^{\alpha}k^{\beta} \tag{8}$$

As the measure of manufacturing firms equals one this is also their total product, and net output is:

$$q = n^{\alpha/\theta}x^{\alpha}k^{\beta} - rk - na - nvx \tag{9}$$

First-order conditions for net output maximization are then:

$$\partial q/\partial k = \partial y/\partial k - r = \beta n^{\alpha/\theta}x^{\alpha}k^{\beta-1} - r = 0 \tag{10}$$

$$\partial q/\partial x = \partial y/\partial x - nv = \alpha n^{\alpha/\theta}x^{\alpha-1}k^{\beta} - nv = 0 \tag{11}$$

$$\partial q/\partial n = \partial y/\partial n - (a + vx) = (\alpha/\theta)n^{\alpha/\theta-1}x^{\alpha}k^{\beta} - (a + vx) = 0 \tag{12}$$

Comparing this with the market equilibrium, we find that (11) and (12) imply a value of x equal to its value under the profit-maximizing equilibrium as determined by (7):

$$x = \theta a/[(1-\theta)v] \tag{13}$$

But this does not apply to the scope of technological infrastructure, n. From (12) and (13) we have at the optimum:

$$\partial y/\partial n = a + vx = a/(1-\theta) \tag{14}$$

While at the profit-maximizing equilibrium, from (6) we have:

$$\partial y / \partial n = vx / \theta^2 = a / [\theta(1-\theta)] > a / (1-\theta) \qquad (15)$$

As y is concave in n this implies that the scope of TI services n at the profit-maximizing equilibrium is less than the scope prescribed by net output maximization. This implies that manufacturing output is less than optimal in the profit-maximizing equilibrium; and as (5) and (10) are identical, the ratio of capital to manufacturing output is the same, implying that the amount of capital employed in production in the profit-maximizing equilibrium is also less than optimal. Thus we have shown:

Proposition 1. If the supply of capital to the national economy is perfectly elastic, TI services and capital are complements in production, and marginal returns to capital and to the scope of TI services are decreasing, then the scope of TI services supplied at the profit-maximizing equilibrium is below the level that maximizes net output, and consequently so is the amount of capital employed and the scale of manufacturing output.

Individual providers of technological services ignore the external benefits, for the economy at large, of extending the scope of TI and attracting more foreign capital.

Subsidization

This under-provision of TI can be rectified by subsidization, *but it is not sufficient to subsidize a, the cost of acquiring new capabilities.* From equations (14) and (15), subsidizing this cost at the rate θ (so that the cost after subsidization is $(1 - \theta)$ will indeed raise the scope of available services to its optimal level. However, equation (7) implies that if only this cost is subsidized the quantity of services provided by each specialist will be less than optimal. In order to maintain the optimal level of x, both a and v must be subsidized at this rate, to offset the monopolistic markup that unique providers of infrastructure services are able to charge: when these services are subsidized at the rate θ users pay v, the marginal cost of providing the service. Of course, this could also be achieved by administrative measures, e.g., by requiring TI firms to provide their services at a reduced cost as a condition of the subsidy they receive in support of creating their capabilities; such arrangements are often observed in practice. When x and n are at their optimal levels, comparing (5) and (6) to (10) and (11) shows that this ensures that k and y will also be optimal. Thus we have shown:

Proposition 2. Subsidizing both the acquisition of new TI capabilities by service providers and their sale to manufacturers at the rate θ rectifies the under-provision of TI services – and its adverse implications for the use of capital and production of output – indicated in Proposition 1.

Subsidizing both the creation of new capabilities and the purchase and sale of the services they provide causes potential providers of these services to internalize their external benefit to the economy in creating new services, and ensures that the price paid by the users of infrastructure services equals the marginal cost of providing them.

Fiscal competition through support of technological infrastructure

Now consider a multi-country setting in which a number of small countries with identical characteristics form an association large enough that their combined effect on the cost of international capital is substantial, i.e., assume that $r = r\left(\sum_{l=1}^{m} k_l\right)$ is upward-sloping, where m is the number of countries. Their total net output equals:

$$Q = \sum_{j=1}^{m} \{n_j^{\alpha/\theta} x_j^{\alpha} k_j^{\beta} - r(\sum_{l=1}^{m} k_l) k_j - n_j a - v n_j x_j\} \tag{16}$$

First-order conditions for net output maximization are then, for $j = 1, \ldots, m$:

$$\partial Q / \partial k_j = \partial y_j / \partial k_j - (r + r` k_j) = \beta n_j^{\alpha/\theta} x_j^{\alpha} k_j^{\beta-1} - (r + r` k_j) = 0 \tag{17}$$

$$\partial Q / \partial x_j = \partial y_j / \partial x_j - n_j v = \alpha n_j^{\alpha/\theta} x_j^{\alpha-1} k_j^{\beta} - n_j v = 0 \tag{18}$$

$$\partial Q / \partial n_j = \partial y_j / \partial n_j - (a + v x_j) = (\alpha / \theta) n_j^{\alpha/\theta-1} x_j^{\alpha} k_j^{\beta} - (\alpha + v x_j) = 0 \tag{19}$$

Now compare this with a Nash equilibrium among the m national governments engaged in fiscal competition, each subsidizing TI in its country at a rate that maximizes its net output taking the actions of other countries as given, and assume it ignores the small effect of its own demand for capital on the rate of interest.[8] From our previous analysis, each country subsidizes TI at the individually optimal rate θ so that the values of k_j, x_j and n_j that are realized satisfy equations (10), (11) and (12). Comparing this to equations (17), (18) and (19) we find that $\partial Q / \partial k_j < 0$ at the Nash equilibrium while $\partial Q / \partial x_j = \partial Q / \partial n_j = 0$, implying that k_j should be reduced while x_j is at its optimal level. But then (19) implies that n_j should also be reduced, as capital and technological services are complements. In summary, when the supply of capital is not perfectly elastic fiscal competition in government support for TI leads to over-provision of the variety of technological services; total net output could be increased through a coordinated reduction of government support for TI:

Proposition 3. Consider a group of identical countries, all drawing investment from a common pool of capital, and assume that they are collectively large enough that they face an upward-sloping supply of capital in the aggregate, but that each is small enough that it ignores its individual effect on the cost of capital. If each of the national governments unilaterally subsidizes domestic TI at a rate that maximizes its own output, taking the actions of other countries as given, then there will be over-provision of TI in Nash equilibrium.

Summary

The formal model of differentiated technological infrastructure presented in this chapter highlights three principal points that bear on earlier analyses of industrial technology policy. First, it shows that there is a formal argument to be made for public support of technological infrastructure in a small open economy. A single country drawing its investment from a perfectly elastic international pool of capital has reason to support its local technological infrastructure. Individual providers of infrastructure services do not take sufficient account of the beneficial economies of scope they generate.

Second, support should be aimed at both the initial acquisition of new capabilities and at the purchase and sale of the services they generate. Supporting only the acquisition of capabilities leads to sub-optimal levels of provision of infrastructure services, echoing Justman and Teubal's (1995) conclusion that both supply and demand for these services should be promoted.

Third, when TI policies are viewed from a broader perspective, the possibility that many governments may adopt similar policies as instruments of fiscal competition must be considered. When this occurs, excessive levels of support may result, signaling a need for international (or inter-regional) coordination. This suggests an enhanced role for supra-national organizations and for agencies of national governments to coordinate the initiatives of lower jurisdictions. In this respect, technological infrastructure policy can be expected to follow a life cycle similar to that of other industrial development policies. It offers governments that move early the opportunity to achieve substantial gains through unilateral policy initiatives, but when such initiatives proliferate all can benefit from some measure of coordinated restraint.

Notes

1 Thanks to Arman Avadikyan, Thierry Burger-Helmchen and the other participants in the BETA conference on "The Economics of Creativity," in Strasbourg, for their helpful comments and suggestions.

2 Of our joint work Zuscovitch and Justman (1995) was more Schumpeterian; Justman and Zuscovitch (2002) more neoclassical. Of course, Ehud had similar collaborations with others, e.g., Yin and Zuscovitch (1998a, 1998b). Among my own work, Justman (1996, 2004) were strongly influenced by Ehud's Schumpeterian perspective.

3 Cohendet's (1996) early analysis of the relation between local and national support for TI, contrasting France and Germany, highlighted the tradeoff implicit in national or regional coordination, between the benefits of tailoring support to specific local needs and the advantages offered by higher-level coordination.
4 Teubal *et al.* (1996) offer a variety of case studies that illustrate different facets of TI, among them: Shapira (1996) and Vickrey (1996).
5 Romer's (1993) blueprint for industry investment boards in the US is an early example. Israeli initiatives are described in Justman *et al.* (1993).
6 Earlier perspectives were more optimistic regarding the scope for more direct government intervention, cf. Johnson (1982).
7 The same analysis could be carried out with a general production function, $f(\int_0^n [x_i(s)]^\theta ds, k_i)$ with f_1, f_2 and f_{12} positive, and f homogenous of degree less than one, and with a negative-semi-definite Hessian matrix.
8 The qualitative argument holds equally (but the exposition is less simple) if each country recognizes its effect on the rate of interest but ignores the adverse effect on other countries of a higher interest rate.

References

Chenery, H.B. (1959) "The interdependence of investment decisions," in M. Abramowitz, (ed.), *The Allocation of Economic Resources*, Stanford, CA: Stanford University Press: 82–120.

Cohendet, P. (1996) "Transfer of technology to SMEs: Conceptual changes and lessons from the two banks of the Rhine," in M. Teubal, D. Foray, M. Justman and E. Zuscovitch (eds.), *Technological Infrastructure Policy*, Dordrecht: Kluwer: 269–84.

Haaland, J. and Kind, H.J. (2006) "Cooperative and non-cooperative R&D policy in an economic union," *Review of World Economics*, **142**: 720–45.

Hirschman, A.O. (1958) *The Strategy of Economic Development*, New Haven, CT: Yale University Press.

Johnson, S. (1982) *MITI and the Japanese Miracle*, Stanford, CA: Stanford University Press.

Justman, M. (1995) "Infrastructure growth and the two dimensions of industrial policy," *Review of Economic Studies*, **62**: 131–57.

Justman, M. (1996) "Swarming mechanics," *Economics of Innovation and New Technology*, **4**: 235–44.

Justman, M. (2004) "Transitional dynamics of output, wages and profits in innovation-led growth," *Structural Change and Economic Dynamics*, **15**: 183–205.

Justman, M. and Teubal, M. (1995) "Technological infrastructure policy: Creating capabilities and building markets," *Research Policy*, **24**:259–81.

Justman, M. and Zuscovitch, E. (2002) "The economic impact of subsidized industrial R&D in Israel," *R&D Management*, **32**: 191–9.

Justman, M., Teubal, M. and Zuscovitch, E. (1993) (eds.), *Technological Infrastructure Policy for Renewed Growth*. In Hebrew. Jerusalem: Jerusalem Institute for Israel Studies.

Justman, M., Thisse, J.F. and van Ypersele, T. (2002) "Taking the bite out of fiscal competition," *Journal of Urban Economics*, **52**: 294–315.

Justman, M., Thisse, J.F. and van Ypersele, T. (2005) "Fiscal competition and regional differentiation," *Regional Science and Urban Economics*, **35**: 848–61.

Neary, J.P. and Leahy, D. (2000) "Strategic trade and industrial policy towards dynamic oligopolies," *Economic Journal*, **110**: 484–508.

Romer, P. (1993) "Implementing a national technology strategy with self-organizing industry investment boards," *Brookings Papers: Microeconomics*, **2**: 345–99.

Shapira, P. (1996) "Modernizing small manufacturers in the United States and Japan: Public technological infrastructures and strategies," in M. Teubal, D. Foray, M. Justman and E. Zuscovitch (eds.), *Technological Infrastructure Policy*, Dordrecht: Kluwer: 285–334.

Spencer, B.J. and Brander, J.A. (1983) "International R&D rivalry and industrial strategy," *Review of Economic Studies*, **50**: 707–22.

Tassey, G. (1991) "The functions of technology infrastructure in a competitive economy," *Research Policy*, **20**: 345–61.

Teubal, M., Foray, D., Justman, M. and Zuscovitch, E. (1996) (eds.), *Technological Infrastructure Policy: An International Perspective*, Dordrecht: Kluwer.

Vickrey, G. (1996) "Modernizing manufacturing: Consultancy, advisory and extension services to improve small business performance," in M. Teubal, D. Foray, M. Justman and E. Zuscovitch (eds.), *Technological Infrastructure Policy*, Dordrecht: Kluwer: 335–73.

Weitzman, M. (1970) "Optimum growth with scale economies in the creation of overhead," *Review of Economic Studies*, **37**: 555–70.

Wildasin, D.E. (2006) "Fiscal competition," in B. Weingast and D. Wittman (eds.), *Oxford Handbook of Political Economy*, Oxford: Oxford University Press.

Yin, X. and Zuscovitch, E. (1998a) "Economic consequences of limited technology transferability," *Australian Economic Papers*, **37**: 22–35.

Yin, X. and Zuscovitch, E. (1998b) "Is firm size conducive to R&D choice? A strategic analysis of product and process innovations," *Journal of Economic Behavior and Organization*, **35**: 243–62.

Zuscovitch, E. and Justman, M. (1995) "Networks, sustainable differentiation and economic development," in D. Batten, J. Casti and R. Thord (eds.), *Networks in Action*, Berlin: Springer Verlag: 269–85.

16 Fits and misfits

Technological matching and R&D networks

Robin Cowan, Nicolas Jonard and Bulat Sanditov

Introduction

Interest in social and economic networks of various types has grown dramatically in the last two decades. The manifold empirical studies emphasize that networks often share two properties. First, they are small worlds – sparse networks with both short distances between pairs of agents, and strong local clustering. Second, they have skewed link distributions, often with a fat right tail. These properties have been observed in a wide variety of networks.[1] A challenge in the economic, mostly game theoretic literature on network formation has been to capture these properties in a sound model. What incentives do (a few) agents have to form shortcuts, and why should they cluster locally? Why do a few agents acquire many links whereas many others seem able to form only a few?

In this chapter we develop a model of R&D networks in which firms seek to innovate, combining theirs and their partners' knowledge in order to produce new knowledge. The key assumption is that the decision to forge a strategic alliance is based on the extent to which the two participating firms' endowments both resemble and complement each other. Two aspects of the model combine to produce small world architectures.

First is the requirement that both similarities and complementarities are present for an alliance to be profitable. The demands of absorptive capacity imply that partnering firms must be similar to each other. As Granovetter (1976) points out, very often such 'similarity' is transitive, and thus in the larger network we should expect clusters of densely interconnected firms. But the fact that firms must also complement each other implies that similarity cannot be 'identity', and so is not perfectly transitive. The simultaneous presence of common and distinct traits will induce correlation in the pairing decision, yielding clustering at the local level. This tendency is strongest when similarities are much more important than complementarities in the innovation process, that is, when what we share matters more than how we differ.

Second is the fact that firms are located within a specific knowledge space. Thus links between firms are implicitly links between points in that space. The interaction between the decision rule, the structure and random location of firms in knowledge space will imply short distances in network space. This creates the second feature of small worlds, namely short pairwise distances.

Within the framework of this model we are able to derive analytically density, degree distribution and clustering, and provide both asymptotic and numerical results on path lengths. The results show that the search for partners with a good technological fit can explain network architecture, and specifically small world properties. Although the search for technological fit is the stated aim for strategic R&D alliance participants, it is often left out of empirical studies. One contribution of this chapter is thus to show that it is essential that technological fit is included. We also contrast two extreme forms of competition. At one extreme every alliance is aimed at a unique innovation, and so if successful gives a positive value to its inventor. At the other extreme we invoke a patent race, in which all alliances are aimed at the same, patentable, innovation, implying that only one success will produce any revenue. We compare these extremes, showing that the strength of competition affects the density of the network but that the basic structures, small worlds with skewed link distributions, are preserved.

Small worlds and degree distributions

Mechanisms underlying a skewed degree distribution are thought to be reasonably well understood. One argument is simply that in most industries firms' size distribution itself is skewed. A larger firm is likely to have more resources with which to create or maintain costly links than does a smaller one. Another popular argument is that skewness arises from a preferential attachment mechanism (Barabasi and Albert 1999): a firm with many links is seen as an attractive partner to other firms either joining the network or revising their partnering decisions. This is enough to create a relatively 'star-ish' network in which a small number of firms acquire many links whereas most have only a few.

A bigger challenge has been to understand the genesis of small worlds. One relevant literature approaches the problem from an algorithmic perspective. Following the initial article by Watts and Strogatz (1998), this literature is contributed to mostly by computer scientists and physicists. It seeks to design algorithms to construct small worlds, asks about the (often asymptotic) properties and functions of small world networks and looks for them in large data sets (see Watts 1999; Amaral *et al.* 2000; Kleinberg 2000). Not surprisingly there are processes that produce small worlds, and not surprisingly their economic underpinnings, if any, are thin (but then, that was not the goal there).

A second literature that provides much more solid foundations is the strategic management literature. Here, network formation and network structure are often discussed in terms of social capital. Clusters emerge through considerations of social control and because link formation often rests on third party referrals. On the other hand, 'structural hole' arguments are used to explain how clique-spanning ties arise, initiated by firms aiming at accessing resources in distant parts of the industry, and at occupying strong brokerage positions. Together, these arguments explain the creation and maintenance of the small world structures that are often observed empirically. (These ideas have been widely discussed: see Coleman 1988; Burt 1992; Kogut *et al.* 1992; Uzzi 1997; Walker *et al.* 1997; Gulati and Gargiulo 1999; Ahuja 2000; Dyer and Nobeoka 2000; Rowley *et al.* 2000; Baum *et al.* 2003.)

In the context of strategic R&D alliances between firms, these arguments seem correct, but incomplete. Two parties to an alliance aimed at innovation of some kind must consider not only social capital but also whether their knowledge and competence complement each other. Indeed one of the standard explanations for why firms engage in (strategic, R&D) alliances, is to access technology or competence that they need but do not possess in-house.[2] We can imagine a space of technological competence, where any firm is located at a point representing its own expertise. If, in its choice of alliance partners, a firm looks to complement its expertise, it will search for partners that are relatively distant in this space. On the other hand though, considerations of absorptive capacity argue that partners to an alliance cannot be too far apart. If they are, their respective absorptive capacities will be too low, and the alliance will fail (Grant 1996 or Nooteboom 2000). This suggests that alliance success should have an inverted-U relationship with distance between the partners in the underlying knowledge space. There are now a small number of studies that seem to support this hypothesis (Mowery *et al.* 1996, 1998; Ahuja and Katila 2001; Schoenmakers and Duysters 2006). This need for potential partnerships to achieve both some degree of commonality and some degree of complementarity is a key feature of the model we develop in this chapter.

The third place where models for the formation of small worlds are provided is the economics literature. There, short pairwise distances tend to arise because indirect benefits decay with distance. And high clustering results from smaller costs of connecting to similar (nearby) nodes. Variants of Jackson and Wolinsky's (1996) communication model all exploit this feature to some degree. A network in which inter-node distances are, on average, long induces some value destruction, which the creation of a shortcut can prevent. Jackson and Rogers (2005) locate nodes on islands. Costs of link formation are low if two nodes are on the same island, and high if they are on different islands. Together with the truncation of benefits, this structure creates arbitrarily dense islands with rare interconnections between islands at equilibrium. Carayol and Roux (2009) explore a framework in which nodes are located on a circle, and the cost of forming a link rises with geographic distance between nodes. Galleoti *et al.* (2006) also have a connections model where players' values and costs are heterogeneous. In a world where society has groups and intra-group links are cheaper to form than inter-group links, they find that interconnected stars are both socially efficient and stable, leading them to claim that centrality, centre-sponsorship and small diameter are robust features of networks. In all of these models shortcuts (which bring down average inter-node distances) arise because a node seeks access to a distant resource, the value of which decays in transmission.

In comparison with the models above, our approach has two important differences. First, firms have richer properties. Each of them is defined as a set of attributes, drawn from a fixed universe (for example firms forming strategic R&D alliances, where attributes are pieces of knowledge, or competence in particular technologies). These attributes only create economic value when combined in the right proportion. Second, we keep the simplest structure with no indirect effects. Value (here, innovation) originates in links, and is not transmitted beyond its inventors. So shortcuts in the equilibrium network generated in our model will not result from firms' desire to shorten their distance to a valuable resource. The combination of firms' heterogeneity and their desire to ally with similar yet complementary partners is enough to induce the formation of small worlds with skewed degree distributions.

Definitions

Networks

The industry consists of a finite population $N = \{1,...,n\}$ of firms. A *network* is a set g of ties (links) between unordered pairs of firms in N. Writing ij to represent the tie between firms i and j, $ij \in g$ indicates that i and j are linked in the network g. The network obtained by adding ij to an existing network g is denoted $g + ij$, whereas the network obtained by deleting ij from g is $g - ij$.

The *neighbourhood* of firm i consists of the firms to whom i is directly connected, denoted $N_i^g = \{j \neq i : ij \in g\}$. The size of the neighbourhood of i is the number of ties held by firm i, also called its *degree*, and is denoted $n_i^g = \#N_i^g$. The total number of links in the network is $A^g = \sum_{i \in N} n_i^g/2$, and thus the density of g is equal to $2A^g/n(n-1)$. A *singleton* is a firm i such that $n_i^g = 0$ i.e. a firm with no partner.

Strategies and equilibrium

Firms play a simultaneous link formation game. A *strategy* for firm i is a list of $(n - 1)$ decisions $(s_{i,1},...,s_{i,i-1},...s_{i,n})$, with $s_{i,j} \in \{0,1\}$. If $s_{i,j} = 1$ then i proposes a partnership to j, whereas no partnership is proposed when $s_{i,j} = 0$. Link formation is bilateral, so in the network $g(s) = g$ induced by the strategy profile s, $ij \in g$ if and only if $s_{i,j}s_{j,i} = 1$, that is, both partners want the alliance and so propose to each other. Denote the payoff to firm i in network g as π_i^g (defined below). The network g is *pairwise stable* (Jackson and Wolinsky 1996) if and only if

$$\min\{\pi_i^g - \pi_i^{g-ij}; \pi_j^g - \pi_j^{g-ij}\} \geq 0, \forall ij \in g$$

and

$$\min\{\pi_i^{g+ij} - \pi_i^g; \pi_j^{g+ij} - \pi_j^g\} < 0, \forall ij \notin g.$$

Links that exist in a stable network g should yield non-negative value to both partners, whereas links yielding negative value to at least one potential partner cannot be part of g.

Assessing small worlds

A small world is defined as a sparse network, displaying both local clustering and short pairwise distances. A sparse network is one in which density is low, that is, the number of edges divided by the number of possible edges is small. In empirical studies of small worlds, densities are typically lower than 5–10 per cent.

There are two classic measures of clustering (see Newman 2003). In the first, clustering is measured by the proportion of neighbours of a node who are neighbours of each other. Computing that proportion for each node and taking the average yields the *clustering coefficient*. An alternative measure, which is more easily determined in our case, is the fraction of transitive triples in the whole network. *Transitivity* is defined as

$$\frac{\sum_i \#\{jk \in g : j,k \in N_i^g\}}{\sum_i \#\{jk : j,k \in N_i^g\}} = \frac{3 \times Nb\, of\, closed\, triangles\, in\, g}{Nb\, of\, connected\, triples\, in\, g}.$$

The factor 3 appears in the numerator because each triangle involves 3 connected triples. On this definition, it is worth noting that the expected value of transitivity is the probability that jk exists conditional on ij and ik existing or, put another way, the probability of closing the connected triple formed by ij and ik by connecting j to k.

Contrasting with clustering, which is a local measure (of redundancy in ties), a measure of global structure is provided by the distribution of pairwise distances between nodes. A *path* of length $l + 1$ in g connecting i and j is a set of distinct nodes $\{i_1,...i_l\}$ such that $i_{i1}, i_1i_2,...,i_lj \in g$. The *distance* separating i from j in g is the length of the shortest path in g between i and j. If no path exists between i and j, then by convention the distance is defined as infinite. Averaging all finite pairwise distances yields the *characteristic path length*. To cope with the issue of disconnected networks (which would introduce at least one infinite distance into the calculation) we restrict attention to distances between 'reachable pairs'.

A network is identified as a small world when clustering significantly exceeds that of a 'comparable' random graph, whereas characteristic path length is of the same order. A benchmark often used in practice is an Erdös and Rényi (1960) random graph (ER), in which a given number of edges are allocated among randomly (and uniformly) selected pairs of nodes. The ER benchmark is made 'comparable' to the graph under scrutiny by taking the same number of nodes and edges.

Innovation

Firms are the carriers of knowledge which they bring together in order to discover new knowledge. Knowledge is modelled as a set or list of discrete elements, where the set of all possible facts is $\{1,...,w\}$. Firm i's knowledge portfolio is a w-position binary vector v_i where $v_i^z = 1$ if i knows fact z and $v_i^z = 0$ otherwise.

In the introduction we argued that for joint innovation, both commonality and difference were important. Define v_{ij} as the commonality of i and j, i.e. the number of facts known to both: $v_{ij} = \#\{z{:}v_i^z = v_j^z = 1\}$. Also define s_{ij} as i's complementing of j, i.e. the number of facts known to i but not to j: $s_{ij} = \#\{z{:}v_i^z > v_j^z\}$. For successful innovation, i and j must both be similar to and complement each other. We model this by assuming success requires a commonality of δ and reciprocal complementarity of γ. The probability that the partnership ij innovates successfully is

$$r_{ij} = \begin{cases} 1 \text{ if } v_{ij} = \delta \text{ and } s_{ij} = s_{ji} = \gamma, \\ 0 \text{ otherwise.} \end{cases} \quad (1)$$

Successful partners must thus have the right technological fit, i.e. share competencies in δ technologies and have their competencies complement each other in γ technologies.[3]

Payoffs

The model represents firms' behaviour in a fast moving industry in which innovation is the motivating goal of firms, and post-innovation competition in the market place is parametrized in a simple manner. Innovative success is affected by the partner firms' technological fit, whereas profits from innovation are affected by the nature of competition. In what follows we examine two extremes of competition.

In the first, the benefits from each successful innovation are fully appropriated by its two inventors, regardless of the amount of R&D in the industry and the success of other partnerships. Here the alliance decision is based entirely on technological fit, and we refer to the resulting network as the knowledge or technology network: t. In the second case, the benefits from each successful alliance have to be shared with all other successful alliances (equivalently, only one alliance will succeed as in a standard patent race). Here market considerations are essential and we refer to the resulting network as the market network: m.

Profits can be written generally as

$$\pi_i^g = \frac{1}{1+\alpha(A^g-1)} \sum_{j \in N_i^g} r_{ij} - cn_i^g,$$

where $g \in \{t,m\}$, $c > 0$ is the cost of forming or maintaining an alliance, paid by each of the partners to it, $\alpha \in \{0,1\}$ captures the intensity of competition for innovative reward ($\alpha = 0$ for the knowledge case, and $\alpha = 1$ for the market case).

Equilibrium networks

In this section we develop the model on the assumption that any second stage competition is not relevant to the alliance decision-making process. In essence, every alliance is aimed at a unique innovation which is not in any way a substitute for any other innovation. This is a strong assumption, but one which is often implicitly made, both in theoretical and empirical network studies.

The expected profit for firm i in a (technology-driven) network t is thus

$$\pi_i^t = \sum_{j \in N_i^t} r_{ij} - cn_i^t. \tag{2}$$

As links convey no indirect benefits, the marginal value of both an existing and a non-existing link ij is simply $\pi_i^t - \pi_i^{t-ij} = \pi_i^{t+ij} - \pi_i^t = r_{ij} - c$. Technological fit being symmetric (see Equation 1), if the incremental value of a link is positive to one partner, it is to the other as well. Proposition 1 below states the results, omitting the straightforward proof.

> **Proposition 1.** *When $c > 1$ the empty network is (uniquely) stable. When $c \leq 1$, , for any values of δ and γ there is a unique equilibrium network, in which all firms with overlap of δ and reciprocal complementarity of γ form an alliance.*

Discussion of general structures

Proposition 1 provides a minimal description of the equilibrium networks. In the sections that follow we present results on global properties of the equilibrium networks. Before turning to the formal derivation of network statistics, however, we can describe heuristically some effects of parameter changes on network architectures.

We model industries in which there is a fixed total number w of relevant facts. One way to differentiate among industries would be by the amount of knowledge (relative to total possible knowledge) that is needed to innovate (that is, $\gamma + \delta$ relative to w). An interpretation is that in young industries, there is a considerable body of knowledge that can or will be useful, but which has not yet been discovered. Firms innovate to find that knowledge, so in our formalization, the amount firms have is small relative to the amount that can be used: $(\delta + \gamma)/w$ is small. In a more mature industry by contrast, less remains to be discovered (there are fewer 'technological opportunities'); most of the knowledge that is, or might be, relevant is known to most firms in the industry (and firms that don't know it cannot compete successfully), hence a higher $\delta + \gamma$ relative to w. Should we then expect in the model that alliances will be more common in a more mature, exploitative industry, or rather that active networking will be a distinct feature of young, innovative industries? This will depend on the relative importance of commonality and complementarity.

Start from the fact that a match between i and j requires that i knows exactly δ of j's $\delta + \gamma$ pieces of knowledge. The probability of this occurring is highest when the probability that i knows any particular fact is $\delta/(\delta + \gamma)$. In this case i would know (in expectation) $w \cdot \delta/(\delta + \gamma)$ facts. But in the model firms know exactly $\delta + \gamma$ facts. So the probability that i and j fit is highest when $(\delta + \gamma)/w = \delta/(\delta + \gamma)$, and decreases as parameter values depart from that equality. If *complementarity* demands dominate ($\delta/(\delta + \gamma)$ small), networks will be dense when innovating firms have little knowledge ($(\delta + \gamma)/w$ also small) so that almost any pair of firms complement each other. Young industries with (relatively) unexplored technological opportunities will be a place of intense technological partnering. On the other hand, when *commonality* demands are high ($\delta/(\delta + \gamma)$ large), dense networks will emerge when total knowledge endowments are high, so that firms are relatively more similar to each other (a larger portion of the knowledge vector is common). Here the requirement of significant technological overlap will only permit intense networking in mature, exploitation-oriented industries, since there, most firms know most of the relevant knowledge and so have much knowledge in common.

Another important feature of alliance networks is transitivity. Consider first an industry in which innovation is driven by creation of novel combinations of existing knowledge, so that we might expect γ, the demand for reciprocal complementarities, to be high. In the extreme, $\gamma = (w - \delta)/2$, and no triangle of firms can form. In this situation, the partners of any firm must all be identical to each other, and so cannot be partners to each other (they would bring no novelty at all to the alliance). In this case however, squares (cycles of period 4) can form: partners of i are mirror images of i

and a mirror image has a cycle of period 2.[4] Note that in this situation firms ally with partners who are identical to each other, and thus possibly redundant from the point of view of innovation. The implication however is that a strong need for complementarity, γ, all else equal, should decrease transitivity in the network. At the other extreme, in an industry where innovation is not driven by creation of novel combinations of existing knowledge, bringing novelty to a partnership may be less important than having a good, easy working relationship. In this case commonality in knowledge would dominate the decision and the demand for reciprocal complementarities γ would be small relative to δ. In the extreme $\gamma = 0$ and the partners of any firm again are all identical to each other, and identical to the focal firm i. This time all possible triangles involving i and its neighbours will form, and the network will fracture into small, isolated cliques of identical firms, each component representing a sub-sector of the industry. Increasing δ in that case would increase the size of the cliques, and the grand coalition (the complete industry network) would obtain when $\delta = w$. The implication here is that when the need for complementarity γ is very small the network will display transitivity. It is important to note that this effect is confounded with changes in density. As γ or δ change, density changes (as discussed in the previous paragraph) and so the potential number of triangles in the network changes. This effect will appear in the transitivity statistic. Rescaling transitivity 'undoes' this effect, so the effect of γ on transitivity just described will be more evident when transitivity is rescaled. (See Figure 16.2 below.)

Density and degree distribution

From Proposition 1, we can derive the properties of the larger network of all alliances. Our interest is in how these properties are affected by the relative importance of commonality (δ) and complementarity (γ) in the innovative process.

The strict condition we place on alliance formation, namely that holding exactly $\delta + \gamma$ ideas is necessary to be part of any alliance, implies that firms endowed with different amounts of knowledge are doomed never to partner. In what follows we focus on a population of firms all of which command the right amount of knowledge. This strategy has two advantages. First, in an arbitrary population endowed with random amounts of knowledge, one effect of changing the parameters γ and δ would be to change the proportion of firms commanding $\gamma + \delta$ technologies. With a fixed total number of firms, this changes the number that can participate in the network. Thus the effects of complementarities and overlap are confounded with changes in the effective, network-participating, population size. Fixing the number of 'alliance-ready' firms over the parameter space removes this effect. The second advantage is that empirical analyses of alliances tend to consider as their population only those firms who have ever had an alliance. Thus they exclude firms that, for whatever reason (perhaps they are too young to have enough knowledge to be attractive; or they command enough knowledge to be self-sufficient) have never had an alliance. We abstract from these issues by considering a population of n firms commanding $\delta + \gamma$ technologies and so which are, even if not participating in any alliances, at least 'alliance-ready'.

The starting element to derive the properties of the network of alliances is the probability q that two firms form a link. As each firm carries $\gamma + \delta$ ideas uniformly distributed over w positions, the probability that i and j form a partnership is simply the hypergeometric probability (of drawing exactly δ of $\delta + \gamma$ existing ideas in $\delta + \gamma$ drawings without replacement from a set of w possible objects), which is written

$$\Pr\{ij \in t \mid a_i = a_j = \delta + \gamma\} = q = \frac{\binom{\delta+\gamma}{\delta}\binom{w-\delta-\gamma}{\gamma}}{\binom{w}{\delta+\gamma}}. \tag{3}$$

Thus we can state the following result.

Proposition 2. *When all firms command* $\delta + \gamma$ *technologies and consider only technological fit,* q *is the (expected) density of the equilibrium network* t*, and the degree distribution of firms in the equilibrium network* t *is binomial with parameters* $n - 1$ *and* q*:* $Pr\{n_i^t = z\} = \binom{n-1}{z} q^z (1-q)^{n-z}$.

Figure 16.1 shows how expected density (q) responds to changes in parameters. Following the discussion in the first section, density is presented as a function of the general knowledge level $(\delta + \gamma)/w$, and the relative importance of commonality versus complementarity, $\delta/(\delta + \gamma)$. In this space, a horizontal move represents the same relative change in δ and γ (so that the ratio $\delta/(\delta + \gamma)$ remains constant), and a vertical move represents changes of identical absolute sizes but opposite directions in δ and γ (so that $\delta + \gamma$ is unchanged).

As suggested earlier, we observe that density is highest along the ray defined by $(\delta + \gamma)/w = \delta/(\delta + \gamma)$. It is worth noting that the probability of matching falls rapidly as we move away, so even though all firms are a priori 'alliance-ready', density remains close to 0 over significant portions of the parameter space.

Transitivity

On the definition given in the section 'Assessing small worlds', the expected value of transitivity (the expected proportion of transitive triples in the equilibrium network) is $\Pr\{jk|ij,ik\}$ the probability that jk forms conditional

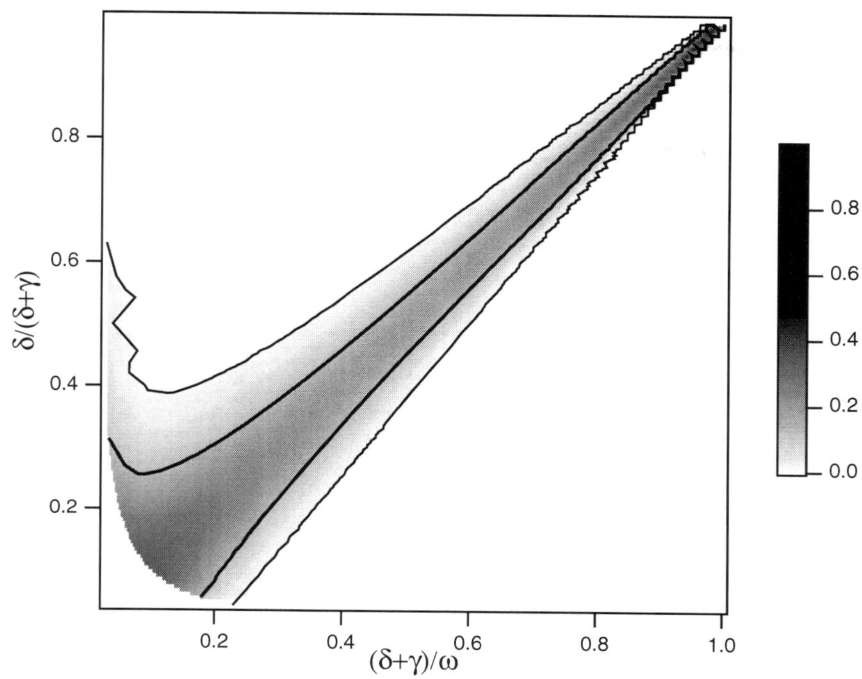

Figure 16.1 Density of the equilibrium network

on *ij* and *ik* existing. From there we want to know whether the model produces *excess* transitivity compared to a random graph of the same density. Noting that in a random graph edges form independently, the probability that the third edge of a triangle forms conditional on two of its edges already existing is simply the probability that an edge forms. This probability in turn is simply q, the density of the random graph. So rescaled transitivity (the extent to which triangles occur relative to a 'comparable' random graph) is written $\Pr\{jk|ij,ik\}/q$, with q defined in Equation 3. A ratio larger (smaller) than 1 indicates the presence (absence) of excess clustering, relative to the random benchmark.

Proposition 3. *When all firms command* $\delta + \gamma$ *technologies, rescaled transitivity in the equilibrium network is equal to*

$$\frac{\sum_{b=\underline{b}}^{\bar{b}} \binom{\delta}{b}\binom{\gamma}{b}^2 \binom{w-\delta-2\gamma}{\gamma-b}\binom{w}{\delta+\gamma}}{\binom{\delta+\gamma}{\delta}^2 \binom{w-\delta-\gamma}{\gamma}^2}$$

where $\underline{b} = max\{0; 3\gamma + \delta - w\}$ *and* $\overline{b} = min\{\gamma; \delta\}$. *Rescaled transitivity is significantly greater than one over much of the (δ,γ) space, and is larger for smaller values of the complementarity parameter γ.*

Proof Obtaining Pr{*ij,ik*} is straightforward: it is the probability of 2 independent events, Pr{*ij,ik*} = q^2. The numerator Pr{*ij*|*ik,jk*} is slightly less direct to compute. This time, write firms *i*, *j* and *k*'s knowledge portfolios as a $w \times 3$ matrix (firm *i* is the first column, etc.) and suppose there are $\delta - b$ rows in the matrix written 111, with $0 \le b \le \delta$. Then the rest of the matrix is perfectly defined. There must be *b* rows of 110, *b* rows of 101 and *b* rows of 011 to ensure the right overlap between all pairs of firms. Then $\gamma - b$ rows of 001, $\gamma - b$ rows of 010 and $\gamma - b$ rows of 100 so that the right complementarities between pairs are obtained. The remaining rows (of which there are $w - \delta - 3\gamma + b$) must be written 000. How many ways there are to array the 1s held by all 3 firms so that $\delta - b$ rows of the matrix are written 111 is a product of binomial coefficients, i.e.

$$\frac{w!}{(\delta-b)!b!^3(\gamma-b)!^3(w-\delta-3\gamma+b)!}.$$

As the 1s held by all 3 firms can be arrayed in $\binom{w}{\delta+\gamma}^3$ ways, the probability that the knowledge matrix has $\delta - b$ rows of 111 is written

$$\frac{(\delta+\gamma)!^3(w-\delta-\gamma)!^3}{(\delta-b)!b!^3(\gamma-b)!^3(w-\delta-3\gamma+b)!w!^2}.$$

Rearranging the factorials, representing them as binomial coefficients and invoking the law of total probability (the knowledge matrix can have from 0 to δ rows written 111) the final result obtains as

$$\Pr\{ij, ik, jk\} = \frac{\binom{\delta+\gamma}{\delta}\binom{w-\delta-\gamma}{\gamma}}{\binom{w}{\delta+\gamma}^2}\sum_{b=\underline{b}}^{\overline{b}}\binom{\delta}{b}\binom{\gamma}{b}^2\binom{w-\delta-2\gamma}{\gamma-b},$$

where $\underline{b} = \max\{0; 3\gamma + \delta - w\}$ and $\overline{b} = \min\{\gamma; \delta\}$, so that all binomial coefficients are well defined. Finally the expected value of transitivity is

$$\Pr\{jk \mid ij, ik\} = \frac{\sum_{b=\underline{b}}^{\overline{b}}\binom{\delta}{b}\binom{\gamma}{b}^2\binom{w-\delta-2\gamma}{\gamma-b}}{\binom{\delta+\gamma}{\delta}\binom{w-\delta-\gamma}{\gamma}}.$$

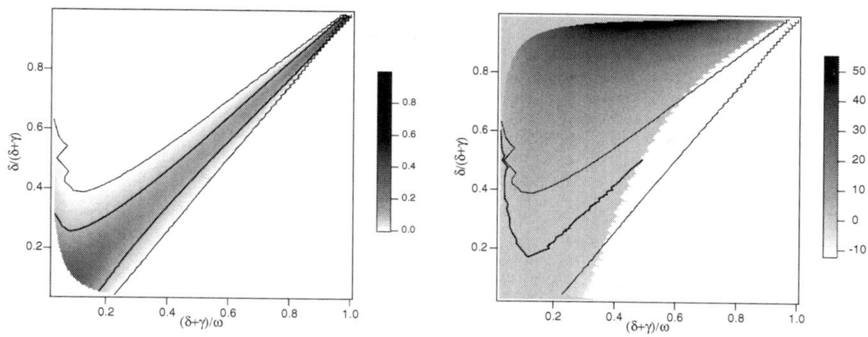

Figure 16.2 Left panel: Expected transitivity: the thin contour level is transitivity = 0.01, the thick contour level is transitivity = 0.1. Right panel: Logarithm of rescaled expected transitivity: the thick contour line is rescaled transitivity = 1; the thin contour line is density = 0.01; in the white area triangles cannot form

From there, rescaled transitivity obtains directly by computing the ratio of $\Pr\{jk|ij,ik\}$ to q, the density of the graph.■

Figure 16.2 displays how transitivity (left panel) and the logarithm of rescaled expected transitivity (right panel) respond to changes in the general knowledge level $(\delta + \gamma)/w$, and the relative importance of commonality versus complementarity, $\delta/(\delta + \gamma)$. One contour of density (density of 0.01) is superimposed on the two images, to show the parts of the parameter space where networks are both reasonably dense and clustered.

The results visible in Figure 16.2 are consistent with the intuitions developed in the section 'Discussion of general structures'. The results are that, first, small values of δ and γ sustain transitivity, as evidenced by the dark region close to the origin in the left panel of Figure 16.2. A move up along the vertical axis starting from the zone of high density corresponds to an increase in δ and the same decrease in γ, the net effect of which is a decline of transitivity. A move down the vertical axis from the high density zone also generates falling transitivity.[5] The effect of density (how many triangles can form in principle) thus dominates.

To detect the presence of excess transitivity, the right panel in Figure 16.2 shows rescaled transitivity (the measure normalized by the amount of transitivity present in an equivalent random graph). Here the main observation is that rescaled transitivity is monotonically increasing with $\delta/(\delta + \gamma)$ (and monotonically increasing with γ, showing the effect described in 'Discussion of general structures' above). Once we correct for the effect of pure randomness on triangle formation, the picture is thus modified. When the dominant consideration in partnering is commonality of knowledge rather than complementarity, the networks that arise are much more locally coherent than equivalent random networks. Rescaled transitivity is greater than one, and often much greater than one.

Distances

The second defining feature of a small world is a short characteristic path length, possibly approaching that of a random graph. The analytical derivation of the distance distribution is not possible in the model; however, we are able to provide the following asymptotic result.

> **Proposition 4.** *As the number of firms in the industry increases, the diameter of the industry network approaches* $\lfloor \delta / \gamma \rfloor + 2$. *When the network is connected, the limit is approached from above.*

Proof Each firm in the industry is located at one point in a discrete knowledge space of size $\binom{w}{\delta+\gamma}$. All possible knowledge configurations are equally likely, so as the number of firms increases, the proportion of occupied nodes in the knowledge space approaches 1. Consider the limiting case of each point in knowledge space being occupied by at least one firm. We can construct a longest shortest path. Consider a source node that has all its knowledge in the first $\delta + \gamma$ positions, and a target node that has its knowledge in the final $\delta + \gamma$ positions. Reserve the first $\delta + \gamma$ positions, then divide the remaining $w - (\delta + \gamma)$ positions in sets of length γ starting from the right end, letting $\xi(<\gamma)$ denote the remainder of the integer division. In Table 16.1 we record the number of 1s in each group of cells as we move from the source node to the target node through the nodes on the path. The movement down the table can be seen as moving $\delta + \gamma$ 1s from the beginning of the vector to the end of it, moving at most γ at a time. Moving exactly γ will permit link formation between adjacent rows of the table.

As a result, the path from source to target has length $\nu + 2$ where ν is the largest integer such that $\delta - \nu\gamma < \gamma$, i.e. $\nu = \lfloor \delta / \gamma \rfloor$, hence the result. The longest pairwise distance is obtained as the integer part of the ratio of knowledge commonality to complementarity increased by 2. This is the limiting case in which there is at least one firm at every point in knowledge space. As population size falls, points in the knowledge space are vacated. Considering any source-target pair, if a point on the constructed path is vacated, the path will be impossible to implement. If this does not disconnect the graph, the shortest path between source and target will increase.■

What we observe here is that the short characteristic path lengths arise from the interaction of the knowledge space with the alliance-formation rule. In contrast to much of the literature, it does not arise from the effects of indirect benefits. In Jackson and Wolinsky's (1996) connections model and its descendants, nodes receive benefits from nodes at all finite distances in the graph, but benefits that decay with distance. This provides an incentive for each node to ensure that it has short paths to all other nodes, implying incentives for at least some nodes to create shortcuts. In our model, there

Table 16.1 Constructing a longest shortest path

Number of ordered position	$\delta + \gamma$	ξ	γ	...	γ	γ	γ	γ	γ	γ
Source firm	$\delta + \gamma$	0	...	...	0	0	...	0	0	0
Firm 2	δ	0	...	...	0	0	...	0	0	γ
Firm 3	$\delta - \gamma$	0	...	...	0	0	...	0	γ	γ
...	continue until $\delta - vy < \gamma$, then									
Firm $v + 2$	$\delta - v\gamma$	0	...	...	0	γ	...	γ	γ	γ
Target firm	0	0	...	0	γ	$\delta - v\gamma$...		γ	γ	γ

are no indirect benefits: benefits to a firm arise only from innovation taking place along a single link.

For the finite case we treat the issue numerically. At each point in the (δ,γ)-(integer)plane, we generate 100 random networks and record the average distance among reachable pairs. Values of δ and γ such that there are no reachable pairs (i.e. when the graph is empty) do not contribute to the set of observations. Using the average density of those 100 networks, we also generate 100 random networks of that expected density, and record for them the average distance among reachable pairs. Re-scaled path length at each point in the (δ,γ)-plane is the ratio of those two path lengths. We then pool the data and build a frequency histogram, as shown in Figure 16.3.

The bulk of the distribution is concentrated around 1, indicating a structure with distances very similar to those of an equivalent random graph over the entire relevant parameter space.

It is worth emphasizing that the asymptotic result predicts a diameter which does not depend on w, the total number of pieces of knowledge in the industry, but solely on the relative importance of commonality and complementarity. Also note that the asymptotic diameter does not scale with n: the existence of a finite limit makes the network a small world in a very strong sense.

Small worlds?

Whereas there is no consensus regarding a measurable definition of a small world, a commonly used statistic is the ratio of rescaled transitivity to rescaled characteristic path length. Ratios significantly larger than one indicate the presence of a small world. In Figure 16.4 we plot the values of this ratio against network density for all relevant (δ,γ) combinations. In addition, we use different markers depending on whether the point lies above or below diagonal in the $(\delta/(\delta + \gamma), (\delta + \gamma)/w)$-plane. We can observe two things from this figure. First, small worlds are only present when the network is not dense. This

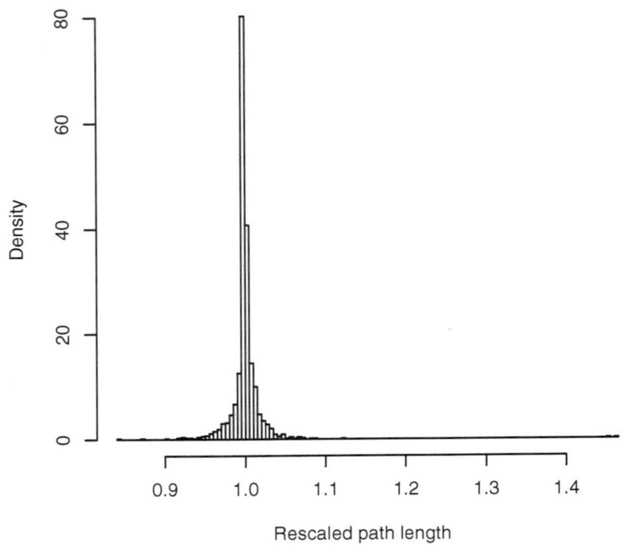

Figure 16.3 Frequency distribution of rescaled distance

is in fact consistent with the general observation that small worlds (and many empirically observed networks) are quite sparse. Second we observe here the importance of complementarity versus commonality in determining network structure. When strong complementarity in partners' knowledge stocks is necessary for an alliance to succeed ($\delta/(\gamma + \delta)$ is small) few triangles form and transitivity is low: small worlds are not present. At the other extreme, when commonality in knowledge is necessary for success ($\delta/(\gamma + \delta)$ is large) small worlds do form, again provided the network is not too dense.

We observe that focusing the alliance decision on technological fit produces networks which are highly clustered relative to equivalent random graphs. These networks also have path lengths that are close to those in equivalent random graphs. The degree distribution does not display power law properties: it is binomial and so is skewed to the right, but with an exponential tail. All together, these results permit us to state that technological fit, which is the stated aim for strategic alliance participants, is a strong explanatory factor for observed, small world, network architectures. More sophisticated network motives, which would involve indirect effects and attempts by firms to manipulate the larger network beyond their immediate set of alliances, are unnecessary.

Equilibrium networks with competition

The model just developed assumed no competition. Partner choice and equilibrium networks were restricted only by knowledge constraints. Now

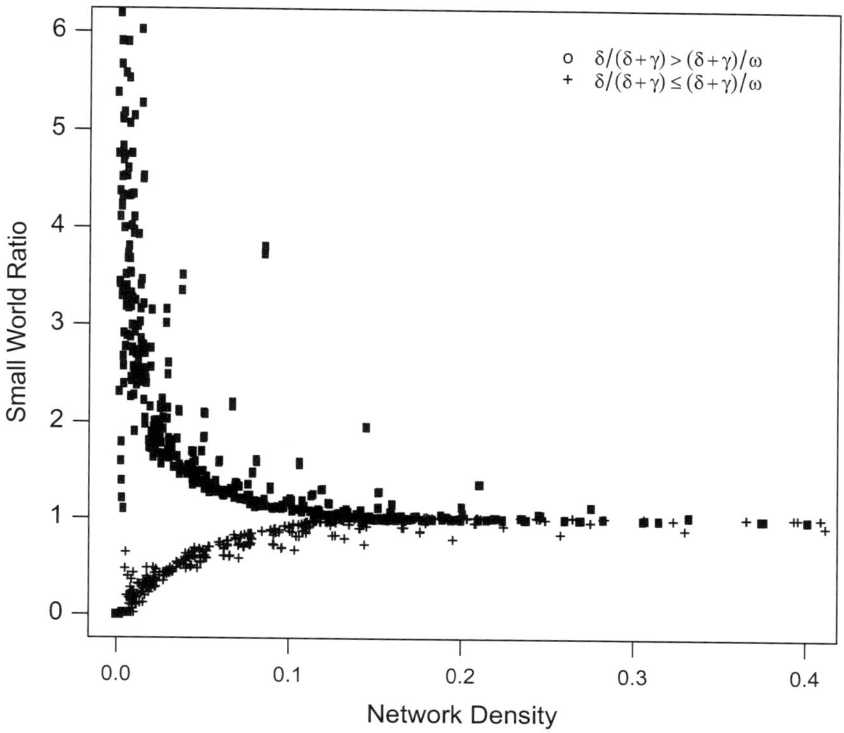

Figure 16.4 Ratio of rescaled transitivity to rescaled average path length as a function of network density

we turn to a case in which market considerations play a role. We examine a situation of extreme competition, in which one and only one alliance will be profitable *ex post*. All firms compete for the same discovery, and the first partnership that innovates obtains a patent and receives monopoly profits, whereas other firms get nothing. The essential issue in this section is how competition affects the equilibrium network, using the no-competition case as a benchmark. In effect, competition will add partner selection constraints, and we examine how they interact with the knowledge constraints discussed earlier, i.e. the relative importance of commonality (δ) versus complementarity (γ) in the innovative process.

Competition in networks

Actors in any alliance network are not only part of a network, they are also part of an industry which may involve inter-firm competition. This observation, although widely acknowledged, seldom appears in the empirical literature on alliances. In some industries, such as pharmaceuticals, the competition between partners in many alliances is quite attenuated (for example an alliance between a dedicated biotech firm and a big pharma firm may involve no competition on any market at all) while in others, such as micro-electronics or automobiles, partners in an alliance are often very direct competitors, even in the market for the outputs of the alliance. The literature on coalition formation tends to include a post-network formation stage of competition (see Bloch 1995; Joshi 2008 and references therein), as does some work on more general R&D networks (Goyal and Moraga-González 2001; Goyal and Joshi 2003, 2006). Typically these articles find that the degree of competition is central in determining which network architectures are observed at equilibrium. Goyal and Moraga-González find that stable networks introducing competition may have only minor effects on the properties of the stable networks. Both absent competition and with Cournot competition, the complete network is always stable. In their example, though, when competition is present a less dense network also becomes stable when global cost-reducing spillovers are weak. Goyal and Joshi (2003) similarly find that under quantity competition the only stable network is the complete network (under fairly general conditions), whereas under price competition, dominant group architectures (networks consisting of a complete sub-component plus singletons) are stable structures, again suggesting that increasing competitive pressures may serve to make the network less dense. We observe that small worlds are not equilibrium networks in either model, but in neither of these articles were small world features the object of study. Goyal and Joshi (2006) mention patent races as examples of 'playing-the-field' network games (games with specific externalities across the links of a given player, and across links of different players). Our model is quite different from these models, due to the constraints imposed by technological fit. In the articles just discussed, an alliance between any pair of firms has the same inherent value: there is no issue that partners must somehow complement each other for an alliance to be valuable. In the model developed below, by contrast, this is the central issue firms face in finding alliance partners. Nonetheless, we also find that although competition often has no effect on the network structure, when it does so, its effect is to make the stable network less dense.

Strategies and equilibrium

We assume that firms are engaged in a patent race with no discounting. In essence, any innovation project will succeed so the project that has a positive return is merely the first one to succeed. We assume that projects undertaken are in every relevant respect identical, so the probability that any particular project is the first to succeed is simply one over the number of projects. Thus the expected profit function for firm *i* in the (market) network *m* is simply

$$\pi_i^m = \frac{n_i^m}{A^m} - cn_i^m .$$

Recall that there are no indirect benefits in the model, thus a link between misfits will never be observed in equilibrium as it creates no value and carries a cost. So $\pi_i^m = (1/A^m - c)n_i^m$, with the understanding that a link only forms between technologically fit players. Denoting A_i the number of alliances not involving i, $A_i^m = A^m - n_i^m$ in the network m, we have $\pi_i^m = f(n_i^m, A_i^m) - cn_i^m$, which depends only on the number of the firm's links and the halved sum of links of other firms in the industry. In Goyal and Joshi's (2006) terminology, the game is a playing-the-field game. As they point out, the classical patent race is also a playing-the-field game. The difference between the game here and the example they consider is that in their case firms are the locus of innovation, whereas in our case it is the partnerships themselves. The aggregate revenue $f(z,A)$ is concave in own links, i.e. $f(z + 1, A) - f(z,A)$ is decreasing with z, and satisfies the strategic substitutes property that $f(z+1,A') - f(z,A') < f(z+1,A) - f(z,A)$ when $A' > A$. The game thus represents a situation in which there is a positive externality across the links of a given firm, and a negative externality across the links of different firms.

If technological fit is not an issue, i.e. firms can form all the partnerships they wish to form, then (as in Goyal and Joshi 2006, Proposition 3, p. 329) a symmetric network[6] with degree $d^* \in \{1,\ldots,n-2\}$ is pairwise stable if

$$\frac{2(n-2)}{n(nd^*+2)} < c < \frac{2(n-2)}{n(nd^*-2)}. \tag{4}$$

In principle, however, other (asymmetric) pairwise stable structures can exist for c in the right interval, and not all values of c fall in an interval which can take the form required by Condition 4.[7] We do not attempt an exhaustive characterization of the equilibrium networks, and will turn instead to numerical experiments. We can however observe that in the full setup in which firms are both constrained by knowledge fit and pairwise stability, the following result obtains readily.

Proposition 5. *The equilibrium market network m is never dense than the technology network t. In addition, if Condition 1 holds so that there exists a symmetric equilibrium network of degree d*, then (a) if* $max\{n_i^t; i \in N\} \le d^*$ *then the technology network t is a market equilibrium network;* (b) *if* $min\{n_i^t; i \in N\} \ge d^*$ *then one (of possibly many) equilibrium market networks is symmetric with each firm having exactly d* alliances.*

Proof All potential links are in the technology network t. Either all are pairwise stable, or there are firms forced by profit maximization to drop a subset of their links in t, in which case there can only be fewer links in total.

Suppose c satisfies Condition 4, and $n_i^t \leq d^*, \forall i \in N$. No pair of firms can add a link. Does a firm want to sever a subset of its links? For any $i \in N, A_i = \left(\sum_{j \neq i} n_j^t - n_i^t\right)/2 < (n-2)d^*/2 = A$. Using the strategic substitutes property and concavity in own links, $\gamma(n_i^t, A_i^t) - \gamma(n_i^t - 1, A_i^t) \geq \gamma(n_i^t, A) - \gamma(n_i^t - 1, A) \geq \gamma(d^*, A) - \gamma(d^* - 1, A) > c$, i.e. no firm wants to drop links.

Suppose c satisfies Condition 1, and $n_i^t \geq d^*, \forall i \in N$. The technology constraints are not binding and so the symmetric equilibrium of degree d^*, which we know to be stable, can be formed by firms.■

Integrating the constraints of pairwise stability and knowledge fit makes the problem very complicated to solve mathematically. The reason is that some firms are bound by the pairwise stability constraint (those who have many partners in the knowledge network t) and some are bound by the knowledge matching constraint (those with few partners in the knowledge network t). We thus resort to a simple, edge-by-edge addition algorithm to randomly generate equilibrium networks starting from the technology network t. An equilibrium market network m will obtain as a subset of the set t of potential links.

Starting with an empty network $m_0 = \{\varnothing\}$, we draw links at random from the technology network t and incrementally add them to the market network. At step s, as m_{s-1} only contains links between fit partners, we pick a random link in $t - m_{s-1}$, check that it is marginally profitable and add it to m_{s-1} to form m_s, the current market network. We then recalculate $A_i^{m_s}, \forall i \in N$. At this stage, we make use of the strict quasi-concavity of the profit function to informally define the 'best response' of firm i as $b_i^{m_s} = b_i^{m_s}(A_i^{m_s}) = -A_i^{m_s} + \sqrt{A_i^{m_s}/c}$. As either $\lfloor b_i^{m_s} \rfloor$ or $\lceil b_i^{m_s} \rceil$ is the optimal number of partners for firm i in m_s, we check that no firm exceeds its best response. If such a firm is found, one of its links is randomly deleted to remove the violation. We repeat this process until no more links can be added. This creates a network that satisfies the constraints of knowledge fit and pairwise stability.

Density and degree distribution

Using the algorithm described above, for each (δ,γ) pair, we generate equilibrium networks and pool the observations to construct our estimated distribution. Industry size is set to $n = 100$, the knowledge universe has size $w = 100$.

Figure 16.5 displays the degree distributions of the technology network and of the market network in 4 different environments. In each of them, we provide the degree distribution in the technology network t, and 3 degree

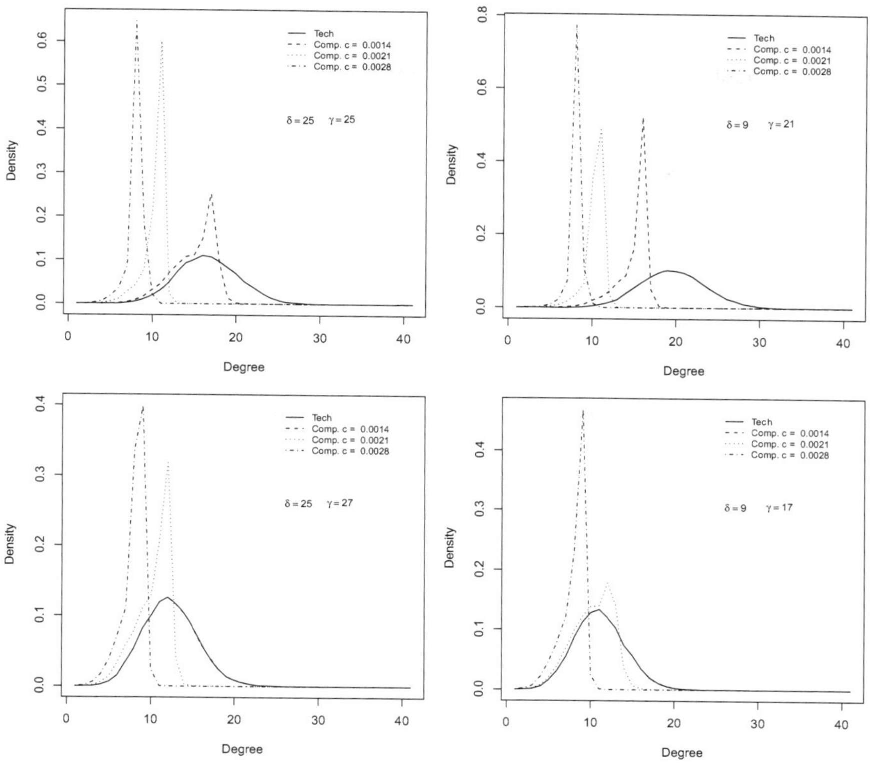

Figure 16.5 Degree distributions for different parameter values. Curves that cannot be seen lie on top of the technology network distribution

distributions obtained for 3 different values of c, to illustrate the influence of cost on the properties of the equilibrium networks. Each distribution represents the average behaviour of 200 instances of the technology network and its attendant market network.

The upper panels of Figure 16.5 correspond to areas of (δ,γ)-plane where the network is dense: $\delta/(\delta + \gamma) \approx (\delta + \gamma)/w$. The left panel corresponds to $(\delta,\gamma) = (25,25)$, a context in which firms know much of the relevant knowledge in the industry, the demand on both commonality and complementarity. The right panel corresponds to $(\delta,\gamma) = (9,21)$, a situation in which knowledge is more scarce and the demand on complementarity is strong relative to that on similarity.

The lower panels in Figure 16.5 correspond to off-diagonal situations and thus sparse industry networks. The lower left panel is a situation in which $\delta/(\delta + \gamma) < (\delta + \gamma)/w$ with $(\delta,\gamma) = (25,27)$ while the right panel depicts a situation on the other side of the diagonal, with $(\delta,\gamma) = (9,17)$.

In each of the panels we observe in principle 3 types of distributions: those whose mass is highly concentrated around a single value; those which mimic the technology distribution almost exactly; and those which mimic the technology distribution (less exactly) up to a certain value, and then concentrate mass at that value. The explanation is that there are in principle 2 types of firms: those whose networking behaviour is constrained by knowledge, and those whose behaviour is constrained by profit maximization. If the technology network is dense, most of the firms will be constrained by profit maximization, and so we will see the mass of the distribution concentrated around the profit-maximizing degree (in all of the 200 realizations, the symmetric equilibrium defined by Equation 4 emerges). If the technology network is sparse, it provides the constraint, and in the extreme case, the competition network is the technology network, and so the degree distributions lie on top of each other: from the perspective of profit maximization, all firms would prefer more partners, but when technology matters, they cannot find more that fit. In the intermediate case, in the same network, some firms are constrained by knowledge fit, and others are constrained by profit maximization. Here we see a combination of the two effects.

The effect of c, the cost per alliance, is to change where the profit maximization constraint binds. That is, higher values of c imply that the optimal number of alliances will fall, so we see a departure from the technology degree distribution at lower values. When c is low, the profit-maximizing number of partnerships is high, so all firms will 'take all the partners they can get', and so the degree network is identical to the technology network.

Small worlds

The technology network showed small world properties over significant parts of the parameter space. Adding competition does not destroy this result. Small worlds remain and, more specifically, small worlds remain in the parts of the parameter space where they emerged when competition was absent. To facilitate comparison with our benchmark case, we show ratios for the statistics in the technology case to those in the network case. For each point in the parameter space we compute the technology to competition ratio, and plot histograms of that ratio. We do this for raw and rescaled transitivity and average pairwise distance, as well as for the small world ratio itself.

The upper right panel of Figure 16.6 shows a frequency distribution of the ratio of transitivity in the two regimes, over the entire (δ,γ)-plane. Most values of this ratio are very close to one, as expected. The reduction in density caused by competition removes some links, and drives transitivity down in a relatively mechanical way. The same obtains with rescaled transitivity, shown in the lower right panel: the ratio of the competition to no

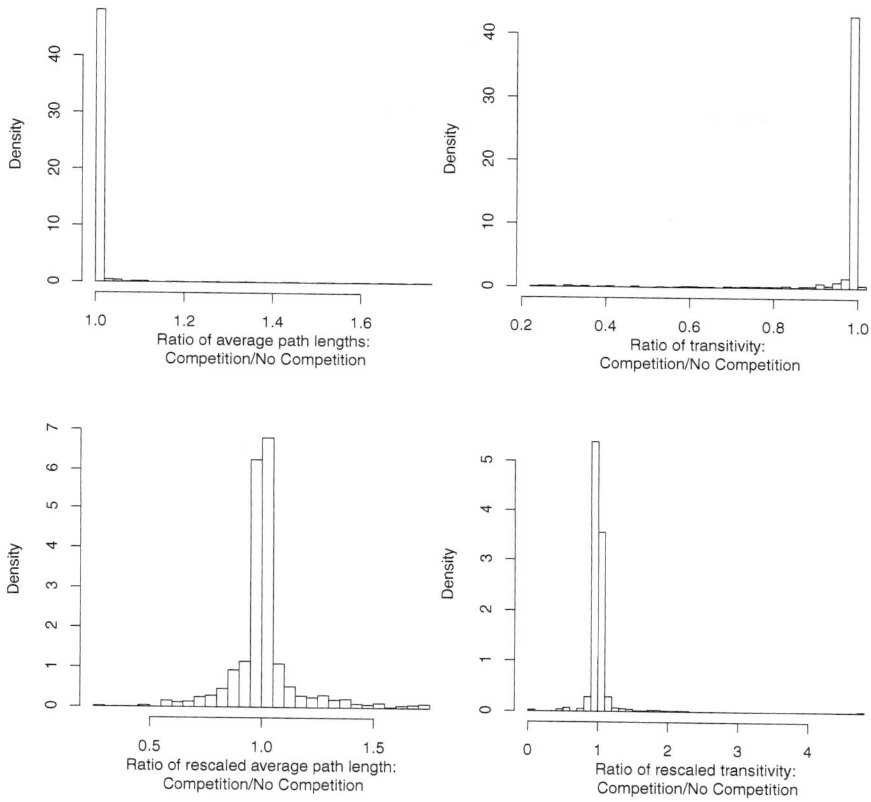

Figure 16.6 Comparing competition with no competition: ratios of path length and transitivity over the parameter space

competition cases is centred around 1, though the distribution is more dispersed than that of the raw measures.

Turning to pairwise distances, we can observe in the upper left panel of the figure that competition also has essentially no effect on average path lengths. The explanation here is that there are many shortest paths between any pair of nodes. Thus the effect of introducing competition, which is to remove edges relative to the no-competition case, does not, except in relatively rare instances, remove all of the shortest paths. Thus competition has essentially no effect on average path length, and any effect it has is to increase path length. The effect would be marked only if competition brought density well below that of the technology network. The same picture obtains when rescaled average pairwise distance is considered in the lower left panel, most of the mass of the distribution lying close to 1.

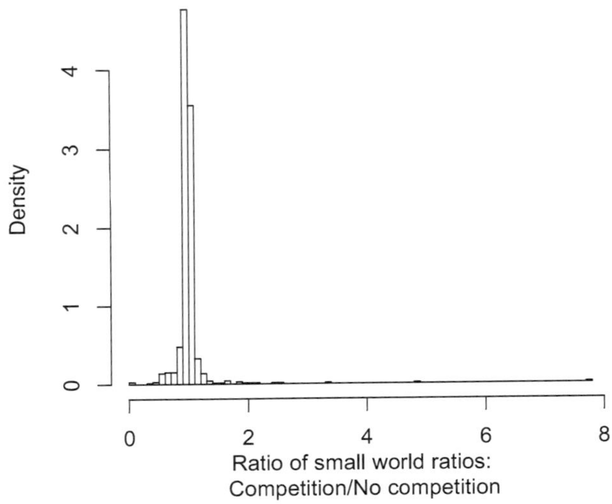

Figure 16.7 Comparing competition to no competition case: ratio of the small world ratio over the parameter space

From this we can conclude that competition, even at its most punishing, does not have a big effect on network architecture insofar as small world aspects are concerned. Density is smaller always, but clustering and short distances are preserved. And this in turn is preserved in Figure 16.7 which computes the ratio of small world ratios in the competition to no-competition cases: again the mass lies mostly around 1.

Conclusion

In this chapter we have proposed and analysed a simple model of alliance formation and joint innovation in which we have assumed that firms must reach a precise balance between commonality and complementarity in order to jointly innovate. Small worlds arise because firms have two essential characteristics. First, firms are heterogeneous. They hold different technological endowments, and seek partners with whom to assemble these technologies in the right portfolio. This will imply random graph features, and specifically short distances, for the equilibrium network. But this is not enough. In addition to holding random knowledge endowments, firms act strategically. They take into account the cost and benefits from each potential alliance, and the behaviour of other firms in the network when it is relevant (when competition is present). This behaviour will create local correlation in the alliance decisions, and thus a tendency for the equilibrium to display excess clustering or transitivity. This tendency is strongest when similarities are much more important than complementarities in the innovation process, that is, when what we share matters more than where we differ. It is this combination of a random process of knowledge assignment and thus location in a knowledge space, and firms' strategic decision to ally only with profitable – here meaning relatively more similar than dissimilar – partners which creates small worlds in equilibrium.

It is worth emphasizing that we obtain small world results without assuming spillovers or any other indirect, possibly long-distance effect. So the mechanism at work to lower the diameter of equilibrium networks is fundamentally different from the more 'traditional' incentive to create a shortcut to capture value created far away in the graph. Also the reason why there is clustering is a very simple and general one of 'transitivity of attractions', a feature suggested but not formalized in Granovetter's (1976) seminal article.

With this model we capture the idea that when a firm makes a strategic alliance it is interested in technological fit with its partners. With respect to the strategic management literature, it is worth emphasizing that we recover the properties of observed alliance networks – small worlds with skewed link distributions – ignoring entirely issues of social capital. This suggests that models of link formation (and hence network emergence) that ignore technology may be missing something important. Again, one thing we observe, though, is that the high values of clustering characteristic of small worlds are not present when γ is high. That is, when the requirement for reciprocal complementarities is very strong, clustered networks do not emerge. One conclusion to draw here is that in empirically observed networks, which tend to have high clustering, the need for this type of complementarity must be weak.

The second part of the model introduces competition following innovation. The basic idea is that of a patent race in which all alliances compete for the patent on a unique innovation. Introducing this aspect, which tends not to be present in empirical studies of network formation, has predictable effects, the most important one being to lower network density. By and large, though, the introduction of competition does not have a big effect on network structure. Here we can conclude that it may be safe to leave this feature out of our investigations without doing much damage to our understanding.

Notes

1 Movie actors, power grids, neural networks (Watts and Strogatz 1998); food webs (Williams *et al.* 2002); academic co-authorship in many disciplines (Newman 2004; Goyal *et al.* 2006), strategic alliances (Verspagen and Duysters 2004; Powell *et al.* 2005); patent co-invention (Balconi *et al.* 2004).

2 See for example Hagedoorn (1993), Mowery *et al.* (1996) or Ahuja and Katila (2001).

3 These are very strong conditions on alliance success. There are two obvious ways they can be relaxed. Overlap could be 'near' optimal overlap and still result in

success. So the overlap rule could be changed to $\delta - \rho \leq v_{ij} \leq \delta + \rho$. Similarly, as firms do not need to share every possible piece of knowledge, the 'reciprocal complements' rule could be changed to $s_{ij} \geq \gamma$ and $s_{ij} \geq \gamma$. (Notice that this also relaxes the strong reciprocity condition.) Both of these generalizations, when made individually, are relatively straightforward, and involve summing the functions given here. Relaxing them simultaneously is more intricate, however.

4 To be more precise, partners of i have δ facts in common with each other and with i, but in the remaining $w - \delta$ places in their knowledge vectors, they are mirror-images of i and identical to each other.

5 Note that to trace the effect of changing γ holding other parameters constant one follows a hyperbola through the $(\delta/(\delta + \gamma), (lta + \gamma)/w)$ space.

6 Note that for a symmetric network g of any degree d^* to be possible, the number of players n cannot be odd: $A^g = nd^*/2$ is guaranteed to be an integer number for all integer value of d^* only if n is even.

7 We should point out that if asymmetric equilibria do exist in fact, they appear to be rare. In several thousand numerically generated networks satisfying Condition 1, asymmetric equilibria never appeared.

References

Ahuja, G. (2000) 'Collaboration networks, structural holes and innovation: A longitudinal study', *Administrative Science Quarterly*, **45**: 425–55.

Ahuja, G. and Katila, R. (2001) 'Technological acquisitions and the innovation performance of acquiring firms: A longitudinal study', *Strategic Management Journal*, **22**: 197–220.

Amaral, L.A., Scala, A., Bartelemy, M. and Stanley H.E. (2000) 'Classes of small-world networks', *Proceedings of the National Academy of Science*: **111**, 49–52.

Balconi, M., Breschi, S. and Lissoni, F. (2004) 'Networks of inventors and the role of academia: An exploration of Italian patent data', *Research Policy*, **33**: 127–45.

Barabasi, A.L. and Albert, R. (1999) 'Emergence of scaling in random networks', *Science*, **286**: 509–12.

Baum, J.A., Shipilov, A.V. and Rowley, T.J. (2003) 'Where do small worlds come from?' *Industrial and Corporate Change*, **12**: 697–725.

Bloch, F. (1995) 'Endogenous structures of association in oligopolies', *Rand Journal of Economics*, **26**: 537–56.

Burt, R.S. (1992), *Structural Holes: The Social Structure of Competition*, Harvard University Press, Cambridge, Mass.

Carayol, N. and Roux, P. (2009) 'Knowledge flows and the geography of networks: A strategic model of small world formation', *Journal of Economic Behavior and Organization*, **71**: 414–27.

Coleman, J.A. (1988) 'Social capital in the creation of human capital', *American Journal of Sociology*, **94**: S95–S120.

Cowan, R. and Jonard, N. (2009) 'Knowledge portfolios and the organization of innovation networks', *Academy of Management Review*, **34**: 320–42.

Dyer, J.H. and Nobeoka, K. (2000) 'Creating and managing a high performance knowledge sharing network: The Toyota case', *Strategic Management Journal*, **21**: 345–67.

Erdös, P. and Rényi, A. (1960) 'On the evolution of random graphs', *Publications of the Mathematical Institute of the Hungarian Academy of Sciences*, **5**: 17–61.

Galeotti, A., Goyal, S. and Kamphorst, J. (2006) 'Network formation with heterogeneous players', *Games and Economic Behavior*, **54**: 353–72.

Goyal, S. and Joshi, S. (2003), 'Networks of collaboration in oligopoly', *Games and Economic Behavior*, **43**: 57–85.

Goyal, S. and Joshi, S. (2006), 'Unequal connections', *International Journal of Game Theory*, **34**: 319–49.

Goyal, S. and Moraga-González, J.L. (2001) 'R&D networks', *Rand Journal of Economics*, **32**: 686–707.

Goyal, S., Van der Leij, M. and Moraga-González, J.L. (2006) 'Economics: An emerging small world', *Journal of Political Economy*, **114**: 403–12.

Granovetter, M. (1976) 'The strength of weak ties', *American Journal of Sociology*, **78**: 1360–80.

Grant, R.M. (1996) 'Towards a knowledge-based theory of the firm', *Strategic Management Journal*, **17**: 109–22.

Gulati, R. and Gargiulo, M. (1999) 'Where do inter-organizational networks come from?' *American Journal of Sociology*, **104**: 1439–13.

Hagedoorn, J. (1993) 'Understanding the rationale of strategic technology partnering: Inter-organizational modes of cooperation and sectoral differences', *Strategic Management Journal*, **14**: 371–85.

Jackson, M.O. and Rogers, B. (2005) 'The economics of small worlds', *Journal of the European Economic Association*, **3**: 617–27.

Jackson, M.O. and Wolinsky, A. (1996) 'A strategic model of social and economic networks', *Journal of Economic Theory*, **71**: 44–74.

Joshi, S. (2008) 'Endogenous formation of coalitions in a model of a race', *Journal of Economic Behavior and Organization*, **65**: 62–85.

Kleinberg, J. (2000) 'Navigation in a small world', *Nature*, **406**: 845.

Kogut, B., Shan, W. and Walker, G. (1992) 'Competitive cooperation in biotechnology: Learning through networks?', in N. Nohria and R. Eccles (eds) *Networks and Organizations: Structure, Form and Action*, Harvard Business School Press, Boston.

Mowery, D.C., Oxley, J.E. and Silverman, B.S. (1996) 'Strategic alliances and interfirm knowledge transfer', *Strategic Management Journal*, **17**: 77–91.

Mowery, D.C., Oxley, J.E. and Silverman, B.S. (1998) 'Technological overlap and interfirm cooperation: Implications for the resource-based view of the firm', *Research Policy*, **27**: 507–23.

Newman, M. (2003) 'The structure and function of complex networks', *SIAM Review*, **45:** 167–256.

Newman, M. (2004) 'Co-authorship networks and patterns of scientific collaboration', *Proceedings of the National Academy of Science*, **101**: 5200–5.

Nooteboom, B. (2000) *Learning and Innovation in Organizations and Economics*, Oxford University Press, Oxford.

Powell, W.W., White, D.R., Koput, K.W. and Owen-Smith, J. (2005) 'Network dynamics and field evolution: The growth of inter-organizational collaboration in the life sciences', *American Journal of Sociology*, **110**: 1132–1205.

Rowley, T. Behrens, D. and Krackhardt, D. (2000) 'Redundant governance structures: An analysis of structural and relational embeddedness in the steel and semiconductor industries', *Strategic Management Journal*, **21**: 369–86.

Schoenmakers, W. and Duysters, G. (2006) 'Learning in strategic technology alliances', *Technology Analysis and Strategic Management*, **18**: 245–64.

Uzzi, B. (1997) 'Social structure and competition in inter-firm networks: The paradox of embeddedness', *Administrative Science Quarterly*, **42**: 35–67.

Verspagen, B. and Duysters, G. (2004) 'The small worlds of strategic technology alliances', *Technovation*, **24**, 563–71.

Walker, G., Kogut, B. and Shan, W. (1997) 'Social capital, structural holes and the formation of an industry network', *Organization Science*, **8**: 108–25.

Watts, D.J. (1999) *Small Worlds: The Dynamics of Networks Between Order and Randomness*, Princeton University Press, Princeton.

Watts, D.J. and Strogatz, S. (1998) 'Collective dynamics of small-world networks', *Nature*, **393**: 440–2.

Williams, R.J., Berlow, E.L., Dunne, J.A., Barabási, A.-L. and Martinez, N.D. (2002) 'Two degrees of separation in complex food webs', *Proceedings of the National Academy of Science*, **99**: 12913–16.

17 Market occupation in a circular market

Product innovation and/or patenting[1]

Murat Yıldızoğlu and Ehud Zuscovitch

Introduction

There is a very clear analytical need to better understand the role that diversity fulfils in advanced industrial economies. The configuration of new technologies that gradually emerges through the current industrial transformation can efficiently sustain a much greater degree of product and skill differentiation because information technologies which are at the heart of the present evolution endow the economic system with much greater flexibility. Scientific and technological progress in areas such as advanced materials and biotechnologies contributes to the understanding of the micro-properties of the matter (whether living or inert) so the potential of new combinations that can feed the industrial system is rapidly increasing as well. Rising living standards in western economies push towards a better matching of these properties to user needs and this much finer adjustment of supply and demand characteristics often creates many micro-markets and areas where standardisation prevailed not long ago. While capital intensity does not seem to decline, information intensity rapidly rises and the ability to transform information and knowledge into market power often takes place through complex strategies of product differentiation.

Such an information-intensive context is a very powerful stimulus to innovative activities both as cause and consequence of frequent changes. The Schumpeterian creative destruction metaphor is of increasing theoretical appeal in such a framework (see Aghion and Howitt 1992) as a very powerful imperfect competition operates through a succession of product configurations and models. No wonder that evolutionary theories gain in importance and that evolutionary and learning mechanisms appear much closer then ever to the hard-core of economics. Under such circumstances one wonders whether price-quantity adjustment is prevailing in economic adjustment or whether competition through product differentiation and qualities fulfils the more important role. Issues of quality and diversity appear now frequently in theories of growth and international trade (Grossman and Helpman 1991).

The interest in product differentiation and its impact on the industry structure and evolution have evolved in two distinct avenues. One is basically the Hotelling tradition that translated the diversity issue into terms of spatial competition.[2] To this tradition, one can naturally relate the Dixit and Stiglitz analysis (1977) of optimal diversity. Works like those of Schmalensee (1978) contributed to further the analysis of brands proliferation and product line strategies. The key feature of such models is that product differentiation appears as a matter of pure strategic concern and no attention is given to technological consideration about the development of the knowledge that is required when moving from one product type to another. In a way all such difficulties are disregarded. Further evolution in the late 1980s and the early 1990s along this line of development tries to relate the issue of product differentiation to the present technological evolution as in the works of von Ungern-Sternberg (1988) and Eaton and Schmitt (1994) that explicitly refer to flexible manufacturing but there again technological constraints do not appear to really affect differentiation decisions.

The second strand of the literature that deals with differentiation is the innovation line. We will not refer here to the relation between innovation and diversity to evolution and growth which is a much wider context but restrict ourselves to industrial organisation considerations. Innovation relates very strongly to product differentiation because the development of knowledge and know-how and their application to micro-markets rather than to an aggregate market is an essential way to secure appropriability of the results of R&D spending. Incorporating the results of search activities in characterising demand in a fine tuning manner to specific product configurations is a way to make much more difficult the action of imitators who will have a much harder job. This view that differentiation secures much better the innovative efforts is something that can be easily identified in major strategic displacements such as the shift of European chemical industries from commodities to specialities. Nesting increasing amounts of experience and knowledge in the diversity of the production line is also an excellent way to regain market shares from newly industrialised countries that are doing much better now with standardised commodities as the basic technology is available to them and their relative efficiency in terms of labour and material costs is overwhelming.

It takes some time for empirical features to end up in theoretical models. Although research portfolios appear earlier, as do concerns with the ways to secure appropriation of innovation (see the Yale inquiry on innovation appropriation) the issue of securing rents through diversity did not really appear in the patent literature until recently with the works of Klemperer (1990), Waterson (1990) and Ordover (1991). Once the breadth of the patent is introduced in these works, a more complete characterisation is offered in further dimensions such as van Dijk (1994) on the influence of the degree of the innovation radicality through the notion of patent height. Breadth and height of patents are the ways in which product differentiation is integrated in the innovation and market structure literature. Yet even there one tends to look for intuition where we have already some light... and strategic spatial considerations are put forward rather than differential returns of scale and scope in the production of technological knowledge. Noticeable exceptions are recent contributions by Luski and Wettstein (1995) and O'Donoghue *et al.* (1995) which both try to relate one innovation to another in a sequential way. Thus the broader project that we pursue is to bring together these two separate lines. We would like to combine strategic 'location' considerations with insights and 'stylised' facts on the relationship between constraints and returns of technological development. Yet the paper presented here is just a first step in this research. We present here a much simplified framework in which the two arguments, those of product differentiation and the breadth of patent protection, are combined in a way that does not yet incorporate increasing returns in relation to the scale and scope of the innovative activity as such so that the theoretical exercise remains somewhat frustrating. Yet we feel that one should start with such a simple framework as a neutral benchmark against which we will then evaluate the results of more satisfactory integration of innovation returns.

Our objective in the present work is therefore more modest. In a model of a horizontally differentiated market, we study the opportunity of these two instruments for an incumbent who aims to protect his market from potential entrants: namely, variety proliferation and patenting.

Our main result shows the superiority of variety proliferation over patenting when the market is horizontally differentiated. In fact, the first strategy enables the incumbent to blockade entry in a rather straightforward manner and, at least, to secure market protection. In fact, the varieties on the front line – facing directly entrants' variety – are mainly those that absorb the shock of entry; the incumbent can use higher prices for protected varieties. In opposition, in the context of horizontal differentiation, only extreme patenting protection can provide absolute entry barriers. Concerning social welfare, we show that, contrary to the usual oligopoly results, in such a market (cf. Tirole (1988, chapter 7), with or without patents, the industry produces too few varieties, even in the proliferation case.

In the next section we present the general structure of the model. The behaviour of an unconstrained monopoly is studied in the third and fourth sections. Our main results concerning variety proliferation and barriers to entry will be presented in the fifth section. The last section is devoted to the study of patenting and the opportunity of such a strategy.

The model

We consider a circular market of unit length with a uniform density of consumers on the circle (following Schmalensee 1978; Salop 1979 and Tirole

1988). Each point on the circle corresponds to one particular variety of the good. Each consumer is located on the circle at the point which corresponds to his ideal variety and the consumers have the same reservation price s. Consequently, the good is horizontally differentiated. Consumers do not necessarily have the possibility to consume their ideal goods. A consumer will accept to consume a variety which is at a distance x from his ideal good and which costs p if and only if:

$$p + tx^2 \leq s \tag{1}$$

where tx^2 represents the quadratic sacrificing cost of the consumer. The higher is t, the more reluctant is the consumer to buy a variety far from his ideal one. Product differentiation is hence increasing with t.

The incumbent (firm 1) is alone on the market at the beginning of the game. The potential entrant (firm 2) will take its decision at the beginning of the second period, after having observed the number of varieties produced by the incumbent.

The incumbent must choose between the following three strategies:

- to produce n^l varieties that would blockade entry in a duopoly and obtain the corresponding monopoly profits;
- to use a sufficient number of patents in order to blockade entry, produce the corresponding number of varieties (n^p) and obtain the corresponding monopoly profit;
- to accommodate entry and produce n^d corresponding to the duopoly number of varieties.

The potential entrant, if he decides to enter, can produce only one variety. If he enters, a price competition takes place in the second period. If he stays out, the incumbent retains the monopoly position.

The cost of producing each variety is given by:

$$C(q) = c.q + f \tag{2}$$

where f represents the cost of installation of the productive capacity corresponding to this variety. These production costs do not depend on the particular characteristics of the variety. This property is not disturbing in a context of horizontal differentiation. It would be much more restrictive in a vertically differentiated market.

Many results of the paper will depend on the ratio $\delta \equiv t / f$. Given that t represents the desire for variety of the consumers and f represents the cost of providing this variety, δ represents a sort of *social desirability* of product variety.

We will study different situations which can arise in the market. First of these is a *mono-variety* monopoly. This case will be followed by an n – *varieties* monopoly of the incumbent. The third case corresponds to a duopoly where the incumbent produces n varieties. Comparing these cases will enable us to evaluate the alternative strategies for the entry blockading strategies. The last case will introduce the possibility of patents by the incumbent and we will study the consequences of such a possibility.

Mono-variety monopoly

If the incumbent is alone in the market, he will face the whole demand. Every consumer will buy his product if the furthest consumer (who is at a distance 1/2) can buy it:

$$p+t\frac{1}{4}\leq s \Leftrightarrow p\leq \frac{4s-t}{4}.$$

The price must also maximise his profits. Hence we must have:

$$\begin{aligned} p^m &= \min\left\{\frac{4s-t}{4}, \arg\max_p (p-c)d(p;s)-f\right\} \\ &= \min\left\{\frac{4s-t}{4}, \frac{1}{3}(2s+c)\right\} \end{aligned}$$

where $d(.)$ is the market demand,

$$d(p)=2\sqrt{\frac{s-p}{t}}\leq 1 \text{ if } s\leq c+\frac{3}{4}t. \tag{3}$$

Assumption A.1: We assume that s is such that the two solutions are identical:

$$s=\frac{3}{4}t+c.$$

This assumption implies that the monopoly problem has an interior solution where the firm serves exactly all the market. (A.1) simplifies our problem and eliminates a parameter from the model without great loss of generality. The following proposition sums up the monopoly solution.

Proposition 1. Under (A.1) the mono-variety monopoly equilibrium is characterised by:

$$p^m = \frac{1}{2}t + c, \quad \Pi_1^m = \tfrac{1}{2}t - f \geq 0 \Leftrightarrow \delta \equiv t/f \geq 2.$$

Proof:
Trivial.
***n*-varieties monopoly**
This case generalises the preceding one to a context where the incumbent can produce n varieties at equidistant locations in the characteristics space (the circle). This is a natural consequence of uniform density distribution of consumers' tastes on the market and the presence of fixed costs. The market share of a variety j covers $1/2n$ at the right and the left of this variety. The price that will be used on this segment of the market is therefore given by:

$$p_j = s - \frac{t}{4n^2}.$$

The corresponding gross profit is:

$$\pi_j = \frac{1}{4}\frac{4sn^2 - t - 4cn^2}{n^3}$$

Total gross profit of the firm is given by:

$$\Pi_g^m = s - c - \frac{t}{4n^2}$$

Under (A.1), profit maximization implies:

$$p_j^* = \frac{1}{4}\frac{n^2(4c + 3t) - t}{n^2} \text{ and } \Pi_g^m = \frac{1}{4}\frac{t(3n^2 - 1)}{n^2},$$

The latter is strictly increasing and concave with n:

$$\frac{\partial \Pi_g^m}{\partial n} = \frac{t}{2n^3} > 0 \text{ and } \frac{\partial^2 \Pi_b^m}{\partial n^2} = -\frac{3t}{2n^4} < 0$$

Therefore, in the absence of fixed production costs for each variety, the optimal monopoly variety becomes infinite: the firm would cover totally the circle producing a variety by consumer (his *ideal* variety).

We introduce fixed production costs depending on the number of varieties produced in order to dismiss this trivial solution:

$$CF(n) = nf \Rightarrow \Pi^m = \frac{1}{4}\frac{t\left(3n^2-1\right)}{n^2} - nf.$$

The following proposition establishes a monopoly solution with fixed costs.
Proposition 2. The monopoly will choose to produce

$$n^m = \frac{1}{2}\left(\frac{4t}{f}\right)^{1/3}$$

varieties and obtain a net profit of

$$\Pi^m = \frac{3}{8}(4t)^{1/3}\left((2t)^{1/3} - 2f^{1/3}\right)$$
$$= f \cdot \frac{3}{4}\left(\delta - (4\delta)^{1/3}\right) > 0 \quad \text{if} \quad \delta > 2.$$

Proof:
Trivial.
Figure 17.1 shows that the number of varieties is increasing with δ but this effect is decreasing when δ increases. It is also interesting to observe that monopoly leads to less variety on the market than the socially desirable one. In fact, the cost minimising number of varieties in this model is given by (cf. Tirole 1988, chapter 7, p. 284):

$$n^s = \frac{1}{2}\sqrt{\delta},$$
$$n^s > n^m \Leftrightarrow \delta > 16.$$

As we can observe in Figure 17.1, this is the case as soon as $n^m > 2$. Therefore the monopoly is socially under optimal concerning the variety creation.

We can now study the role of the entry on the market equilibrium.

Entry and duopoly

We will now consider the entry of a mono-variety producer. To this end, we will rename the n varieties of the incumbent such that the entry will take place between the nth variety and the first one (cf. Figure 17.2).

The entry, if it is chosen, will be followed by a price competition between n varieties of the incumbent and the variety of the entrant. If we do not limit

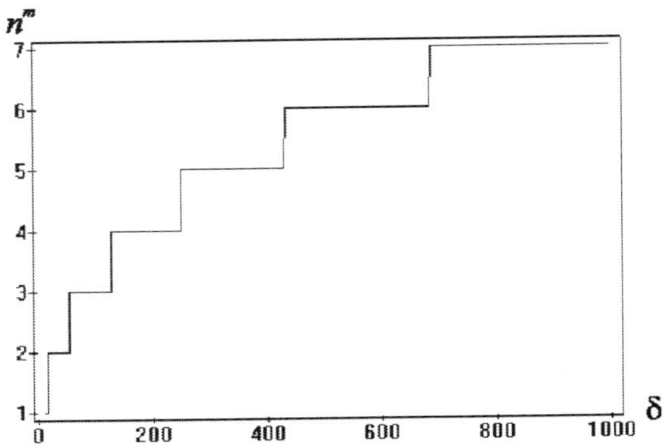

Figure 17.1 Optimal number of varieties in monopoly

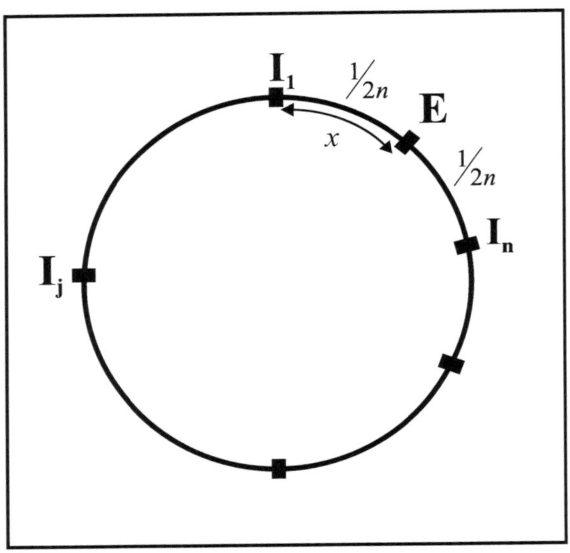

Figure 17.2 Circular city and entry

our analysis to symmetrical price equilibria of this game, a solution of this sub-game is quite difficult to obtain for any n in general.

However, we can compute this solution analytically for each particular level of n. And this solution implies in fact different prices for the varieties of the incumbent.

We will compute backwardly the **sub-game perfect equilibrium** of this game.

Price competition and location of the entrant

The incumbent fixes his prices in order to maximise his total profit over n varieties. We represent the price vector of the incumbent by p_I and the price of the entrant by p_e. The reaction functions of the firm(s) will be given by:

$$p_I^*(p_e;n,x) = \arg\max_{p_1 \cdots p_n} \sum_{j=1}^{n} \pi_j(p_I, p_e; n, x)$$

$$p_e^*(p_I;n,x) = \arg\max_{p_e} \Pi_e(p_I, p_e; n, x)$$

The price equilibrium (p_I^*, p_e^*) must be the solution of this system of $(n + 1)$ equations. It should be noted that the size of the system depends on n. But we can compute the equilibria for different values of n.

Proposition 3. For each value of n and x, a unique price equilibrium (p_I^*, p_e^*) can be established.

Proof:

cf. appendix.

When deciding on location, the entrant must take into account this price equilibrium which will follow his decision. If $p_1 = p_n$, the entrant will choose to locate equidistantly to varieties n and 1. This distance will then be equal to $1/2n$.

Proposition 4. In equilibrium $p_1^* = p_n^*$ and $x^* = 1/2n$.

Proof:

cf. appendix.

The resolution of this model for different values of n shows that this is the only equilibrium.

This location will be followed by a price equilibrium where the incumbent will choose the lowest prices in the neighbourhood of the entrant and his prices will be increasing with the distance of varieties to the entrant product. Therefore, the varieties n and 1 will first *absorb* the shock of entry and this shock will propagate to other varieties at decreasing intensity.

An example

We present here the analytical solutions for $n = 5$. The best reply functions of the firms are given by:

$$p_I^*(p_e)=\begin{pmatrix}p_1^*(p_e)\\p_2^*(p_e)\\p_3^*(p_e)\\p_4^*(p_e)\\p_5^*(p_e)\end{pmatrix}=\begin{pmatrix}\frac{1}{2}tx(1-tx)+\frac{1}{2}(p_e+c)\\ \frac{1}{2}tx\left(\frac{3}{5}-tx\right)+\frac{2}{25}t+\frac{1}{2}(p_e+c)\\ \frac{1}{2}tx\left(\frac{1}{5}-tx\right)+\frac{3}{25}t+\frac{1}{2}(p_e+c)\\ -\frac{1}{2}tx\left(\frac{1}{5}+tx\right)+\frac{3}{25}t+\frac{1}{2}(p_e+c)\\ -\frac{1}{2}tx\left(\frac{3}{5}+tx\right)+\frac{2}{25}t+\frac{1}{2}(p_e+c)\end{pmatrix},$$

$$p_e^*(p_I)=\frac{1}{2}tx\left(\frac{1}{5}-tx\right)+\frac{1}{2}c+\frac{5}{2}x(p_5-p_1)+\frac{1}{2}p_1.$$

It is not surprising to observe that these functions are increasing in the competitor's prices and that the entrant reacts only to p_1 and p_5 Prices are strategic complements. Solving the previous linear system, we can compute equilibrium prices:

$$p_1^*=tx\left(\frac{13}{15}-\frac{7}{3}tx\right)+c,$$

$$p_2^*=tx\left(\frac{10}{15}-\frac{7}{3}tx\right)+c+\frac{2}{25}t,$$

$$p_3^*=tx\left(\frac{7}{15}-\frac{7}{3}tx\right)+c+\frac{3}{25}t,$$

$$p_4^*=tx\left(\frac{4}{15}-\frac{7}{3}tx\right)+c+\frac{3}{25}t,$$

$$p_5^*=tx\left(\frac{1}{15}-\frac{7}{3}tx\right)+c+\frac{2}{25}t,$$

$$p_e^*=11tx\left(\frac{1}{15}-\frac{1}{3}tx\right)+c.$$

Equilibrium profits are given by:

$$\Pi_I^*(x)=\frac{1}{450}t\left(181x+18-905x^2\right)-nf,$$

$$\Pi_e^*(x)=\frac{121}{450}xt(1-5x)-f,\ \frac{d^2\Pi_e^*(x)}{dx^2}=-\frac{121}{45}t<0,$$

$$\frac{d\Pi_e^*(x)}{dx}=0\Leftrightarrow x^*=\frac{1}{10}=\frac{1}{2n},$$

Therefore, the entrant will choose to produce an equidistant variety. Equilibrium prices and profits then follow:

$$p_1^* = \frac{19}{300}t + c = p_5^*,$$
$$p_2^* = \frac{37}{300}t + c = p_4^*,$$
$$p_3^* = \frac{43}{300}t + c, \quad p_e^* = \frac{11}{300}t + c,$$
$$p_e^* < p_1^* = p_5^* < p_2^* = p_4^* < p_3^*,$$

$$\Pi_I^* = f \cdot \left(\frac{541}{9000} \delta - n \right) \quad > \quad \Pi_e^* = f \cdot \left(\frac{121}{9000} \delta - 1 \right).$$

We observe that the varieties on the *front line* (1 and 5) absorb the shock of entry and the incumbent can use higher prices for protected varieties. These results are robust and we obtain similar analytical results for other values of n.

Optimal variety choice in duopoly

We normalise all the values in order to obtain $f = 1$. The resolution of the problem for different values of n shows that the optimal profit functions of the incumbent and the entrant have the following structure:

$$\Pi_I(n) = \gamma(n)\delta - n,$$
$$\Pi_e(n) = \lambda(n)\delta - 1$$
with $\gamma' < 0, \gamma'' > 0, \lambda' < 0$ et $\lambda'' > 0$

Figure 17.3 gives a representation of these profits for $\delta = 200$.

Two consequences ensue:

- the incumbent's profit is decreasing with n and hence in duopoly the incumbent prefers a minimal number of varieties;
- the incumbent can effectively blockade the entry choosing to produce n^l varieties such that $\Pi_e(n) \leq 0$. In this case, he obtains his monopoly profits corresponding to n^l

For each δ we can compute a unique n^l which blockades the entry.

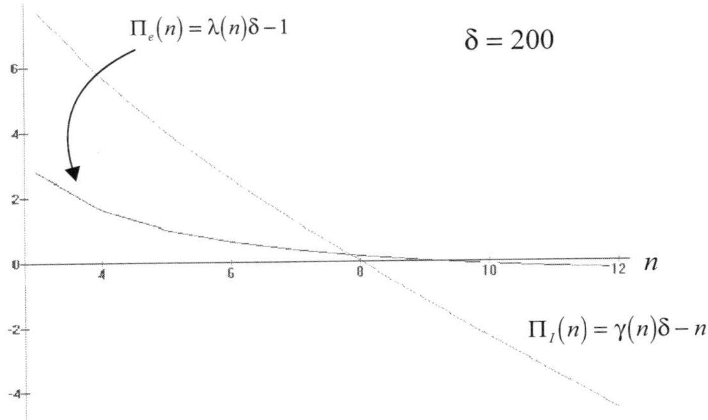

Figure 17.3 Duopoly profits and variety

Variety proliferation and entry blockading

If he chooses to blockade the entry, the incumbent will obtain his monopoly profits corresponding to n^l varieties: $\Pi^m(n^l)$. In this case n^l must be the root of the following equation:

$$\Pi_e(n^l) = 0 \Longleftrightarrow \lambda(n^l) = 1/\delta.$$

for any δ. It is clear that n^l is increasing in δ (see Figure 17.4): when the variety desire increases, the incumbent has to install more variety in order to cover all the market.

But the incumbent will only choose to blockade entry if his monopoly profits are higher than his duopoly profits with one variety (optimal number in duopoly).

We observe, in Figure 17.5, that for any δ the blockaded entry monopoly profit is higher than the duopoly profit of the incumbent: entry will be blockaded. Moreover, this is the best choice because the unconstrained monopoly equilibrium implies necessarily free entry ($n^m < n^l$).

Patents and entry blockading

Variety proliferation is not the only strategy for occupying a market. A well known alternative is given by patenting. If the incumbent uses patents to blockade the entry, he has to cover the whole market segment in this model. If the legally permitted breadth of a patent under a given patent law (cf. Ordover 1991 and van Dijk 1994) is l, the incumbent must obtain at least $1/l$ patents to blockade the entry. In this case he

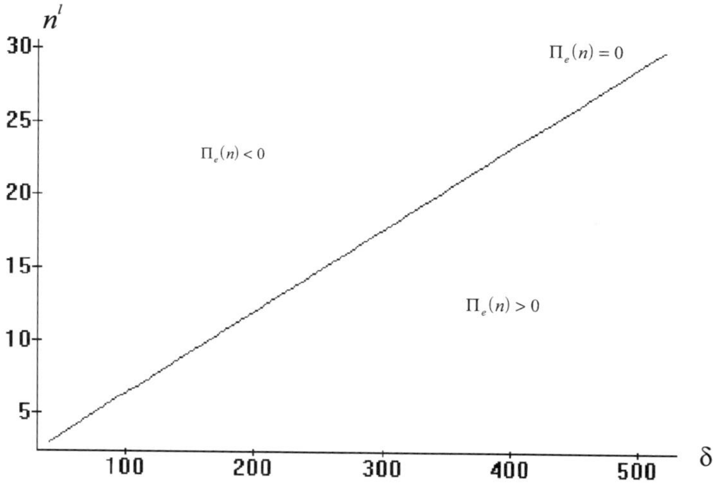

Figure 17.4 Variety proliferation and entry blockading

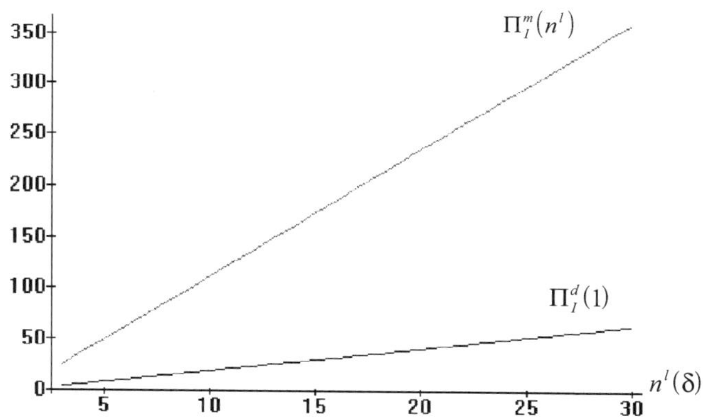

Figure 17.5 Entry blockading: incumbent's profits

can produce optimal monopoly variety and obtain his monopoly profits minus patent costs: *P*. In this setting, the role of the patent system is very limited.

In fact, the solution of this problem is quite trivial: if patent cost is low, patent and produce n^m; if not produce n^l.

But if the patent development includes product innovation costs, the patent costs must include some of the fixed production costs *f* and patenting can modify the optimal varieties of monopoly and hence the structure of the problem. Let us assume that a fraction $(1 - \alpha)$, $0 \le \alpha \le 1$, of fixed costs is in

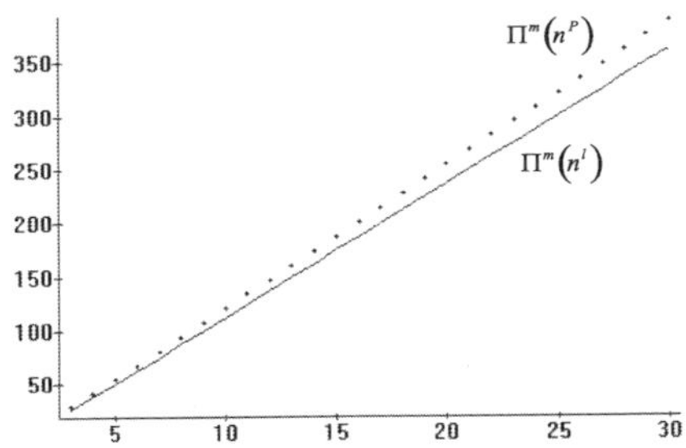

(a)

α = 0.2

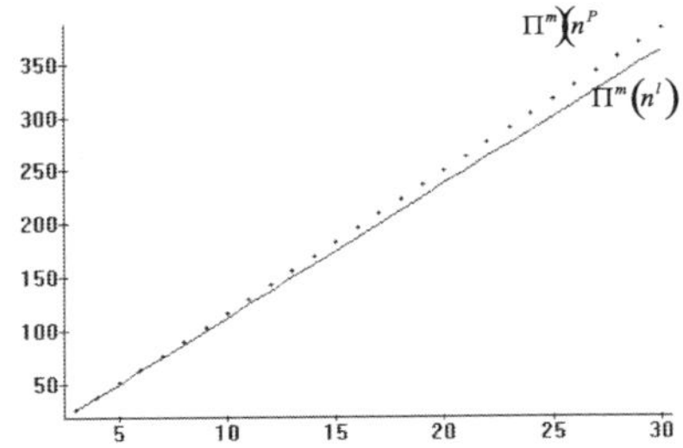

(b)

α = 0.99

Figure 17.6 Gross profits: patent or not? Two cases, one result…

fact product innovation costs. Therefore, a patenting incumbent pays only αf for each supplementary variety. In this case, the profits of the incumbent are given by:

$$\Pi_P^m = \frac{1}{4}\frac{t\left(3n^2-1\right)}{n^2} - n\cdot\left(\alpha f\right).$$

The optimal number of monopoly varieties becomes then:

$$n^P = \frac{1}{2}\left(\frac{4t}{\alpha}\right)^{1/3} = \frac{1}{\alpha^{1/3}}n^m \geq n^m$$

Unsurprisingly, the patenting monopoly produces more variety than the ordinary monopoly.

Given this result, we must reconsider the incumbent's optimal strategy. Variety proliferation or patenting? In order to answer this question, we must compare again the monopoly profits corresponding to patenting with n^P varieties and to variety proliferation with n^l varieties.

We can easily make this comparison for $P = 0$.

Figure 17.6-(a) corresponds to a case where an important fraction of the fixed production costs is development costs and they must be incurred in the patenting process.

Figure 17.6-(b) corresponds to a case where patenting absorbs only 1.

Consequently, if we take into account the patenting cost P, the incumbent's choice will be in favour of the variety proliferation, even in case (a).

Conclusion and extensions

We showed in a simple model of horizontal differentiation that variety is a superior strategy when entry blockading is the key consideration. More generally, this strategy provides indubitable market protection.

Even if this model does not tackle all our initial questions, it provides us with simple benchmark cases on the role of diversity in the determination of market structure and entry. To fully understand the role of diversity, one must extend this model in several directions. The most immediate is the combination of these two instruments in order to obtain an optimal protection of the market. A symmetrical consideration of the incumbent and the potential entrant allowing multiple varieties for the latter should not modify our results in a meaningful way. More radical extensions should include sequential variety creation and scale (or scope) economies in the production of those varieties.

Notes

1 Note by Murat Yıldızoğlu: The initial idea of this work came from a question Ehud asked me at the cafeteria of 38, Bld d'Anvers. Ehud was already living between Strasbourg and Beer Sheva, and advancing on this paper had not been easy. A somewhat complete first draft of the paper was presented during the 1996 Stockholm Schumpeter Conference. Our last real interaction on this paper took place at the end of 1996 (December) and the article is proposed here with its final state at this date.

2 See for instance Anderson *et al.* (1992).

Appendix A: Proof of Proposition 3

The market share of each variety j of the incumbent depends on the price p_j but also on prices p_{j-1} and p_{j+1} Consequently, this market share can be broken up into two segments: one market share preceding the variety (y_j^p), which takes place between $j - 1$ and j and one following it (y_j^f) and taking place between j and $j + 1$. These market shares extend from the position of the variety j to the position of the *limit consumer*:

$$p_j + t \cdot \left(y_j^p\right)^2 = p_{j-1} + t \cdot \left(\frac{1}{n} - y_j^p\right)^2$$

$$p_j + t \cdot \left(y_j^f\right)^2 = p_{j+1} + t \cdot \left(\frac{1}{n} - y_j^f\right)^2.$$

The market share of the variety is therefore given by

$$d_j\left(p_j, p_{j-1}, p_{j+1}; n\right) = y_j^p\left(p_j, p_{j-1}; n\right) + y_j^f\left(p_j, p_{j+1}; n\right).$$

We have two exceptions for varieties 1 and n: if $j = 1$, $y_j^p\left(p_j, p_{j-1}; n\right) = y_1^p\left(p_1, p_e; n, x\right)$ and if $j = n$, $y_j^f\left(p_j, p_{j+1}; n\right) = y_n^f\left(p_n, p_e; n, x\right)$

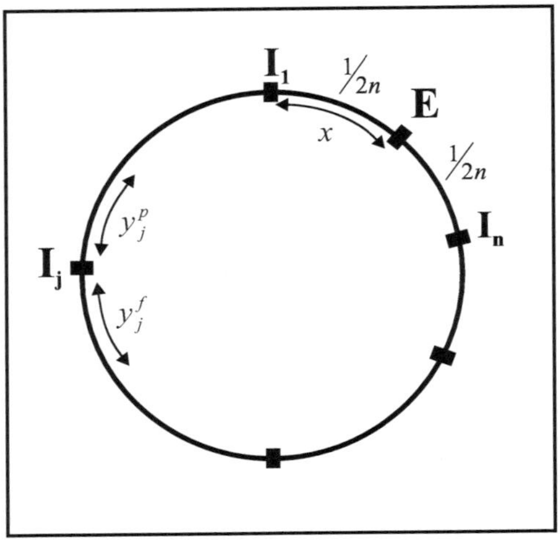

Figure 17.7 Circular city and entry

where x is the distance of the entrant to the variety 1. The total demand of incumbent is therefore given by:

$$D_I(p_I, p_e; n, x) = y_1^p(p_1, p_e; n, x) + \sum_{j=2}^{n} y_j^p(p_j, p_{j-1}; n)$$
$$+ \sum_{j=1}^{n-1} y_j^f(p_j, p_{j-1}; n) + y_n^f(p_n, p_e; n, x).$$

For the potential entrant, we have simply:

$$D_e(p_e, p_n, p_1; n, x) = y_e^p(p_e, p_n; n, x) + y_e^f(p_e, p_1; n, x).$$

Using these results, we can compute profits:

$$\Pi_I(p_I, p_e; n, x) = \sum_{j=1}^{n} (p_j - c) \cdot d_j(p_j, p_{j-1}, p_{j+1}; n, x)$$
$$= \sum_{j=1}^{n} \pi_j(p_j, p_{j-1}, p_{j+1}; n, x)$$
$$\Pi_e(p_I, p_e; n, x) = (p_e - cight) \cdot D_e(p_e, p_n, p_1; n, x).$$

The price sub-game equilibrium is given by $(p_I^*(n,x), p_e^*(n,x))$ such that

$$\frac{\partial \Pi_I(p_I, p_e; n, x)}{\partial p_1}\left(p_I^*(n,x), p_e^*(n,x)\right) = 0,$$

$$\vdots$$

$$\frac{\partial \Pi_I(p_I, p_e; n, x)}{\partial p_n}\left(p_I^*(n,x), p_e^*(n,x)\right) = 0,$$

$$\frac{\partial \Pi_e(p_I, p_e; n, x)}{\partial p_e}\left(p_I^*(n,x), p_e^*(n,x)\right) = 0.$$

For a given n and x, we have a system of $n + 1$ linear equations in prices. Moreover, p_j appears only in $\pi_{j-1}(\cdot)$, $\pi_j(\cdot)$ and $\pi_{j+1}(\cdot)$. Therefore, these equations are linearly independent and hence, this system possess a unique solution. It can easily be shown that these conditions are also sufficient because $\partial^2\Pi_I(\cdot)/\partial p_j^2 < 0$.

Appendix B: Proof of Proposition 4

If $p_1^* = p_n^* > p_e^*$ then the best reply of the entrant is to produce an equidistant variety $x^* = 1/2n$. In fact, the profit of the entrant is given by:

$$\Pi_e(p_I, pe; n, x) = \frac{(p_e - c)\left(tnx^2 + np_e - tx - p_n n^2 x - p_1\left(n - xn^2\right)\right)}{2xnt(xn-1)}$$

If $p_1 = p_n = p$, this profit becomes:

$$\Pi_e(p_I, pe; n, x) = \frac{(p_e - c)\left(tnx^2 + np_e - tx - np\right)}{2xnt(xn-1)},$$

$$\forall p, n, \quad \frac{\partial \Pi_e}{\partial x} = 0 \Leftrightarrow x^* = \frac{1}{2n}.$$

If the entrant chooses to produce such a variety, the incumbent has no reason to use different prices for varieties 1 and n. Therefore, $p_1^* = p_n^*$ and $x^* = 1/2n$ is an equilibrium.In fact this is the only equilibrium of our game.

References

Aghion, P. and Howitt, P. (1992) 'A Model of Growth through Creative Destruction', *Econometrica*, **60**: 323–51.

Anderson, S., de Palma, A. and Thisse, J.-F. (1992) *Discrete Choice Theory of Product Differentiation*, MIT Press, Cambridge, MA.

Dixit, A. and Stiglitz, J.E. (1977) 'Monopolistic Competition and Optimum Product Diversity', *American Economic Review*, **67**: 297–308.

Eaton, B.C. and Schmitt, N. (1994) 'Flexible Manufacturing and Market Structure', *American Economic Review*, **84**: 875–88.

Grossman, G. and Helpman, E. (1991) *Innovation and Growth in the Global Economy*, MIT Press, Cambridge, MA.

Klemperer, P. (1990) 'How Broad Should the Scope of Patent Protection Be?' *Rand Journal of Economics*, **21**: 113–30.

Luski, I. and Wettstein, D. (1995) 'An Optimal Patent Policy in a Dynamic Model of Innovation', Monaster Center Discussion Paper, 95–3, Ben Gurion University.

O'Donoghue, T., Scotchmer, S. and Thisse, J.-F. (1995) 'Patent Breadth, Patent Life and the Pace of Technological Improvement', Working Paper, 95–242, University of California at Berkeley.

Ordover, J. (1991) 'A Patent System for Both Diffusion and Exclusion', *Journal of Economic Perspectives*, **5**: 43–60.

Salop, S. (1979) 'Monopolistic Competition with Outside Goods', *Bell Journal of Economics*, **10**: 141–56.

Schmalensee, R. (1978) 'Entry Deterrence in the Ready-to-eat Breakfast Cereal Industry', *Bell Journal of Economics*, **9**: 305–27.

Tirole, J. (1988) *The Theory of Industrial Organisation*, MIT Press, Cambridge, MA.

van Dijk, T.W.P. (1994) *The Limits of Patent Protection*, Universitaire Pers Maastricht, Holland.

von Ungern-Sternberg, T. (1988) 'Monopolistic Competition and General Purpose Products', *Review of Economic Studies*, **55**: 231–46.

Waterson, M. (1990) 'The Economics of Product Patents', *American Economic Review*, **80(4)**: 860–9.

18 Creativity and uncertainty in the act of work

The contribution of the viability theory

Francis Munier

Introduction

The neoclassical approach considers labor as a "negative" act, an effort as a source of disutility. We consider that this assumption is very simplistic. It is not a fair representation of reality and it does not provide a relevant framework in order to understand the notion of creativity in the act of work in economics (Cates, 1979; Rubenson and Runco, 1992; Pannells and Claxton, 2008; Menger, 2009; Rasulzada and Dackert, 2009), while others disciplines analyze this topic with some greed, such as psychology (Lubart, 2003), and more recently management science.

This research attempts to understand precisely the notion of creative work that, like Menger (2009) is meaningful and exists only in situations of uncertainty and disequilibrium. Therefore, it is necessary to define precisely the nature and logic of creativity, and to specify the forms of uncertainty according to this process. We show that the standard approach does not provide a relevant theoretical framework to explain the creative act in the sense of working in uncertainty. Conversely, given our assumptions and definitions, we suggest that the mathematical approach provided by the viability theory (Aubin, 1997, 2010) appears as a more adequate corpus of theory and much more promising. The first part is devoted to a review of the literature on the theme of creativity in order to show the vacuity of the term within standard economics. The second part consists of proposing a draft of implementation of creative work in a situation of radical uncertainty in the context of viability theory.

Creativity

We criticize the orthodox economic model by highlighting the crucial role of radical uncertainty in the dynamics of creativity. We also criticize the teleological and deterministic characteristics of the orthodox economic approaches. We emphasize the dichotomous nature of creativity in order to claim that the theoretical corpus of viability seems promising.

The dichotomous nature of creativity and the art of reconciling oppositions

Defining creativity is not easy because of the polysemous meanings of this concept (Aleinikov *et al.*, 2000; Haddad, 2011). According to Haddad (2011), historically, it is the engineer Louis Armand who was the first, in 1970, to encourage the introduction of the term creativity in the *Supplément* to the *Grand Dictionnaire Analogique* of Paul Robert.

This concept is quite recent, and we notice that originally creativity was in fact considered as synonymous with inventiveness (Haddad, 2011). This meaning, more akin to the idea of mechanistic and deterministic behavior, is closer to what can be found in economics, though in our opinion it is not relevant. Therefore to define creativity in economics seems still more difficult because it is easy to consider that the *Homo œconomicus* is really the opposite of a creative being. Its standard behavior, which is to maximize an objective function in a probabilistic context, whose ultimate goal is to maximize his satisfaction, curbs, or even annihilates, any form of creativity. The *Homo œconomicus* is intrinsically conservative, rigid, and chilly against all forms of deviation from his only behavior that is maximizing the satisfaction. The fact that "the desire to be happy is more powerful than the desire to be," as Helvetius (1758) already said (Laval, 2007), induces the idea that *Homo œconomicus* is jailed in a steel cage (Weber, 2000).

Of course, much research attempts to reconsider this scheme. Some reconsider the assumption of homogeneity of labor. The heterogeneity constitutes a potential of utility (and talent) with respect to, precisely, the variety of jobs (Doeringer and Piore, 1971; Ashenfelter and Layard, 1987). Others introduce the variable training / learning and interpret it as a source of differentiation resulting from the accumulation of human capital (Mincer, 1993). We can also mention the approach based on matching (Jovanovic, 1979) which considers that individuals are unequally productive, that jobs are varied according to intrinsic characteristics, and that the quality of the union between employment and the employee is a priori unknown.

Notwithstanding these advances, the assumption of substantive rationality remains for these approaches the core of the behavior. However, it implies that the individual is by nature homogeneous, and omniscient, while paradoxically omniscience is the antidote to creativity. *Homo œconomicus* does not learn, he has no doubt, he asks no question, and he does not rebel against established ideas, the mainstream! He is rather favorable to consensus, to stability and equilibrium.

Nevertheless, as acknowledged by Sternberg and Lubart (1999, p. 3) "Creativity is the ability to produce work that is both novel (i.e. original, unexpected) and appropriate (i.e. useful, adaptative concerning task constraints)." This seminal definition allows us to point precisely to an essential feature of creativity: its dichotomous nature where both elements of chance and necessity are intimately related (Monod, 1970), and where rupture is still viable.

Indeed, creativity requires more rupture, instability, and disequilibrium, although, as noted by Munier (1999) and Haddad (2011), a certain amount of rationality (in the sense of reasonable reason) is required in the process of creativity. This dual need also corresponds to the two hemispheres of the brain. The left hemisphere is related to rationality, consciousness, logical reasoning, while the right hemisphere is more concerned with creativity and imagination.

This dichotomous logic seems to be the engine of creativity in which notions such as oxymoron, cons-behavior, and creative destruction are all concepts that help understand the phenomenon. Chance and necessity, useless and useful, rational and irrational, *pensée magique* (Piaget, 1987) and reason, immobility and development, statics and dynamics, movement and stationary equilibrium are all founding couples of the creative process within the meaning of a dazzling madness channeled according to constraints of viability (Brenot, 1997). This dichotomy is explained by different elements that we propose to develop below.

Some definitional elements of creativity: slow hunch, warranted assertions, possible adjacent, "bricolage," and uncertainty

Sternberg (2006, p. 87) proposes "*that creative people are those who are willing and able to 'buy low and sell high' in the realm of ideas*". We share this definition but in the same way (at the same time) we consider that it is important to explore the process between the act of buying and selling. For that, we introduce some concepts outside the economic field, but differently from Sternberg (2006) who focuses more on the psychological qualities of a creator.

In *Science et Méthode, Livre I, Chapitre III* (1908), Henri Poincaré wrote:

> *A ce moment, je quittai Caen, que j'habitais alors, pour prendre part à une course géologique entreprise par l'École des Mines. Les péripéties du voyage me firent oublier mes travaux mathématiques ; arrivés à Coutances, nous montâmes dans un omnibus pour je ne sais quelle promenade ; au moment où je mettais le pied sur le marche-pied, l'idée me vint, sans que rien de mes pensées antérieures parut m'y avoir préparé, que les transformations dont j'avais fait usage pour définir les fonctions fuchsiennes sont identiques à celles de la Géométrie non-euclidienne. Je ne fis pas la vérification ; je n'en aurais pas eu le temps, puisque, à peine assis dans l'omnibus, je repris la conversation commencée, mais j'eus tout de suite une entière certitude. De retour à Caen, je vérifiai le résultat à tête reposée pour l'acquit de ma conscience.*

Does it mean that the *Eureka* would be considered as the only explanation of the creative act, and that it would be a lonely activity that is random? In fact, Poincaré (1908) himself considers the fact that the discovery takes place in four phases: preparation, incubation, illumination, verification. This framework is also proposed by Wallas (1926), which is considered as the father of a real theory of creativity (Gallate *et al.*, 2012).

Recent research, related to innovative "milieux" or open science (Cribb and Hartomo, 2010), endeavor instead to prove the opposite of Eureka. We can also include the highlighting of a process that is called *the slow hunch* (Johnson, 2010) in order to explain this idea of the connectionist nature of creativity. One explanation is based on the nature of the cognitive process of creativity. The philosopher C. Peirce (1893) was the first to introduce the notion of *warranted assertions* which are the source of creativity. Thus, in a holistic sense, creativity emerges through redundancies of sets of knowledge which, through precisely this redundancy, allow a sense of novelty within a given corpus of knowledge. This induces the idea that contingent uncertainty is an important element of creativity, as a viable rupture, which can be radical without any form of LUCA (Last Universal Common Ancestor), although some researchers consider that creativity is a path dependent phenomenon. This idea was quite well developed by the notion of the *possible adjacent* introduced by Kauffman (1993). Creativity always takes place at the margin, at the border of the existing, the established codes. Dogan and Pahre (1991) introduced the concept of creative marginality in order to explain this idea in social sciences.

Moreover, as suggested by Valery (1894) when he attempted to understand the genius of Leonardo da Vinci, creativity is also a process that recombines multidisciplinary knowledge. As we said before, this is the notion of the possible adjacent (Kauffman, 1993) which considers that inventions come always from a handset of existing knowledge. In this, creativity is therefore especially not endeavoring to achieve optimal solutions, as deducted logic, but more as warranted assertions.

In this sense, creativity is neither a deterministic nor a teleological process (Gigerenzer *et al.*, 1990), moreover in a situation of radical uncertainty. It is more akin to a kind of "*bricolage*" in the sense of Levi Strauss (1962).

This is why we consider that the standard economic model is inappropriate to introduce and understand the creative process. We suggest that the analysis of the creative process should be based on a non-probabilistic uncertainty. Menger (2009) shares the idea of an existing relationship between uncertainty and creativity. Wallerstein (1998) develops the same hypothesis that uncertainty and creativity are encompassed concepts, and that the belief in certainty is a mistake. We try to go away by introducing a specific type of uncertainty. To overcome the uncertainty of events without statistical regularity, which deals only with averages, Peirce (1893) proposed the notion of tychastic uncertainty (Aubin, 2010). Creativity is then opportunistic, seizing tychastic uncertainty to find something it did not seek. This is the notion of serendipity, a term introduced in 1754 by Horace Walpole, inspired by the title of a Persian tale entitled *The Three Princes of Serendip* (Sri Lanka in Persian) (Bourcier and van Ancel, 2011; Haddad, 2011).

To conclude this first part, it is essential to recall that one of the main motivations was to prove that the standard economic approaches are not able to provide a relevant framework to understand creativity. One of these arguments is based on the fact that creativity is the opposite of the deterministic and teleological behavior. Then, we propose a new corpus (implementation of the viability theory) in order to take into account these stylized facts of creativity. We consider that it is more relevant and original compared to the work of Menger (2009), who still keeps the epistemological aspects of the standard approaches, both in sociology and economy.

Creativity as a viable behavior

Some glimpses on the viability theory

The main object of the viability theory (Aubin, 1997) is to explain the evolution of a system described by given nondeterministic dynamics and viability constraints, to reveal the concealed regulation laws that allow the system to be regulated, and provide selection mechanisms for implementing them. A common definition of viability is something that can be adapted to constraints in order to be able to live. The viability theory rejects therefore the teleological principle of optimization to focus on viability constraints, on the adaptation of the system. At each moment, a state of the system must respect the viability constraints.

A system is described by state variables and regulation variables. The former correspond to components of the state of the system and can evolve under the impulse of agents. The states evolve according to the regulee, defined as a regulating control or a message detected by the system.

A dynamic law associated with this state and with all regulees defines the velocity of evolution of the state. A second law, called the *a priori retroaction*, describes the constraints to which the regulees are subjected according to the current or past state of the system. It determines all the available regulees, thus clarifying a form of diversity, in the framework of a situation whereby a contingent uncertainty prevails.

The viability theory therefore constitutes a study of the dynamic evolution of a system beyond contingent uncertainty and under viability constraints. The set of states of the system which comply with viability constraints is referred to as a *constrained set*. A viable state is an element of this set. A viable evolution is an evolution, which verifies, at each moment, the viability constraints.

This consistency between the dynamics of the system and the viability constraints is important. The initial question, to which the viability theory provides some answers, can be expressed as follows: with each viable state of the macrosystem as an initial state (element of the constrained set), is there at least one viable evolution? (cf. definition 4.1.2 [Aubin, 1997, p. 143]). The mathematical resolution of compatibility of constraints and of dynamics of a system is made using the *regulation law*. This law results from the basic viability theorem, which consists of considering that a viable evolution is governed by a viable regulee, obtained through the intermediary of the *regulation map*.

The system is also characterized by three basic principles. On the one hand, its evolution is nondeterministic in the sense that the system is subjected to a contingent uncertainty, and that the existence of a multitude of evolutions according to the different regulees is possible. It also illustrates the idea of flexibility if one considers that it is possible to develop different possible regulees where the environment does not provide for this. On the other hand, the system is subject to viability constraints, which the state of the system must respect "on pain of death." These constraints limit the evolution of the system and determine the domain of constraint in which this evolution must imperatively occur. The third basic principle suggests that the evolution of the system complies with a principle of inertia according to which the regulee evolves only if viability conditions are questioned. If the evolution happens as slowly as possible, this is called the *principle of heavy evolution*.

The notion of the *viability domain* is introduced in order to avoid a situation where all the viable regulees are empty in at least one state of the system. The distinction between a *constrained set* and a *viable domain* is crucial in the sense that it allows the notion of consistency between a dynamics and the constraints to which it is confronted. The theorem of the theory of viability is based on this necessary compatibility: a constrained set is compatible with a macrosystem if and only if it is a viability domain of that macrosystem (cf. theorem 4.1.5 [Aubin, 1997, p. 145]).

The mathematical theory of viability uses tools from set-valued analysis such as set-valued maps. It consists in using a set-valued map $\mathbf{C}$, where we associate $\mathbf{x} \in \mathbf{E}$ with a set (possible empty) $\mathbf{C(x)} \subset \mathbf{F}$. On the contrary, in the single-valued analysis, each element $\mathbf{x} \in \mathbf{E}$ is a unique element $\mathbf{f(x)} \in \mathbf{F}$.

This approach enables the treatment of nondeterministic and non-stochastic dynamic models and can also take into account regulation parameters. Its relevance is to widen the spectrum of the study for a solution of the differential equation by replacing it with the notion of differential inclusion.

The advantage of using differential inclusion is thus, on the one hand, to allow the treatment of certain random forms, in particular the random form defined by nonstochastic uncertainty. On the other hand, the differential inclusions fit the case of complex systems, which differ from each other by the absence of control and a variety of possible dynamics. These are short-sighted systems that do not attempt to attain teleological objectives (*satisfacing behavior*). In contrast to the optimization, only minimal satisfaction requirements are to be respected. The differential inclusion is described by $\mathbf{x'(t)} \in \mathbf{F(x(t))}$, where when at each moment $\mathbf{t}$ the state of the system is $\mathbf{x(t)}$, the set $\mathbf{F}$ determines the set of possible velocities for this state.

Creativity and viability: conceptual framework

We apply the main tenets of the viability theory (Aubin, 1991) to the underlying assumption that creativity could appear during a crisis of viability according to a tychastic evolution within a viable domain.

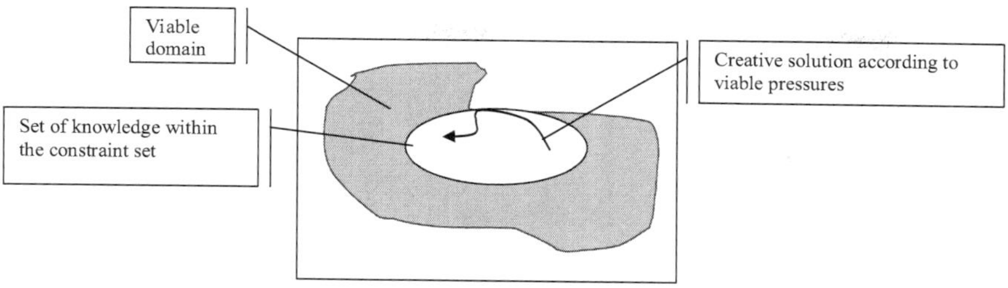

Figure 18.1 Viable evolution of creative acts

Evolution of viable creative works

The evolution of a set of knowledge, among a non-stochastic variety (*warranted assertions*), appears as a creative evolution at a critical moment, on the border of the viable domain (*possible adjacent*). The creative acts are defined in terms of the regulee within a set of knowledge accepted by society, academia, etc. We find again the core idea that creativity is the art of reconciling oppositions.

The regulee is one part of the couple, as an engine of creativity, i.e., the necessity part. The other one is contingent hazard (tychastic uncertainty). Even, viability is based on nondeterministic logic, some creators can behave according to their sets of knowledge and the evolution is not a myopic evolution in terms of goals. The difference with a teleological approach is based on the fact that the objectives are not strictly speaking the only one criterion to optimize. In other words, the viable evolution of creative works (within the meaning of dynamic evolution) must meet the viability constraints represented by the set of knowledge accepted. We represent here the notions of *slow hunch*. Figure 18.1 illustrates this situation.

The evolution of creative works "accepts" the pressure of the existing set of knowledge to the extent that autonomy and respect are viable for them in the creative process.

Viable crisis, heavy and creative solution

In the case where the pressure is too high, a viability crisis occurs; the evolution of creative works is not viable. The set of knowledge is now in crisis. The set of knowledge may refuse to accept – the viable crisis continues. However, the set is forced to adjust its knowledge and to evolve according to a more appropriate evolution. In this case, creativity is accepted and is viable. Figure 18.2 below shows the situation for different pressures.

The evolution of creative acts corresponding to pressures P_0 and P_1 represents the crisis of viability. Pressure P_2 is a viable evolution, but within a finite time; this evolution also reaches the boundary of the viable domain

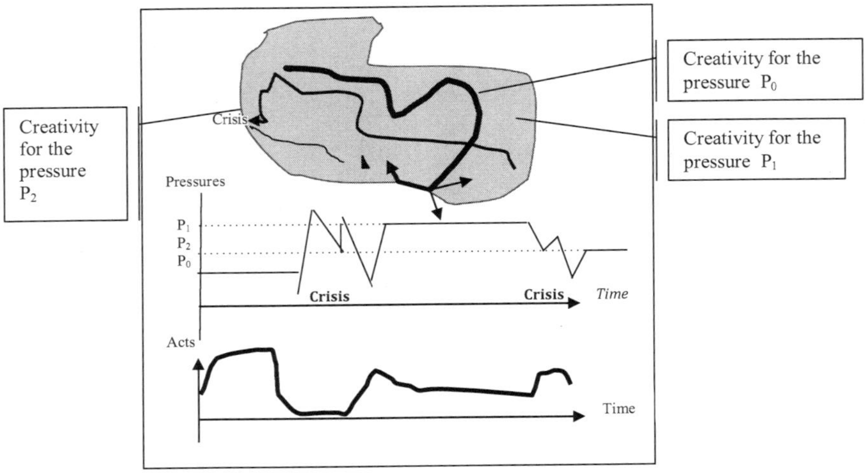

Figure 18.2 Creative evolution and crisis of viability

and induces a crisis. To avoid this crisis, it is assumed that the set of knowledge behaves as a "closed behavior" with respect to novelty.

The principle of inertia reflects this behavior. Rather than behaving in a creative evolution, which could create a crisis again, the set of knowledge chooses behavior based on a work evolving as little as possible (we find here a phenomenon of "oxymoron").

A heavy solution is one that is characterized by a behavioral evolution as constant as possible. This situation arises when the evolution of creative works is located in a niche of viability (see Figure 18.3).

By definition, a niche of viability is related to the dynamics when the regulee is fixed. In the figure, the work (which is no longer creative) is a kind of creative lock-in in the niche of viability for pressure P_2 in which the behavior within the set of knowledge remains viable but is similar to misoneist behavior. The creative process is thus annihilated, as the pressure is too heavy and "unacceptable" according to the set of knowledge.

Viable creation or how to resolve the antinomy of the couple: tychastic uncertainty and necessity

The creative process could be the following: a creation is an achievable capacity making it possible according to the needs of the context. It stores knowledge and evolves according to the evolution of the context, in particular the environment (socio-cultural system) that exerts pressures (necessity).

We suppose then that the creative acts could be viable and be represented as a connection between evolution of the set of knowledge and the

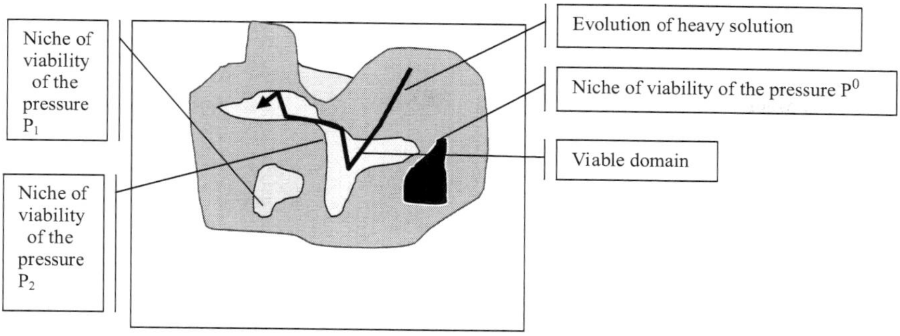

Figure 18.3 Heavy solution and non-creativity

socio-cultural system, and between this system and its institutional environment. The evolution is explained according to socio-cultural codes within a system which itself undergoes pressures of its institutional environment that could be the mainstream in so many fields.

These creative acts result from the practice of each evolution confronted with the economic, social, institutional, and cultural conditions of its environment. Creation thus constitutes a set of problem-solving of need in a particular environment. Creation appears according to codes that one can describe from this socio-cultural system. This system undergoes in its turn pressures of the global environment of the creator.

In other words, creators belong, generally in a very abstract way, even unconsciously, to a socio-cultural group where socio-cultural codes are important.

The dynamic units in this system represent the creations belonging to this group (viable domain). It is supposed that if an individual does not behave according to a set of knowledge (which itself obeys socio-cultural codes of its socio-cultural system), his creation is not viable, within the meaning of this system. In our logic, the external pressure of environment influences the nature of the system and consequently the behaviors of the creators.

The interactions between the creative acts of the system and its environment adjust and shape the viable domain. Consequently, an interface appears between the system and its environment as a border of the system. It is estimated that this interface is a function of regulation, which is indeed a combined force of environment regulation and self-regulation (adaptations and adjustments) of the system. Behaviors, which evolve/move out of this border, are not viable because they do not respect the principles of the socio-cultural system. The creation is not accepted.

It is necessary to specify that the system we propose is an open system. Its evolution is endogenous whereas that of its environment is supposed to be exogenous. We can consider that the constraints of viability (socio-cultural codes) to which the system is confronted articulate the interactions between the system and its environment. Location of these interactions defines the border of viability. These interactions define the concept of macrosystem structured in two subsystems: the socio-cultural group and the environment. These two subsystems are interdependent and thus form a hierarchical macrosystem, where the creative acts could emerge.

We suppose that creation and its adoption modify the existing individual creative acts. Then the whole of creative acts, including new creative acts, tend to modify the nature of the system. Creation brings new possibilities to solve problems of need.

Creation can produce effects on creators' cognition and behavior and consequently on individual creative acts within a given system. It is allowed that influence of different creations on evolution of the socio-cultural system could be different according to the inertia of the system. In our illustration, thickness of the interface represents resistance of the system compared to change. There appears a confrontation between two opposed pressures. In other words, the impact of creation will be neutralized by the system if it offers a significant resistance to reject the new idea (the interior pressure overrides the external one). In the same manner, the system can be open (the interior pressure is low) and allows a fast introduction of the creation.

Let us suppose that a creation is introduced into the environment of a socio-cultural system. An external pressure is exerted. Because of differences of creative acts (e.g., structure of knowledge, cognitive capacity, deviation of imitation, etc.) with limited rationality, incomplete information and knowledge, creators behave differently vis-à-vis a creation and with their "attachments" to socio-cultural codes.

Because of interior and external resistances, certain changes of creative acts are carried out in the field permitted by the interface between a socio-cultural system and its environment. Creation induces emergence of new creative acts. Parts of old creative acts are eliminated. Creation thus induces a mechanism of selection of creative acts. More precisely, the pressures coming from modification of the socio-cultural system and interactions from the global environment eliminate individuals' creative acts. This elimination is the consequence of adjustments/adaptations, and these latter could be simultaneous with the global change by individual members of the system, or just follow it. The elimination can relate to the "too excessive" or "too delayed" creative acts compared to new socio-cultural codes.

Conclusion

Our conceptualization of the creative acts supposes that creators behave according to constraints, which emanate from the socio-cultural system and more generally from its environment. According to the main aspects of definition of creativity, it is essential to propose a new framework in economics in order to be able to include uncertainty, slow hunch, warranted assertions, and possible adjacent. We think that the viability theory provides this framework.

References

Aleinikov, A.G., Kackmeister, S. and Koenig, R. (2000) *Creating Creativity – 101 Definitions (What Webster Never Told You)*, Alden B. Dow Creativity Center, Michigan.

Ashenfelter, O. and Layard, R. (1987) *Handbook of Labor Economics*, Elsevier Edition, Amsterdam.

Aubin, J.-P. (1991) *Viability Theory*, Birkhäuser, Berlin.

Aubin, J.-P. (1997) *Dynamic Economic Theory: A Viability Approach*, Springer-Verlag, Berlin.

Aubin, J.-P. (2010) *La mort du devin, l'émergence du demiurge, Essai sur la contingence, la viabilité et l'inertie des systèmes*, Éditions Beauchesne, collection Miroir des savants, Paris.

Bourcier, D. and van Ancel, P. (2011) *La sérendipité. Le hasard heureux*, Éditions Hermann, Paris.

Brenot, P. (1997) *Le Génie et la folie*, Plon éditeur, Paris.

Cates, C. (1979) "Beyond Muddling: Creativity," *Public Administration Review*, Vol. **39**, No. 6.

Cribb, J. and Hartomo, T.S. (2010) *Open Science Sharing Knowledge in the Global Century*, CSIRO Publishing, Collingwood, Victoria.

Doeringer, P.B. and Piore, M.J. (1971) *Internal Labor Markets and Manpower Analysis*, D.C. Heath, Lexington, Mass.

Dogan, M. and Pahre, R. (1991) *L'innovation dans les sciences sociales: La marginalité créatrice*, Presses universitaires de France, Paris.

Gallate, J., Wong, C., Ellwood, S., Roring, R.W. and Snyder, A. (2012) "Creative People Use Nonconscious Processes to Their Advantage," *Creativity Research Journal*, Vol. **24**, Issue 2–3: 146–51.

Gigerenzer, G., Swijtink, Z., Porter, T., Daston, L., Beatty, J. and Kruger, L. (1990) *The Empire of Chance: How Probability Changed Science and Everyday Life*, Cambridge University Press, Cambridge.

Haddad, G. (2011) *Les défis de la créativité*, Inaugurazione dell'Anno Accademico 2010–2011, Università degli Studi di Bergamo, UNESCO.

Helvetius, C.-A. (1758) *De l'esprit*, A Paris, Chez Durand, Libraire MDCCLVIII.

Johnson, S. (2010) *Where Good Ideas Come from: The Natural History of Innovation*, Riverhead Books, Éditions, Paris.

Jovanovic, B. (1979) "Job Matching and the Theory of Turnover," *Journal of Political Economy*, Vol. **87**, Issue 5: 972–90.

Kauffman, S. (1993) *The Origins of Order: Self-Organization and Selection in Evolution*, Oxford University Press, New York.

Laval, C. (2007) *L'Homme économique: Essai sur les racines du néolibéralisme*, Gallimard, coll. Nrf essays, Paris.

Levi Strauss, C. (1962) *La pensée sauvage*, Plon Editeur, Paris.

Lubart, T. (2003) *Psychologie de la créativité*, Armand Colin, Paris.

Menger, P.-M. (2009) *Le travail créateur*, Hautes Etudes, Gallimard Seuil, Paris.

Mincer, J. (1993) *Studies in Human Capital: Collected Essays of Jacob Mincer*, Edward Elgar, Cambridge.

Monod, J. (1970) *Le hasard et la nécessité. Essai sur la philosophie naturelle de la biologie moderne*, Éditions du Seuil, Paris.

Munier, F. (1999) "Taille de la firme et innovation: approches théoriques et empiriques fondées sur le concept de compétences," Thése de doctorat, Université Louis Pasteur.

Pannells, T.C. and Claxton, A.F. (2008) "Happiness, Creative Ideation, and Locus of Control," *Creativity Research Journal*, Vol. **20**, Issue 1: 67–71.

Peirce, C. (1893) "Evolutionary Love," *The Monist*, Vol. **3**: 176–200.

Piaget, J. (1987) *Six études de psychologie*, Folio Essai, Paris.

Poincaré, H. (1908) re-edition of *Science et Méthode, Philosophia Scientice*, Cahier spécial 3, July 2011, Editeur Kime, Paris.

Rasulzada, F. and Dackert, I. (2009) "Organizational Creativity and Innovation in Relation to Psychological Well-Being and Organizational Factors," *Creativity Research Journal*, Vol. **21**, Issue 2&3.

Rubenson, D.L. and Runco, M.A. (1992) "The Psychoeconomic Approach to Creativity," *New Ideas in Psychology*, Vol. **10**: 131–47.

Sternberg, R.J. (2006) "The Nature of Creativity," *Creativity Research Journal*, Vol. **18**, Issue 1: 87–98.

Sternberg, R.J. and Lubart, T.I. (1999) "The Concept of Creativity: Prospects and Paradigms," in R.J. Sternberg (ed.), *Handbook of Creativity*, Cambridge University Press, Cambridge.

Valery, P. (1894) "Introduction à la méthode de Léonard de Vinci," réédition de 1992, Folio Gallimard, Paris.

Wallas, G. (1926) *The Art of Thought*, Harcourt Brace, New York.

Wallerstein, I. (1998) "Uncertainty and Creativity," *American Behavioral Scientist*, Vol. **42**, No. 3: 320–22.

Weber, M. (2000) *L'éthique protestante et l'esprit du capitalisme*, Champs Flammarion, Paris.

19 What makes movies tick

Ivory tower insights, studio views and research directions

S. Abraham Ravid

Introduction

The well-known screenwriter William Goldman famously described the modus operandi of the movie industry as "nobody knows anything." This implies that the movie industry is unpredictable, and therefore gut feeling should be at least as good as data-based decision making. This chapter will try to summarize work that others and I have done over the last decade and a half towards debunking the "nobody knows anything" mantra which research seems not to support.

This chapter is not a comprehensive review of all the many papers written on the topic of film profitability. McKenzie (2012) provides a good survey of these papers.[1]

The movie industry is indeed characterized by a somewhat disappointing distribution of revenues and returns (more about that later). However, by no stretch of the imagination is movie production the only industry to have uncertain outcomes and an "irregular" distribution. The pharmaceutical industry is very similar and almost all entrepreneurial ventures are also characterized by many misses and a few big hits. Furthermore, when you crunch the numbers, you see that decision makers in the movie industry seem to act rationally and respond to incentives perhaps more so than in other industries, due to the quick turnover in Hollywood and the project by project nature of production which provides detailed data about every project undertaken.

Finally, audiences respond to the products (films) in ways that we would expect. In fact, research supports the view that star worship, which seems to be the media's characterization of audience behavior, is all but a myth – audiences respond to well written stories and well executed movies. Stars can still sell magazines, but not movies.[2]

Profitability

Art De Vany and David Walls (1996, 1999, 2002) were among the first researchers to put a statistical face on movie profitability. The important message from their work is the characterization of movie revenues as a function with "fat tails," where there are a few large hits and many misses. More specifically, De Vany and Walls suggest that the function that best fits the distribution of movie revenues is Bose Einstein (a general function used in particle physics) with a variance which may be infinite. If indeed the variance is infinite, obviously one cannot predict very well. However, you can make statements regarding the distribution. For example, one can argue that stars increase the median much more than the mean and make the revenue distribution less skewed (De Vany and Walls, 1999 p. 300). Such statements can be made with respect to any choice variable. The characterization of revenues as a function with fat tails and possibly an infinite variance is perhaps why even academics adhere to the famous Goldman quote. A function with an infinite variance where average is not very meaningful is not a good guideline for decision making based upon past data. Most academics, however, have a more tempered view of predictability in the industry, and even De Vany and Walls suggest, as noted earlier, that some predictive statements can be made.[3]

One point to mention is that De Vany and Walls, and in fact many of the researchers in the field, use only US theatrical revenues. The share of this source of revenues has been declining over the years, whereas the share of international revenues as well as home entertainment income has been increasing. By 2011, the US and Canada brought in only 31 percent of the global movie revenues (www.MPAA.org). For the average movie, home entertainment revenues are the largest component of income. In general, additional revenues tend to make the distribution less skewed. While there still are fat tails (big hits can be big hits everywhere), movies that are not doing very well in the US market can be "salvaged" by international or DVD sales and vice versa. For example, the *Titanic* re-release in 3D in 2012 made about 58 million dollars during the first 6 weeks of its release in the US, but international box office was 283 million. On the other hand, *21 Jump Street* made in 9 weeks 135 million dollars in the US and only 49 million dollars overseas (*Variety*, May 21–27, 2012).

My study (Ravid, 1999) was, to my knowledge, the first to include all sources of income in the revenue and profit equation and indeed it is immediately evident that total revenues are less variable than US theatrical revenues. The standard error (SE) of the domestic revenue regression is 1.54, whereas the SE of the total revenue regression, including all sources, is only 0.94. The inclusion of all sources of revenues and the comprehensive set of independent variables allow for an interesting regression analysis which can identify factors contributing to revenues and returns. First, it turns out that budgets drive revenues, but not necessarily rates of return (in fact, as far as we can tell, lower budgets lead to higher rates of return). The importance of budgets has been evident in all studies since. Interestingly, the only factor that was consistently statistically significant in the return equation was classification as a family movie (G and PG rating) and sequel status (more about sequels later).[4] De Vany and Walls (2002) also find that G and PG movies have a higher rate of return for each quartile of the distribution.[5] These conclusions are quite intuitive on further reflection since as you increase the potential audience, obviously revenues increase. Similarly, sequels follow a successful original movie.

At the time of the publication of the Ravid (1999) study industry seemed to find the conclusions instructive and the paper made some headlines.[6] A logical follow up would have been a world full of children's movies and yet, as De Vany and Walls lamented (2002), Hollywood kept, at least for a while, making "too many" R rated movies. Several issues came up in subsequent research. First, G rated movies have indeed the highest rate of return but they may be riskier than other films. Ravid and Basuroy (2004) question the production of very violent movies but find that these movies may increase revenues and, in particular, tend to lose money less often. In other words, at least part of the "R rating puzzle" (too little family friendly fare) can be attributed to risk reduction. However, we need to consider not only the number of films but also the allocation of budgets and financing dollars in the industry. Palia *et al.* (2008) consider single financed and co-financed films and show that, in support of many existing alliance theories, studios tend to farm out the riskiest part of their portfolio. All studios diversify to some extent. Even Disney produces most years some PG-13 rated or unrated movies. However, regardless of how risky their portfolio is, studios always co-finance the riskiest movies. The paper finds that PG rated movies and sequels tend to be less risky and thus are almost never co-financed. The paper can also support some other ideas, i.e., that big budget films are co-financed more often. The lesson that family friendly sequels are the way to go, first pointed out in Ravid (1999), seems to have been learned by now.

Out of the ten top grossing movies of 2011 nine were sequels or franchises. Only one movie had an MPAA rating of R (the sequel to *The Hangover*). For 2010 the list included many sequels, none of the top ten was R rated and five films were rated PG or G. The world chart is similar (source: www.the-numbers.com). We note that the percentage of G and PG rated (family) films in the general population of movies is much lower (3 percent for G rated movies and 11 percent for PG rated movies in the period 1995–2012 according to www.the-numbers.com).

Stars

Stars (defined in Ravid, 1999 as prior academy award winners and nominees) did not make much of a difference to a film's financial success as measured in that study. This conclusion seemed to be quite a surprise to academics and to industry, and much research in the years that followed tried to find out why stars are employed and make millions of dollars per movie if they do not add much to the financial success of the project. We should note that in Ravid (1999) stars do not have a negative sign in any of the regressions run, in other words, they do not make movies fail either. One interpretation offered at the time was that stars received their marginal product. This interpretation is supported by quick adjustment of star salaries as their market value fluctuated.

De Vany and Walls (1999) also suggest that the star myth is just a myth. Their sample included 2015 movies out of which 326 had a star (a person listed in the Premier list of that year's 100 most powerful people). A total of 19 stars had a significant impact on the probability of success in the past; however, De Vany and Walls note: "None of these estimates guarantee that a particular star will make a movie successful. In fact, they assure that no star can guarantee any outcome because the variance is infinite."

One of the more interesting subsequent papers in this area is by Elberse (2007). She finds that the announcement of star inclusion in a movie increases the expected revenue in that movie in the Hollywood stock exchange by 3 million dollars. This figure varies by the type of project and star. Given that the average movie in 2006 cost 65 million dollars excluding marketing,[7] 3 million is, relatively speaking, not much. Indeed Elberse (2007) also shows that a star casting announcement does not move the stock prices of media companies. In summary, there is no convincing economic rationale for (or against) employing stars in movies.

Why is it then that studio executives keep signing these huge star contracts?[8] On a philosophical level one can refer back to Rosen (1981) and Adler (1985) who extends the Rosen model.[9] Adler (1985) has perhaps the more startling prediction, showing that stars need not have more talent than any other performer. This is because in equilibrium, consumers find it less costly to "consume" performers that they can discuss with others and thus are willing to pay to see these performers more even if they do not bring more talent to the table. In our context we may conjecture that there is a large pool of available talented people who can competently perform in any given movie. Stars are performers that attract attention independent of their talent.

Another possible answer is provided in Basuroy *et al.* (2003). We split our sample into films that were panned by reviewers and others that succeeded. Stars provide a revenue bump to the failed films. In other words, stars may serve as insurance policy for failed movies. In an industry where a big failure is much more dreaded than a big success is wished for, insurance is worth its weight in gold, or in eight figure salaries. In Ravid and Basuroy (2004) this theme of risk aversion is developed much further (see later discussion).

It is the story, stupid

Beyond some predictable characteristics, there is some new work that shows that the most boring and traditional elements, i.e., good stories and good directors, are indeed the most important determinants of the success of a movie project. Let me start with some interesting statistical facts, cited in Goetzmann *et al.* (2012) and Harris *et al.* (2012).

Table 19.1 Percentage of nominations and awards in different categories for the Oscar winning film over 40 years (1971–2010) and Razzies from their inception, 1981–2010

Category	*Oscar nominated (Best)*	*Oscar won (Best)*	*Razzies nominated (Worst)*	*Razzies won (Worst)*
Writing	98	75	100	82
Directing	98	83	97	67
Editing	93	50		
Sound (combined)	68	43		
Best Actor	63	33	67	33
Best Actress	38	20	55	39

Table 19.1 shows the frequency of Oscars and nominations in different categories awarded to members of the team of the "best picture" each year. A similar tabulation is presented for the Razzies, which are the semi-humorous awards presented to the worst movies of the year. For example, the top left cell shows that in 98 percent of the cases the writer for the best picture was also nominated for the best writing. The next cell to the right shows that in 75 percent of the cases that writer also won an Oscar for writing. Similarly, all writers for the worst film were nominated for the worst screenplay and 82 percent "won."

The conclusion from 40 years of Oscars and the entire history of the Razzies presented in this table is very clear. Best pictures are highly correlated with best writing. Worst pictures are highly correlated with bad writing. In only one case (*Titanic* in 1997) did the academy not nominate the writer who wrote the best picture, and the writer of the "worst picture" has been nominated for a Razzie every year since the inception of the awards.

We should note that the correlation between acting awards and the best picture Oscar is much lower. At the low end of the spectrum, only 20 percent of the "best actress" winners act in the best picture. This illustration seems to support the view that it is the screenplay (and the director[10]) that drive the success of the movie rather than the stars.

This simple tabulation is only indicative because obviously awards do not necessarily correlate with revenues. Yet Eliashberg *et al.* (2007) and Goetzmann *et al.* (2012) support the view that writing quality translates into revenues too. In Goetzmann *et al.* (2012) we discuss the pricing of screenplays and show that screenplays are priced lower if the story (as approximated by the pitch, the summary which is used to promote and sell the screenplay) is fuzzier. We interpret fuzzier pitch as softer information, which is harder to evaluate (see Petersen and Rajan, 2002; Stein, 2002). Information regarding the screenwriter has the opposite effect – the more experienced and well known is the writer, the higher the price (for a discussion of contractual forms in screenplays see Harris *et al.*, 2012). Yet one finding is harder to interpret and supports the casual evidence in Table 19.1. When we run regressions of revenues and returns of the resulting films (obviously a biased sample of presumably the best scripts in the sample) on prices paid for the screenplays and numerous control variables, we find that prices paid correlate with revenues and rates of return of the resulting films. The latter finding is even more surprising because, as noted, it is very difficult to predict returns. The fact that screenplay prices correlate with returns suggests that spec screenplay purchases are motivated by some *ex ante* evaluation of the value of the film, which is already evident in the screenplay. We should note that the sample in Goetzmann *et al.* (2012) is of spec screenplays, i.e., does not include screenplays commissioned by studios, which may be based upon well known brand names. Eliashberg *et al.* (2007), one of the two other papers, to our knowledge, to consider how screenplay sales[11] affect the success of the resulting projects, reaches similar conclusions. Using a sophisticated textual analysis technique, Eliashberg *et al.* (2007) try to evaluate what elements in the text of a "spoiler," i.e., an extensive summary of the screenplay, lead to high returns of the resulting movies. Part of the sample is used as a training ground for the methodology called "bootstrap aggregated classification and regression tree" and the other part is used to test the predictions. Eliashberg *et al.* conclude (2007, p. 883):

> Based on a holdout-prediction test, we find that our model is able to capture signals from stories contained in spoilers and use them to differentiate between profitable and unprofitable stories for movies. Compared to randomly choosing movie scripts, making production decisions based on our forecast will significantly increase ROI … This improvement is especially significant considering the fact that the stories in scripts represent only one of the many factors that can affect a movie's ROI.

Again, this progression from screenplay content to film profitability seems to be there. This may not surprise industry insiders. Here is a quote from Eliashberg *et al.* (2007, p. 882):

> Ridley Scott, a famous director of motion pictures, once pointed out, "any great film is always driven by script, script, script" (Silver-Lasky, 2003, p. 108). Of course, what are in the script are the stories. Peter Gruber, the producer of Batman, suggested the same: "At the end of the day when you get done with all the fancy production design … what's up on the screen is the script. Plain, old-fashioned words. It all starts there and it all ends there" (Silver-Lasky, 2003, p. 108).

Our simple tabulation seems to indicate that directors are important as well. In a recent study (John *et al.*, 2012) we try to see whether directors are hired and fired "correctly" from an economic point of view. The findings support job matching theories (such as Jovanovic, 1979). We show that directors' careers are short, and they are re-evaluated at every project they do. The hiring decision is based upon the total career path. In spite of industry beliefs to the contrary you are better (or worse) than your last movie. In fact, it is fair to say that you are as good as your average project. However, perhaps more interesting is the "director factor." In line with recent literature that finds a "CEO factor" (see Bertrand and Schoar, 2003 and Bennedsen *et al.*, 2010, 2011) we find that directors who make a total of four films or more (which we interpret to mean that they have higher, market proven ability) have a large positive, and statistically significant, effect on both the rate of return of their movies as well as on the ratio of favorable reviews. The impact on favorable reviews is significant (but lower) for directors with two and three movies as well. There are various ways of looking at the data. Tests of differences show that the difference between the highest group (career length of five or more films) and those who made less than four films over their entire career is very significant. Longer careers are thus associated with higher ability directors. Further, the difference in the director fixed effect between directors who make only one film versus those who go on to make five or more films is significant, showing that "better" directors have a better and more significant outcome. It is then the story and the leader of the project that make a difference, rather than the stars.

Other issues

There are many other issues that determine the success of motion pictures. For example, release dates and windows matter a great deal (see Einav, 2007; Hennig-Thurau *et al.*, 2007). Furthermore, it seems that the stock market views any change in release dates as a negative sign (see Einav and Ravid, 2009). The importance of the allocation of screen space and competition by other movies is analyzed by Krider and Weiberg (1998) and in a simultaneous framework by Elberse and Eliashberg (2003). There is the question of the optimal allocation of movies to screens (see Eliashberg *et al.*, 2009) and the allocation of movies to theaters within an area (see Chisholm and Norman, 2006). There is also an intriguing question of how resources should be optimally used in the context of motion pictures (see Miller and Shamsie, 1996). Finally, in the context of directors and actors, teams may matter too (see Shamsie and Mannor, 2012). However, it seems that without a great script and a skillful and creative director all is for naught.

Can audiences be influenced?

Eliashberg and Shugan (1997) were the first to consider the role of movie critics, more specifically, whether movie critics were "influencers" or "predictors." In other words, whether good reviews attract people to movies or whether it is that critics predict, using "formal or informal" techniques, how the movie would fare and translate it into an informative review. Eliashberg and Shugan (1997) find that weekly revenues are correlated with critical reviews in later but not earlier weeks, and conclude that critics are predictors rather than influencers. If they influenced movie going, the critical reviews would be correlated with early outcomes as well. However, Basuroy *et al.* (2003) find correlations between critical reviews and revenues in all weeks, and conclude that critics are both influencers and predictors, or at least you cannot tell the two apart. They discover what seems to be a negativity bias; in other words, negative reviews are worse for movies than positive reviews help. As noted, budgets and stars can moderate negative reviews; perhaps this is another reason for star studded films.

Morris Holbrook (1999) analyzes consumers' (viewers') preferences collected by HBO and compares them to critical reviews of the same films. The paper considers quite a few dimensions of films – such as genres and even length. He finds that audiences prefer more run of the mill movies whereas reviewers go for the edgier stuff. There are several findings, however, that agree with the thrust of what we have discussed so far. First, audiences like family movies; in fact, it is one of the highest coefficients in his tests. The paper also shows that audiences like stars but reviewers are not influenced by stars either way. This disagrees with earlier findings; however, the definition of stars in this paper is quite different. Perhaps most interesting are two findings – first, even though critics and "the masses" have different views of various film characteristics, there is still a positive and significant correlation between their choices, consistent with the previous discussion. Also, out of 19 characteristics one of the very few which affects popular appeal and expert judgment in the same direction and in a statistically significant manner (positive) is a good director defined in the same manner as in our paper (the set of directors who have directed many films + Steven Spielberg).

In recent years there has been a lively debate regarding the question of whether internet reviews (so called word of mouth, although user reviews is probably a more accurate description) have taken the place of professional reviews in influencing consumer choice in the movies. Here the jury is still out.

Liu (2006) finds that the volume of internet postings affects revenues in each of the first eight weeks of a run. On the other hand, the valence does not seem to matter. Similar results are found at Duan *et al.* (2008). However, Chintagunta *et al.* (2010) and Basuroy and Ravid (2012) suggest that there may be endogeneity issues. For example, there can be unobservable variables that affect the word of mouth and movie revenues or there can be reverse causality – i.e., high revenue movies cause more buzz on the Internet. At any rate, measuring the impact of internet postings on movie revenues is much less clean than measuring the impact of professional reviews, as reviews come out at a specific time and before the opening, whereas internet buzz comes in all the time.

Chintagunta *et al.* (2010) use instrumental variable techniques and show that once one accounts for endogeneity it is the valence rather than the volume that matters. Basuroy and Ravid (2012) also show that the significance of volume disappears once endogeneity is taken into account. However, Chintagunta *et al.* (2010) find that professional critics do not matter, whereas Basuroy and Ravid (2012) find that they are important for the success of movies, even in the presence of internet postings, supporting earlier work. The two studies, however, use different measures of critical acclaim.

Two final points on critics – first, in a new study which analyzes the impact of the content of ads on the success of movies and estimates the impact of specific content elements (Basuroy *et al.*, 2012) we find that the best content item one can put in an ad is a favorable quote from a reputable reviewer. Also, in another study (Ravid *et al.*, 2006) we show that critics can be statistically biased, but their recommendations still matter (statistically) to the success of films.

Conclusion

In conclusion, there is still a way to go in understanding the elusive nature of success in motion pictures. We know that the revenue distribution is skewed, with many failures and a few large hits, but this is also the case for many idea based products such as drugs or software. We know that sequels and family movies do better than other types of movies and we also know that by now industry knows this – after all, they rarely agree to co-finance them.

The star culture is still a puzzle, but a reasonable way of thinking about the issue is that there is a large pool of very talented people who can make a movie a success. The "stars" are "stars" for various other reasons. Whereas stars are a debatable proposition, there seems to be increasing evidence that screenplays and directors are indispensable.

Most work assigns a role to the opinions of expert critics. The debate over the role of the Internet continues – one of the difficult issues is that whereas technology evolves quickly, scientific work takes years to complete and thus findings lag behind the fast evolving reality. However, I would tend to agree with Joe Morgenstern, the movie critic of the *Wall Street Journal*, who writes, "Far from worrying that my supposed power will be diminished by the recent democratization of criticism, I find encouragement in the change, as any sensible person should" (Morgenstern, 2006).

Notes

1 Vogel (2010) is a standard reference on the industry.
2 Conversations with industry insiders suggest that very low budget films for niche audiences or direct to DVD productions may receive financing based upon attaching a star to the project. There is no real research about that issue due to lack of data.

3 Without getting too technical, under many circumstances right skewed or left skewed distribution can still yield meaningful results in regression analysis.
4 In some runs of the rate of return regression (in particular if we included small films) the budget was negative and significant and the number of reviews mattered as well.
5 See also studies by Simonoff and Sparrow (2000) and Fee (2002).
6 See, for example, *Wall Street Journal*, front page, August 13, 1998; *New York Times*, March 5, 1999 or *Variety*, March 22, 1999 and many others.
7 Elberse's sample period is 2001–2005. MPAA stopped collecting budgets in 2006 and this number is available at www.the-numbers.com/.
8 Contracting is an under-explored area, due to lack of available data. However, Chisholm (1997) provides some interesting insights based upon a painstakingly collected sample.
9 See also MacDonald (1988).
10 See John *et al.* (2012).
11 In a recent working paper, Luo (2011) provides a theoretical model and uses a similar database to Goetzmann *et al.* (2012) to predict whether a writer should sell a pitch or a completed screenplay.

References

Adler, M. (1985) "Stardom and Talent," *The American Economic Review*, **75** (1): 208–12.

Basuroy, S. and Ravid, S. A. (2012) "Has Word of Mouth Swamped Critical Reviews? Evidence from the Motion Pictures Industry," Working paper.

Basuroy, S., Chatterjee S. and Ravid, S. A. (2003) "The Impact of Critics on the Success and Failure of Motion Pictures Influencers vs. Predictors," *Journal of Marketing*, **67**, October: 103–17.

Basuroy, S., Rao, V. and Ravid, S. A. (2012) "What is Advertising Content Worth? Evidence from the Motion Pictures Industry," Working paper.

Bennedsen, M., Perez Gonzales, F. and Wolfenzon, D. (2010) "Do CEOs Matter?" Working paper, Columbia University.

Bennedsen, M., Perez Gonzales, F. and Wolfenzon, D. (2011) "Estimating the Value of the Boss: Evidence from CEO Hospitalization Events," Working paper, Stanford University.

Bertrand, M. and Schoar, A. (2003) "Managing with Style: The Effect of Managers on Firm Policies," *Quarterly Journal of Economics*, **118** (4): 1169–208.

Chintagunta, P. K., Gopinath, S. and Venkataraman, S. (2010) "The Effects of Online User Reviews on Movie Box Office Performance: Accounting for Sequential Rollout and Aggregation Across Local Markets," *Marketing Science*, **29** (5): 944–57.

Chisholm, D. C. (1997) "Profit Sharing vs. Fixed Payment Contracts – Evidence from the Motion Pictures Industry," *Journal of Law, Economics and Organization*, **13**: 169–201.

Chisholm, D. C. and Norman, G. (2006) "When to Exit a Product: Evidence from the U.S. Motion-Picture Exhibition Market," *The American Economic Review*, **96** (2): 57–61.

De Vany, A. and Walls, D. (1996) "Bose Einstein Dynamics and Adaptive Contracting in the Motion Pictures Industry," *Economic Journal*, **439** (106): 1493–1514.

De Vany, A. and Walls, D. (1999) "Uncertainty in the Movies: Can Star Power Reduce the Terror of the Box Office?" *Journal of Cultural Economics*, **23** (November): 285–318.

De Vany, A. and Walls, D. (2002) "Does Hollywood Make Too Many R-Rated Movies? Risk, Stochastic Dominance, and the Illusion of Expectation," *Journal of Business*, **75** (3): 425–51.

Dhar, T., Guanghi Sun and Weinberg, C. B. (2012) "The Long Term Box Office Performance of Sequel Movies," *Marketing Letters*, **23** (1): 13–30.

Duan, W., B. Gu and A. B. Whinston (2008) "The Dynamics of Online Word-of-mouth and Product Sales: An Empirical Investigation of the Movie Industry," *Journal of Retailing*, **84** (2): 233–42.

Einav, L. (2007) "Seasonality in the U.S. Motion Picture Industry," *Rand Journal of Economics*, **38** (1): 127–45.

Einav, L. and Ravid, S. A. (2009) "Stock Market Response to Changes in Movies' Opening Dates," *Journal of Cultural Economics*, **33** (4): 311.

Elberse, A. (2007) "The Power of Stars: Do Star Actors Drive the Success of Movies?" *Journal of Marketing*, **71** (4): 102–20.

Elberse, A. and Eliashberg, J. (2003) "Demand and Supply Dynamics Behavior for Sequentially Released Products in International Markets: The Case of Motion Pictures," *Marketing Science*, **22** (3): 329–54.

Eliashberg, J. and Shugan, S. M. (1997) "Film Critics: Influencers or Predictors?" *Journal of Marketing*, **61** (April): 68–78.

Eliashberg, J., Hui, S. K. and Zhang, Z. J. (2007) "From Story Line to Box Office: A New Approach to Green Lighting Movie Screens," *Management Science*, **53** (6): 881–93.

Eliashberg, J., Hegie, Q., Ho, J. et al. (2009) "Demand-Driven Scheduling of Movies in a Multiplex," *International Journal of Research in Marketing*, **26** (2): 75–88.

Fee, E. (2002) "The Costs of Outside Equity Control: Evidence from Motion Picture Financing Decisions," *Journal of Business*, **75** (4): 681–711.

Goetzmann, W., Ravid, S. A. and Sverdlove, R. (2012) "The Pricing of Soft and Hard Information: Economic Lessons from Screenplay Sales," *Journal of Cultural Economics, forthcoming.*

Harris, M., Ravid, S. A. and Basuroy, S. (2012) "Contingency Without Moral Hazard: A Theory of Intellectual Property Contracts and Evidence from Screenplay Sales," Working paper, University of Chicago, Booth School of Business.

Hennig-Thurau, T., Henning, V., Sattler, H., Eggers, F. and Houston, M.B (2007) "The Last Picture Show? Timing and Order of Movie Distribution Channels," *Journal of Marketing*, **71** (4): 63–83.

Holbrook, M. B. (1999), "Popular Appeal versus Expert Judgements of Motion Pictures," *Journal of Consumer Research*, **26** (September): 144–55.

John, K., Ravid, S. A. and Sunder, J. (2012) "Performance and Managerial Turnover: Evidence from the Career-Paths of Film Directors," Working paper, New York University.

Jovanovic, B. (1979) "Job Matching and the Theory of Turnover," *Journal of Political Economy*, **87** (5): 972–90.

Krider, R. E. and Weinberg, C. B. (1998) "Competitive Dynamics and the Introduction of New Products: The Motion Picture Timing Game," *Journal of Marketing Research*, **35** (1): 1–15.

Liu, Y. (2006) "Word-of-Mouth for Movies: Its Dynamics and Impact on Box Office Revenue," *Journal of Marketing*, **70** (3): 74–89.

Luo, H. (2011) "When to Sell Your Idea: Theory and Evidence from the Movie Industry," Working paper, Harvard University.

MacDonald, G. M. (1988) "The Economics of Rising Stars," *The American Economic Review*, **78** (1): 155–66.

McKenzie, J. (2012) "The Economics of Movies: A Literature Survey," *Journal of Economic Surveys*, **26** (1): 42–70.

Miller, D. and Shamsie, J. (1996) "The Resource-Based View of the Firm in Two Environments: The Hollywood Film Studios from 1936 to 1965," *Academy of Management Journal*, **39** (3): 519–43.

Morgenstern, J. (2006) *Wall Street Journal*, April 29, p. 6.

Palia, D., Ravid, S. A. and Reisel, N. (2008) "Choosing to Co-finance: An Analysis of Project Specific Alliances in the Film Industry," *Review of Financial Studies*, April: 483–511 (lead article).

Petersen, M. A. and Rajan, R. (2002) "Does Distance Still Matter? The Information Revolution in Small Business Lending," *Journal of Finance*, **57**: 2533–70.

Ravid, S. A. (1999) "Information Blockbusters and Stars: A Study of the Motion Pictures Industry," *Journal of Business*, October: 463–92.

Ravid, S. A. and Basuroy, S. (2004) "Executive Objective Function, the R-Rated Puzzle and the Production of Violent Films," *Journal of Business*, **77** (2): 155–92.

Ravid, S. A., Wald, J.K. and Basuroy, S. (2006) "Distributors and Critics: Does it Take Two to Tango?" *Journal of Cultural Economics*, **30** (3): 201–18.

Rosen, S. (1981) "The Economics of Super-stars," *American Economic Review*, **71**: 845–58.

Shamsie, J. and Mannor, M. (2012) "Looking Inside the Dream Team: Tacit Knowledge as a Driver of Performance," Working paper.

Silver-Lasky, P. (2003) *Screenwriting in the 21st Century*. New York: Sterling Publishing Company Inc.

Simonoff, J. and Sparrow, I. (2000) "Predicting Movie Grosses: Winners and Losers, Blockbusters and Sleepers," *Chance*, **13**: 15–24.

Stein, J. (2002) "Information Production and Capital Allocation: Decentralized vs. Hierarchical Firms," *Journal of Finance*, October: 1891–1921.

Vogel, H. (2010) *Entertainment Industry Economics*. 8th edn. Cambridge: Cambridge University Press.

Index

acquisitions 208
advertising 210
Air Liquide 115
Aldrich, H. E. 25
ALMIM 168
Amabile, T. M. 89, 140
appropriation strategies 151
AREVA 114
automobile industry 27, 29, 298
Avadikyan, A. 110, 112, 278

Bach, L., 3, 7, 14, 17
Baumol, J. B. 9
biotechnology 198, 263
biotechnology industry 214
Bootz, J. P. 55, 73, 89, 90, 91, 92, 93, 95
Bunel, M. 226
Burger-Helmchen, T. xxv, 71, 258

Callon, M. 139, 157, 255
Canon 181
CATI database 198
Cavallucci, D. 41
Chesbrough, H. 136, 151, 164
circular market 311
CIS data 233
C-K theory 48
cluster 12, 78, 83, 114, 194, 281
CNRS 255
cognition 91
Cohendet, P. 13, 24, 71, 72, 93, 95, 131, 151, 209, 252
commercialisation of research 253
commitment 95, 193
communities 8, 60–73, 78, 89–105, 133, 252
competences 93
competitiveness 6, 34, 84, 98, 135, 271
contracts 257
Cowan, R. 73, 90, 194, 281
creative economy 3, 85, 144, 252
creative system 67
creativity 4, 30, 49, 60, 80, 83, 89, 110, 131, 226, 231, 266, 271, 327
cultural industries 5, 78, 137, 334
Cyert, R. M. 23

demand side 10, 171
density 289
design 4, 24, 41–55, 60, 80, 117, 132, 159, 169, 194, 226, 272, 344
digital camera industry 169
division of labour, 155
Dosi, G. 151
dynamic capabilities 131, 226
Dupouët, O. 71, 73, 89, 96, 99, 104

economies of scale 42, 218
economies of scope 271
EDF 66, 114
Electricite de Strasbourg EDF 34, 66, 99, 114
electronic design automation industry 208
entrepreneurship 17, 73, 102, 110, 138, 209, 247, 255, 339
entrepreneurial university 252
Eparvier, P. 226
equilibrium 81, 168, 213, 274, 283, 287, 314, 327
Ermine, J. L. 58, 67, 68, 69, 71, 74
EURODRIVE 33
evolution 43
expert 89

finance 234
Florida, R. 12, 252
forgetting 20

Frank, L. 226
Freyermuth, J. 151
Fujifilm 181

GDF 114
Guittard, C. 58, 71

Harvey, J. F. 104, 131
Heraud, J. A. 78
high-technology industries 208
Hussler, C. 226

ideas 12, 71, 79–86, 91, 131–45, 246, 328
innovation 3, 21, 41, 61, 78, 98, 110, 131, 151, 167, 194, 208, 226, 267, 271, 281, 286, 309
INSERM 255
intangible resources or assets 135, 154, 231, 246
intellectual property IPR 4, 67, 154

Jonard, N. 281
Justman, M. 271

Kern, F. 71, 73, 89
knowledge 83, 192
knowledge management 49, 58, 96
Kodak 181

laboratories 99, 115, 252
Lambert, G. 20, 36, 38
law of evolution 41
Lerch, C. 38, 41
licensing 110, 153, 167
Lièvre, P. 58, 62, 63, 70
Llerena, P. 24, 72, 192
Lou, W. 167

manager 73
market 116, 309
market size 173, 244
March, J. 20, 98, 141
MASK 67
Matt, M. 7, 252
media 5, 339
movie 10, 305, 339
MPAA 340
Muller, E. 72, 82, 83
Munier, F. 327

national innovation systems 110
Nelson, R. R. 24, 109, 134, 154
network 78, 114, 156, 162, 281
network externalities 7
network position 192
Nikon 181

Olympus 181
Ozman, M. 192
open innovation 1, 131, 209
open source 164
opportunity 58, 135, 144, 195, 227

paradigm 8, 25, 48, 142, 153, 159, 252
Paraponaris, C. 58
patent 46, 110, 151, 198, 208, 246, 252, 257, 282, 309–10
paths of evolution 47
Penin, J. 71, 151, 153, 154, 155
policies 3, 78, 110, 230, 252, 271
product innovation 309
product line 41, 118
production 4, 29, 42, 80, 96, 110, 139, 152, 169, 208, 232, 253, 271, 311, 339
public economics 271

Ravid, S. A. 339
rationality, 20, 32, 48, 64, 336
Renault 28
research productivity 253
routine 21, 60, 64, 95, 132

Sanditov, B. 281
Schenk, E. 41, 55, 89, 90, 92, 93, 95
Schumpeterian perspective 41, 111, 132, 155, 195, 271, 309
sectoral system of intellectual property SSIP 151
semiconductor industry 25, 216
service sector 4, 34, 80, 117, 134, 226, 257, 272
Schaeffer, V. 252
SEW USOCOME 32
Simon, L. 131
small worlds 282
social dimension 92
social network 16, 141
software 32, 152, 347
Sony 181
strategic alliances 192
supply side 10, 172, 208

tangible resources 246
teaching 252
technological infrastructure 271
technological matching 281

technological regime 111, 151
technologies 116
technology 43
TEEXMA 66
territories 78
TQM 30
transitivity 290
TRIZ 42

uncertainty 327
university 252
University Louis Pasteur 255

Valeo 27
value chain 11, 35, 62, 112, 139
viability theory 327
video game 10

Wagner, M. 208, 222, 223
Winter, S. G. 27, 154

Yildizoğlu, M. 309

Zenker, A. 78
Zuscovitch, E. 3, 20, 151, 158, 271, 309

Printed in the USA/Agawam, MA
December 6, 2013

582791.177